Lecture Notes in Computer Science 3846

Commenced Publication in 1973
Founding and Former Series Editors:
Gerhard Goos, Juris Hartmanis, and Jan van Leeuwen

Editorial Board

David Hutchison
  Lancaster University, UK
Takeo Kanade
  Carnegie Mellon University, Pittsburgh, PA, USA
Josef Kittler
  University of Surrey, Guildford, UK
Jon M. Kleinberg
  Cornell University, Ithaca, NY, USA
Friedemann Mattern
  ETH Zurich, Switzerland
John C. Mitchell
  Stanford University, CA, USA
Moni Naor
  Weizmann Institute of Science, Rehovot, Israel
Oscar Nierstrasz
  University of Bern, Switzerland
C. Pandu Rangan
  Indian Institute of Technology, Madras, India
Bernhard Steffen
  University of Dortmund, Germany
Madhu Sudan
  Massachusetts Institute of Technology, MA, USA
Demetri Terzopoulos
  New York University, NY, USA
Doug Tygar
  University of California, Berkeley, CA, USA
Moshe Y. Vardi
  Rice University, Houston, TX, USA
Gerhard Weikum
  Max-Planck Institute of Computer Science, Saarbruecken, Germany

H. Jaap van den Herik   Yngvi Björnsson
Nathan S. Netanyahu (Eds.)

Computers and Games

4th International Conference, CG 2004
Ramat-Gan, Israel, July 5-7, 2004
Revised Papers

 Springer

Volume Editors

H. Jaap van den Herik
Universiteit Maastricht
Institute for Knowledge and Agent Technology, IKAT
6200 MD, Maastricht, The Netherlands
E-mail: herik@cs.unimaas.nl

Yngvi Björnsson
Reykjavik University
Department of Computer Science
Ofanleiti 2, IS-103 Reykjavik, Iceland
E-mail: yngvi@ru.is

Nathan S. Netanyahu
Bar-Ilan University
Department of Computer Science
Ramat-Gan 52900, Israel
E-mail: nathan@cs.biu.il

Library of Congress Control Number: 2006920436

CR Subject Classification (1998): G, I.2.1, I.2.6, I.2.8, F.2, E.1

LNCS Sublibrary: SL 1 – Theoretical Computer Science and General Issues

ISSN   0302-9743
ISBN-10   3-540-32488-7 Springer Berlin Heidelberg New York
ISBN-13   978-3-540-32488-1 Springer Berlin Heidelberg New York

This work is subject to copyright. All rights are reserved, whether the whole or part of the material is concerned, specifically the rights of translation, reprinting, re-use of illustrations, recitation, broadcasting, reproduction on microfilms or in any other way, and storage in data banks. Duplication of this publication or parts thereof is permitted only under the provisions of the German Copyright Law of September 9, 1965, in its current version, and permission for use must always be obtained from Springer. Violations are liable to prosecution under the German Copyright Law.

Springer is a part of Springer Science+Business Media

springer.com

© Springer-Verlag Berlin Heidelberg 2006
Printed in Germany

Typesetting: Camera-ready by author, data conversion by Scientific Publishing Services, Chennai, India
Printed on acid-free paper   SPIN: 11674399   06/3142   5 4 3 2 1 0

Preface

This book contains the papers of the 4th International Conference on Computers and Games (CG 2004) held at the Bar-Ilan University in Ramat-Gan, Israel. The conference took place during July 5–7, 2004, in conjunction with the 12th World Computer-Chess Championship (WCCC) and the 9th Computer Olympiad.

The biennial Computers and Games conference series is a major international forum for researchers and developers interested in all aspects of artificial intelligence in computer-game playing. After two terms in Japan and one in North America, the fourth conference was held in Israel.

The Program Committee (PC) received 37 submissions. Each paper was initially sent to two referees. Only if conflicting views on a paper were presented, was it sent to a third referee. With the help of many referees (see list after this preface), the PC accepted 21 papers for presentation and publication after a post-conference editing process. For the majority of the papers this implied a second refereeing process.

The PC invited Brian Sheppard as a keynote speaker for CG 2004. Moreover, Dr. Sheppard was Guest of Honour at the 9th Computer Olympiad and recipient of the 2002 ChessBase Award for his publication "Towards Perfect Play of Scrabble." Dr. Sheppard's contribution "Efficient Control of Selective Simulations" was taken as the start of these proceedings and as a guideline for the order of the other contributions. Brian Sheppard's contribution deals with Scrabble, Poker, Backgammon, Bridge, and even Go. So, his contribution is followed by papers on these games if presented at the conference. Otherwise the international and varied nature of the papers of CG 2004 would be difficult to order owing to their diversity of backgrounds and their many different views on games and related issues. This diversity, however, makes the book attractive for all readers.

Dr. Sheppard's contribution is followed by a Poker contribution, viz., "Game-Tree Search with Adaptation in Stochastic Imperfect-Information Games" by Darse Billings et al. and a Backgammon contribution, viz., "*-MINIMAX Performance in Backgammon" by Thomas Hauk et al. Since the paper on Backgammon uses *-MINIMAX Search, we decided to let it be preceded by "Rediscovering *-MINIMAX Search" by the same authors. Then four papers on Go follow. The remaining papers are on Chinese chess (two papers), and thereafter one paper for each of the following games: Amazons, Arimaa, Chess, Dao, Gaps, Kayles, Kriegspiel, Loa, and Sum Games. The book is completed by three contributions on multi-player games.

We hope that our readers will enjoy reading the efforts of the researchers. Below we provide a brief characterization of the 22 contributions in the order given above. It is a summary of their abstracts, yet it provides a fantastic three-page overview of the progress in the field.

"Efficient Control of Selective Simulations" by Brian Sheppard describes a search technique that estimates the value of a move in a state space by averaging the results of a selected sample of continuations. Since exhausted search is ineffective in domains characterized by non-determinism, imperfect information, and high branching factors, the prevailing question is: can a selective search improve upon static analysis? The author's answer to this question is affirmative.

"Game-Tree Search with Adaptation in Stochastic Imperfect-Information Games" is written by Darse Billings, Aaron Davidson, Terence Schauenberg, Neil Burch, Michael Bowling, Robert Holte, Jonathan Schaeffer, and Duane Szafron. It deals with real-time opponent modelling to improve the evaluation-function estimates. The new program called VEXBOT is able to defeat PSOPTI, the best poker-playing program at the time of writing.

"Rediscovering *-MINIMAX Search" by Thomas Hauk, Michael Buro, and Jonathan Schaeffer provides new insights into the almost forgotten STAR 1 and STAR 2 algorithms (Ballard, 1983) by making them fit for stochastic domains.

"*-MINIMAX Performance in Backgammon" also by Thomas Hauk, Michael Buro, and Jonathan Schaeffer presents the first performance results for Ballard's (1983) *-MINIMAX algorithms applied to Backgammon. It is shown that with effective move ordering and probing STAR 2 considerably outperforms EXPECTIMAX. Moreover, empirical evidence is given that today's sophisticated evaluation functions do not require deep searches for good checker play in Backgammon.

"Associating Shallow and Selective Global Tree Search with Monte Carlo for 9×9 Go" by Bruno Bouzy continues to advocate that Monte-Carlo search is effective in examining search trees. An iteratively-deepening min-max algorithm is applied with the help of random games to compute mean values. The procedure is stopped as soon as one move at the root is proved to be superior to the other moves. Experiments demonstrate the relevance of this approach.

"Learning to Estimate Potential Territory in the Game of Go" is a contribution by Erik van der Werf, Jaap van den Herik, and Jos Uiterwijk. It investigates methods for estimating potential territory in the game of Go. New trainable methods are presented for learning to estimate potential territory from examples. Experiments show that all methods described are greatly improved by adding knowledge of life and death.

"An Improved Safety Solver for Computer Go" by Xiaozhen Niu and Martin Müller describes new, stronger search-based techniques including region merging and a new method for efficiently solving weakly dependent regions. In a typical final position, more than half the points on the board can be proved safe by the current solver. This result almost doubles the number of proven points compared to the earlier reported percentage of 26.4.

"Searching for Compound Goals Using Relevancy Zones in the Game of Go" is written by Jan Ramon and Tom Croonenborghs. A compound goal is constructed from less complex atomic goals, using standard connectives. Compound-goal search obtains exact results. A general method is proposed that uses relevancy zones for searching for compound goals.

"Rule-Tolerant Verification Algorithms for Completeness of Chinese-Chess Endgame Databases" by Haw-ren Fang attempts to verify a conjecture, viz., that the rule of checking indefinitely has much more effect on staining the endgame databases than other special rules. It turned out that three endgame databases, KRKCC, KRKPPP, and KRKCGG are complete with the Asian rule set, but stained by the Chinese rules.

"An External-Memory Retrograde Analysis Algorithm" by Ping-hsun Wu, Ping-Yi Liu, and Tsan-sheng Hsu gives a new sequential algorithm for the construction of large endgame databases. The new algorithm works well even when the number of positions is larger than the number of bits in the main memory computer. A 12-men database KCPGGMMKGGMM is built. It has 8,785,969,200 positions after removing symmetrical positions. The construction process took 79 hours. The author found the largest DTM and DTC values currently known, viz., 116 and 96.

"Generating an Opening Book for Amazons" is by Akop Karapetyan and Richard Lorentz. The authors discuss a number of possible methods for creating opening books. They focus mainly on automatic construction and explain which seem best suited for games with large branching factors such as Amazons.

"Building a World-Champion Arimaa Program" by David Fotland describes a new two-player strategy game designed by Omar Syed. The game is difficult for computers. Omar offers a $10,000 prize to the first program to beat a human top player. BOT-BOMB won the 2004 computer championship, but failed to beat Omar for the prize. The article describes why this is so.

"Blockage Detection in King and Pawn Endgames" by Omid David Tabibi, Ariel Felner, and Nathan S. Netanyahu is the only contribution on chess. A blockade detection method with practically no additional overhead is described. The method checks several criteria to find out whether the blockage is permanent.

"Dao: a Benchmark Game" is written by Jeroen Donkers, Jaap van den Herik, and Jos Uiterwijk. The contribution describes many detailed properties of Dao and its solution. The authors conclude that the game can be used as a benchmark of search enhancements. As an illustration they provide an example concerning the size of a transposition table in α-β Search.

"Incremental Transpositions" by Bernard Helmstetter and Tristan Cazenave deals with two single-agent games, viz., Gaps and Morpion Solitaire. The authors distinguish between transpositions that are due to permutations of commutative moves and transpositions that are not. They show a depth-first algorithm which can detect the transpositions of the first class without the use of a transposition table. In a variant of Gaps, the algorithm searches more efficiently with a small transposition table. In Morpion Solitaire a transposition table is not even needed.

"Kayles on the Way to the Stars" by Rudolf Fleischer and Gerhard Trippen provides a solution for a previously stated open problem in proving that determining the value of a game position needs only polynomial time in a star of bounded degree. So, finding the winning move — if one exists — can be done in linear time based on the data calculated before.

"Searching over Metapositions in Kriegspiel" is written by Andrea Bolognesi and Paolo Ciancarini. It describes the rationale of a program playing basic endgames. It is shown how the branching factor of a game tree can be reduced in order to employ an evaluation function and a search algorithm.

"The Relative History Heuristic" is authored by Mark Winands, Erik van der Werf, Jaap van den Herik, and Jos Uiterwijk. The authors propose a new method for move ordering. It is an improvement of the history heuristic. Some ideas are taken from the butterfly heuristic. Instead of only recording moves which are best in a node, moves which are applied in the search tree are also recorded. Both scores are taken into account in the relative history heuristic. So, moves are favored which on average are good over moves which are sometimes best.

"Locally Informed Global Search for Sums of Combinatorial Games" by Martin Müller and Zhichao Li describes algorithms that utilize the subgame structure to reduce the runtime of global α-β Search by orders of magnitude. An important issue is the independence of subgames. Important notions of a subgame are temperature or its thermograph. The new algorithms exhibit improving solution quality with increasing time limits.

"Current Challenges of Multi-Player Game Search" by Nathan Sturtevant focuses on the card games Hearts and Spades. The article deals with the optimality of current search techniques and the need for good opponent modelling in multi-player game search.

"Preventing Look-Ahead Cheating with Active Objects" by Jouni Smed and Harri Hakonen first discusses a lockstep protocol. This requires that the player starts at announcing a commitment to an action and thereafter announces the action itself. Since the lockstep protocol requires separate transmissions, it slows down the turns of the games. Another method to prevent look-ahead cheating is the use of active objects. It relies on parameterizing the probability of catching cheaters. The smaller the probability, the less bandwidth and transmissions are required.

"Strategic Interactions in the TAC 2003 Supply Chain Tournament" is a joint effort of Joshua Estelle, Yevgeniy Vorobeychik, Michael P. Wellman, Satinder Singh, Christopher Kiekintveld, and Vishal Soni. The authors introduce a preemptive strategy designed to neutralize aggressive procurement, perturbing the field to a more profitable equilibrium. It may be counterintuitive that an action designed to prevent others from achieving their goals actually helps them. Yet, strategic analysis employing an empirical game-theoretic methodology verifies and provides insight into the results.

Acknowledgements

This book would not have been produced without the help of many persons. In particular we would like to mention the authors and the referees. Moreover, the organizers of the events in Ramat-Gan have contributed quite substantially by bringing the researchers together. A special word of thanks goes to the Program Committee of CG 2004. Moreover, the editors gratefully acknowledge the expert assistance of all our referees. Additionally, the editors happily recognize the generous sponsors. Finally, we would like to express our sincere gratitude to Martine Tiessen and Jeroen Donkers for preparing the manuscript in a form fit for the Springer publication.

September 2005
Jaap van den Herik,
Yngvi Björnsson,
Nathan Netanyahu,

Maastricht, Reykjavik, and Ramat-Gan

Organization

Executive Committee

Co-chairs: H. Jaap van den Herik
             Yngvi Björnsson
             Nathan S. Netanyahu

Program Committee

Michael Buro   Martin Golumbic   Hitoshi Matsubara
Ken Chen   Ernst A. Heinz   Jontahan Schaeffer
Jeroen Donkers   Hiroyuki Iida   Takenobu Takizawa
Ariel Felner   Richard Korf   Jos Uiterwijk
Aviezri Fraenkel   Shaul Markovich

Referees

Darse Billings   Ariel Felner   Jack van Ryswyck
Adi Botea   Ernst A. Heinz   Jonathan Schaeffer
Bruno Bouzy   Markian Hlynka   Pieter Stone
Mark Brockington   Hiroyku Iida   Nathan Sturtevant
Yngvi Björnsson   Graham Kendall   Gerald Tesauro
Neil Burch   Akihiro Kishimoto   Jos Uiterwijk
Michael Buro   Jens Lieberum   Clark Verbrugge
Tristan Cazenave   Richard J. Lorentz   Erik van der Werf
Ken Chen   Shaul Markovitch   Mark Winands
Jeroen Donkers   Martin Müller   Ren Wu
Markus Enzenberger   Xiaozhen Niu   Ling Zhao

Sponsors

The City of Ramat-Gan
Intel Israel
Israel Ministry of Tourism
Aladdin
Mercury
IBM Israel
Pitango
PowerDsine
Israeli Chess Fedaration
Golan Heights Winery
Rimonim Hotels
ChessBase

Table of Contents

Efficient Control of Selective Simulations
   Brian Sheppard 1

Game-Tree Search with Adaptation in Stochastic Imperfect-Information Games
   *Darse Billings, Aaron Davidson, Terence Schauenberg,
Neil Burch, Michael Bowling, Robert Holte, Jonathan Schaeffer,
Duane Szafron* 21

Rediscovering *-MINIMAX Search
   Thomas Hauk, Michael Buro, Jonathan Schaeffer 35

*-MINIMAX Performance in Backgammon
   Thomas Hauk, Michael Buro, Jonathan Schaeffer 51

Associating Shallow and Selective Global Tree Search with Monte Carlo for 9×9 Go
   Bruno Bouzy 67

Learning to Estimate Potential Territory in the Game of Go
   *Erik C.D. van der Werf, H. Jaap van den Herik,
Jos W.H.M. Uiterwijk* 81

An Improved Safety Solver for Computer Go
   Xiaozhen Niu, Martin Müller 97

Searching for Compound Goals Using Relevancy Zones in the Game of Go
   Jan Ramon, Tom Croonenborghs 113

Rule-Tolerant Verification Algorithms for Completeness of Chinese-Chess Endgame Databases
   Haw-ren Fang 129

An External-Memory Retrograde Analysis Algorithm
   Ping-hsun Wu, Ping-Yi Liu, Tsan-sheng Hsu 145

Generating an Opening Book for Amazons
   Akop Karapetyan, Richard J. Lorentz 161

Table of Contents

Building a World-Champion Arimaa Program
  David Fotland 175

Blockage Detection in Pawn Endings
  Omid David Tabibi, Ariel Felner, Nathan S. Netanyahu 187

Dao: A Benchmark Game
  *H. (Jeroen) H.L.M. Donkers, H. Jaap van den Herik,
  Jos W.H.M. Uiterwijk* 202

Incremental Transpositions
  Bernard Helmstetter, Tristan Cazenave 220

Kayles on the Way to the Stars
  Rudolf Fleischer, Gerhard Trippen 232

Searching over Metapositions in Kriegspiel
  Andrea Bolognesi, Paolo Ciancarini 246

The Relative History Heuristic
  *Mark H.M. Winands, Erik C.D. van der Werf,
  H. Jaap van den Herik, Jos W.H.M. Uiterwijk* 262

Locally Informed Global Search for Sums of Combinatorial Games
  Martin Müller, Zhichao Li 273

Current Challenges in Multi-player Game Search
  Nathan Sturtevant 285

Preventing Look-Ahead Cheating with Active Objects
  Jouni Smed, Harri Hakonen 301

Strategic Interactions in the TAC 2003 Supply Chain Tournament
  *Joshua Estelle, Yevgeniy Vorobeychik, Michael P. Wellman,
  Satinder Singh, Christopher Kiekintveld, Vishal Soni* 316

Author Index 333

Efficient Control of Selective Simulations

Brian Sheppard

Sheppard Company, Inc,
Concord, MA, USA
sheppardco@aol.com

Abstract. Selective simulation is a search technique that estimates the value of a move in a state space by averaging the results of a selected sample of continuations. The value of selective sampling has been demonstrated in domains such as Backgammon, Scrabble, poker, bridge, and even Go. This article describes efficient methods for controlling selective simulations.

1 Introduction

The domains dealt with are characterized by three issues: (1) non-determinism, (2) imperfect information, and (3) high branching factors. In such domains, exhaustive algorithms achieve a shallow search depth, which is frequently insufficient. The ineffectiveness of exhaustive search seemingly leaves static analysis as the only option for such domains. Searching for an alternative, the question arises: can a selective search improve upon static analysis?

In some domains it is possible to create effective selective-search policies. This is not true in every domain, but it is true in many domains that are of current research interest, such as Backgammon [9], Scrabble [11], poker [3], bridge [10], and even Go [5,7]. The selective-search methods described in this paper employ *selective sampling* of possible continuations. The distinction between selective sampling and Monte Carlo is subtle and maybe not even well defined, but here is an attempt at the distinction: in selective sampling we try to *choose* continuations that make the distinctions between alternatives clearer, whereas in Monte Carlo we select samples in the same distribution as the branching structure of the domain. This difference should not obscure the goal of selective simulation, which is the same goal as that of a Monte-Carlo simulation: to find the best move We just want to find the best move more quickly.

For example, assume that the player to move is playing a bear-off in Backgammon. The player considers two options. After either option the opponent can roll a doublet to bear off all his men, so the two moves work out equally in that case. In a selective simulation framework, the player to move can cease considering doublets, since any difference between the plays must be revealed in the other rolls. By avoiding doublets, the player reduces CPU usage to 5/6 of a Monte-Carlo simulation, but still arrives at the best move. In this example, the payoff is tiny. In other situations, the payoff of selectivity can be arbitrarily large.

The course of the article is as follows. Section 2 provides the basic simulation framework. In Section 3 three illustrative domains are introduced: Hold'em poker,

Scrabble, and Backgammon. The emphasis is on Hold'em poker, since for that game the data used are specifically collected for this article. The data for the other games are from the literature. Section 4 focuses on generating plausible moves. The essence of this article is in Section 5: Selecting a sample of continuations. Section 6 briefly discusses time control. In Section 7 the issue of selecting a sample move (in relation to a continuation) is considered. Section 8 provides methods on how to evaluate the end result of a simulation. In Section 9 a summary is given.

2 Basic Simulation Framework

The following pseudo-code outlines the process of simulation.

1. Generate plausible moves.
2. Select a sample of continuations (an arbitrary process that can generate state changes).
3. While time remains for further simulation:
   - select a continuation,
   - select a move to simulate,
   - follow up that move with the selected continuation,
   - evaluate the end result,
   - average with all previous results for this plausible move.
4. The move with the highest average is the best.

The choices that a player makes throughout this framework determine the effectiveness of the search policy. The article discusses some options for each step in the framework. Among the options that fit the framework is a pure Monte-Carlo approach.

3 Illustrative Domains

The article contains several examples, most from three domains: Hold'em poker, Scrabble, and Backgammon. The Scrabble [12] and Backgammon [13] examples mostly summarize data from existing literature. The Hold'em data was collected specifically for this article, and does not appear elsewhere.

The game Hold'em poker is described by Billings, Papp, Schaeffer, and Szafron [2]. For this purpose, it suffices to note that each player has 2 hole cards, and there are 5 community cards. The winner is the player that makes the best 5-card hand by combining his hole cards and the community cards.

The Hold'em engines used in these experiments were rudimentary. They are nothing special in themselves, and this article is not about them. It suffices to note that they play at an intermediate level, and exhibit the variability of the domain. They are typical of the programs that a developer might struggle to improve.

Hold'em can be played with up to 10 players, but in this article we will use the 2-player game. Each experiment measures the difference between two players by playing 1600 games of Hold'em. The difference between the players is reported in the units of "small bets per hand," a standard measure of skill. We will perform each experiment 400 times, and measure the standard deviation of the mean measured by each experiment. The goal of the article is to quantify the advantage of selecting a sample in a systematic way, rather than simply using Monte Carlo.

The baseline Monte-Carlo simulation used no selectivity at all to compare the players. That is, a sample of 1600 random hands was played out, and the results were summarized. The standard deviation of the measurement was 0.251 small bets. The true value of the difference between the players in the experiment was known to be about 0.025 small bets per hand.

This poses a huge obstacle to progress. To quantify the problem, consider that an advantage of 0.025 small bets per hand is considered significant in Hold'em. Yet, the standard deviation of a 1600-game Monte-Carlo simulation is ten times as large. To reduce the standard deviation of a Monte-Carlo sample to only 0.025 would require about 160,000 trials. This is feasible, but time consuming. In addition to the time, there is a 16 per-cent chance that the weaker program would have better simulation results. Drawing the wrong conclusion because of statistical noise can set back a research agenda for months.

Now consider a more typical case, where the players differ by a small amount because of some tweak. Assume that the difference is 0.005 small bets per hand. To detect such a difference with 84 per-cent confidence would require 4,000,000 trials! Such differences are very expensive to detect, yet a difference of 0.005 is a worthwhile improvement. If a player wishes to produce a program that is aware of the subtleties of the game, then the player would have to detect such small differences.

4 Generating Plausible Moves

The selective simulation framework starts with generating a list of moves to simulate (i.e., at the route of the search). Below we briefly discuss four related topics concerning the list of plausible moves: the need for selectivity (in 4.1), the flexibility (in 4.2), a breadth-versus-depth decision (in 4.3), and metrics (in 4.4).

4.1 Need for Selectivity

In the ideal case a player can consider all legal moves. For example, in 9 × 9 Go, there are never too many legal moves, and a program can consider all of them [7]. Another example is poker, where the options are either Bet, Call, or Fold. The number of legal moves in Bridge is also quite manageable, so a program can consider all of them.

Other domains are not so simple. In Backgammon, for instance, it is possible to have hundreds of legal moves, particularly if one rolls small doublets such as 1-1 or 2-2. In Scrabble, the number of legal moves averages over 700. Moreover, one can imagine war games where the number of legal moves is so large that the complete list cannot even be generated. In such cases a program needs to generate plausible moves by a static analysis.

4.2 Flexibility

One valuable feature of a simulation framework is that the plausible-move selection process does not need to be particularly discriminating. A program is not relying on the plausible-move generator to determine the best play. The program simply wants the best move sorted sufficiently high that it makes a cut. This can be helpful. For example, if a program wants to select good Scrabble moves by static analysis then it should include some evaluation of access to triple word squares. But that is not necessary if the moves will be simulated. Triple word squares are neither significant enough to deny a move its place on the list of moves, nor to assure a move its place.

Plausible-move generation need not be an "all or nothing" gamble. It is possible to simulate a few moves, then look at the outcomes and decide to add a few more moves. On the basis of simulations it is possible to discover that a move that was initially discarded should be considered. The Go proverb to "move where your opponent wants to move is an example of a similar heuristic." In Scrabble, for example, a simulation may tell you where the opponent obtains big scores, and then you can generate moves that block those locations. MAVEN [12] does not actually do this, as we have not yet seen a case where it is necessary, but the possibility is attractive.

4.3 A Breadth-Versus-Depth Decision

Simulating more moves means that there is less chance of overlooking the best play. But there is a downside: more moves means that the program either uses more time, or can simulate fewer continuations. Having fewer continuations means that there is a greater chance of missing the best play because of statistical noise.

Each program faces the breadth-versus-depth decision after a time limit for the simulation has been fixed. From the time limit, the program can estimate how many trials it will be able to do in the available time, and from that it can figure how many moves it wants to spread those trials over. There is no hard and fast rule for how to do this. In general there is a tradeoff between the chance that the Nth move is best versus the chance that the best move will not be selected because of less precision. In judging this, keep in mind that the number of trials has to be quadrupled in order to halve the standard deviation.

4.4 Metrics

Each programmer should use a metric to evaluate the effectiveness of a plausible-move generator. This subsection briefly discusses two metrics. One is well known and has been described in the literature. The other was created for the MAVEN project, but has not yet been described in the literature.

A common metric is to count how often the generated moves include the moves actually selected by expert human players. This metric is frequently used in Go, and has also been used in Chess [1]. The metric has the advantage of being quick to evaluate. However, it has the disadvantage of providing no information about the quality of second-best moves.

A more robust metric is the average number of points lost. This metric is more complicated to calculate, and may be impossible to calculate, but it is more useful when it can be achieved. For example, assume the time control for a Scrabble program is set such that the program has time to simulate only 10 moves. Assume further that we fix a policy P that chooses 10 moves. During development, we can simulate 50 moves instead of 10. When the simulation selects a move that P ranks in the top 10, then the loss due to P is 0. When the simulation selects a move that P does not rank in the top 10, then the loss due to P equals the evaluation of the selected move minus the evaluation of the best move that P selected. This metric is useful when you do not have a large supply of expert games, or when the program plays better than humans.

5 Selecting a Sample of Continuations

A peculiar feature of the basic simulation framework described in Section 2 is seen in the lines 1 and 2: the selection of continuations is listed as a separate step from the generation of moves. Hence the question arises: Does the framework require that a domain's continuations are independent of the moves?

If the continuations of a domain are intricately related to the moves, then selective simulation might not be a good search engine. For example, I would speculate that simulation could not play a game such as Reversi, where the legality of continuations depends heavily on every legal move.[1] But when the domain has less dependence on the initial position, then simulation can be a better match. For example, the legality of continuations in a Go position has very little dependence on the first move. Brügmann [7] and Bouzy and Helmstetter [6] showed that simulations are surprisingly effective in Go.

Simulation is at its best when there is hidden information, because then the first move of the continuation is to select hidden information for the opponent. Such information is necessarily independent of the move choices, so the framework described above is completely correct in that case. If the domain additionally has randomization, then simulation is an almost perfect match.

When selecting continuations, the goal is to achieve convergence of the simulated results faster than simple random selection. Ideally, this should be done without biasing the results. The techniques described below fall into two broad categories. The first category attempts to reduce the variance of sampling. The techniques used in this category are: explore parallel continuations (see 5.1), adjust for non-uniformity (see 5.2), and enforce uniformity (see 5.3). The second category attempts to amplify differences by focusing on continuations in which outcomes differ (see 5.4).

5.1 Explore Parallel Continuations

The simplest approach to achieve convergence and to reduce the variance of sampling is to use parallel continuations. That is, apply each continuation after each plausible move. This is a very simple technique for controlling variance, since it means that any biases that exist in the sample are the same for all moves. For example, in Hold'em

---

[1] Of course, there are good deterministic algorithms for Reversi.

simulations, you can evaluate the Bet and Call options after each possible pair of opponent's hole cards. This way, it is impossible for the opponent's hole cards to be better for either option. Below we elaborate on reducing the variance by duplicating trials (in 5.1.1) and by synchronizing random generators (in 5.1.2).

5.1.1 Duplicate Trials in Hold'em

We can reduce variance by playing each deal twice, once with each player moving first. This simple change uses only half as many deals, but the luck of the draw is lower because the cards exactly balance. Using duplicate deals is a trick taken from bridge tournaments, and first applied to poker by Billings et al. [2].

Duplicate trials appear to be interdependent, since a duplicate experiment is actually 800 trials each of which measures the *difference* between the players. However, this does not change the statistical formulas. The observed standard deviation of the mean in duplicate trials is 0.075. Because the number of trials required to achieve a specific level of accuracy is a quadratic function, the reduction in variance from 0.251 to 0.075 is equivalent to an increase in speed of a factor of 11.

Not every such experiment will produce such drastic gains. The structural similarity of the poker players used in this experiment makes the gains more drastic, since there is a strong tendency for play to cancel across pairs of games. The degree of improvement that a program may experience in its own simulations may vary. The same comment applies to all of the experiments in this article.

5.1.2 Synchronizing Random Generators

The fundamental decision unit of poker AI is the "strategy triple". It is the triple (F, C, R) of positive real numbers such that $F + C + R = 1$. A poker AI program generates such a triple at every decision point, to represent the decision to Fold with probability F, Call (or Check) with probability C, and Raise (or Bet) with probability R.

It is natural that the probability triples of well-tuned players will have a large amount of overlap, which means that the programs will often play the same strategy. In the event that the programs play the same strategy to the end of a game then the difference between the programs is zero. When using the duplicate sampling technique (cf. 5.1.1), we can synchronize the random state so that the same sequence of numbers is generated on the duplicate plays of a deal. This reduces the variance of a duplicate sample from 0.075 to 0.048. (Synchronizing random generators makes no difference without also duplicating trials.) This reduction in variance is equivalent to a speedup of 2.4 times.

5.2 Adjust for Non-uniformity

Sometimes it is easier to remove the impact of non-uniformity after it has happened. For example, in a poker simulation a player must have some notion about the distribution of the opponent's hole cards. Rather than generating the hole cards in the assumed distribution, the player can generate a uniform distribution, and then weight the outcomes according to the presumed distribution. This works well in poker when the opponent's hole cards are easily enumerated, as they are in Hold'em. It would not work so well in 5-card draw, where the opponent has one of C(47, 5) possible hands.[2]

One downside to adjusting after the fact is a loss of information. We take a simplified example. Assume that the domain has two semantic categories that should occur in ratio 9 to 1. If the program makes a sample containing 1000 of each and scales them down to 9 to 1 ratios, then the effective sample size is only 1000 + 1000/ 10 = 1100, whereas the program has taken 2000 samples. In effect, each sample that the program has taken from the minority option counts as only 1/10 of a sample. This loss can be viewed as inefficiency, or as increase in variance, or as a loss of information, but it is a bad thing in any way the program looks at it.

For this reason, such manipulations are a poor design choice unless the programmer has no alternative. But there is one situation where the manipulations make a lot of sense: namely, if a program has simulated a sample under one distribution, and then the program wishes a simulation under a different distribution. For example, in Scrabble the program can do simulations assuming that the opponent's tiles are uniformly random. But then the program might conclude on the basis of the opponent's last turn that the opponent was unlikely to have kept the Q. It would cost a lot to do a new simulation under this hypothesis. But it is inexpensive to reweigh the outcomes under the new distribution. If necessary, after reweighing the program can add trials to categories that are very important in the new distribution and that did not occur often under the original. Below we give an example of class weighting in Hold'em.

5.2.1 Class Weighting in Hold'em

The goal of class weighting is to eliminate the variance that arises from unequal distributions of hole cards. Though there are C(52,2) = 1326 different pairs of hole cards, there are only 169 essentially different pairs after symmetries are accounted for. The 169 equivalence classes are as follows:

1. 13 hands consist of a pair of cards of identical rank ("Pocket pairs").

2. C(13,2) = 78 hands have different rank and the same suit ("Suited").

3. C(13,2) = 78 hands have different rank and different suits ("Off-suit").

The number of symmetries of each hand is as follows:

1. Pocket pairs occur with weight C(4,2) = 6.

2. Suited hands occur with weight C(4,1) = 4.

3. Off-suit hands occur with weight P(4,3) = 12.

The implementation of class weighting is to recombine the samples according to the true frequencies of each class, as follows:

    Sum = Count = 0;
    for (each equivalence class) {
        if (there are trials in this class) {

---

[2] Categorize hole cards at a higher level of aggregation (e.g., ace-high, two-pairs, ...) to apply this method to 5-card draw.

```
        v = total in this class / number of trials in this class;
        Sum += v * equivalence class weight;
        Count += equivalence class weight;
    }
}
Mean = Sum / Count;
```

In Hold'em, the two hole cards account for the entire difference between the hands of the players, so attacking the variance of hole card distribution is important. An interesting aspect of Hold'em is that a program can reweigh the classes based on the hole cards of either player. In my actual implementation, I apply class weighting to both positions, and then average the means.

In the Hold'em experiment, class weighting had essentially the same result as Monte Carlo. We might have expected a reduction, but this is actually not a bad result. One of the problems with class weighting is that scaling increases the variance of classes that have a relatively small sample size. Accordingly, class weighting "gives back" some of its variance reduction.

Despite the outcome of this experiment, class weighting is a valuable technique because it allows us to estimate an unbiased mean even when trials have an imbalanced distribution. That capability is important to other selective sampling techniques.

5.3 Enforce Uniformity

When a program has a strong model of the distribution of continuations, then it can sample in a way that strictly matches the known distribution. We call this enforcing uniformity. Below we provide two examples of enforcing uniformity, viz. in Scrabble (5.3.1) and in Hold'em (5.3.2).

5.3.1 Tile Distribution in Scrabble

In Scrabble, the opponent's rack can be assumed to come from a uniform distribution.[3] Maven draws the first rack randomly, and for subsequent racks it begins each rack by placing the one tile that has been most underrepresented in the sampling this far. This policy very strongly pushes the sample towards the desired distribution, but does not bias the outcome.

Table 1 provides some data that illustrates the benefits. It is taken from a MAVEN simulation that contained 2718 continuations. The table shows that blanks were expected to occur in 7.53% of all racks, and actually occurred in 7.47% of all racks, a difference of only 0.06%. For comparison purposes, if the same experiment were conducted using a pure Monte-Carlo approach with no controls on sampling, then we would have a standard deviation of 0.51%. If you look down the column labeled Difference, it is clear that the sample has much greater conformance to the expected distribution than a Monte-Carlo sample would.

---

[3] Alternatively, you can draw inferences from the opponent's recent moves and bias the distribution of tiles accordingly. This makes implementation more difficult, but does not change the essence of the idea.

The benefit of this policy is to eliminate biases that relate moves to continuations. In the case of Scrabble, an illustrative example is the following: if the sample included more Qs than expected then it would reduce the evaluation of any move that pays off to an opponent's Q.

Table 1. Data on tile distribution

Tile	Expected	Actual	Difference
?	7.53%	7.47%	0.06%
A	67.74%	67.92%	0.18%
B	7.53%	7.65%	0.13%
C	15.05%	14.94%	0.12%
D	30.11%	30.10%	0.01%
E	67.74%	67.73%	0.01%
F	15.05%	14.97%	0.08%
G	22.58%	22.59%	0.01%
H	15.05%	14.97%	0.08%
I	67.74%	67.59%	0.16%
J	7.53%	7.54%	0.02%
K	7.53%	7.51%	0.02%
L	30.11%	30.06%	0.05%
M	15.05%	14.97%	0.08%
N	45.16%	45.03%	0.13%
O	52.69%	52.61%	0.08%
P	15.05%	15.38%	0.33%
Q	7.53%	7.62%	0.09%
R	37.63%	37.93%	0.30%
S	30.11%	30.02%	0.09%
T	45.16%	45.14%	0.02%
U	30.11%	29.99%	0.12%
V	15.05%	14.94%	0.12%
W	15.05%	15.27%	0.21%
X	7.53%	7.58%	0.05%
Y	15.05%	14.97%	0.08%
Z	7.53%	7.51%	0.02%

5.3.2 Hole Cards in Hold'em

When a program has control over the distribution of trials, it is natural to ensure that every equivalence class occurs with the proper frequency. An experiment that samples every equivalence class at the predicted frequency is called a *stratified sample*. The technique is used in public-opinion polling, for example. This goes beyond class weighting, which merely scales results. Stratified sampling makes certain that every class is appropriately represented, which avoids the drawback of class weighting while achieving the same result.

The application to Hold'em has a few technical fine points that illustrate how to adapt a technique to a domain. It is natural to arrange the 1326 distinct hole cards in a sequence that will repeat as long as there are trials, but we can do slightly better. Note that the weights of the equivalence classes are 6, 4, and 12, which are all even numbers. It follows that a cycle of length 1326 / 2 = 663 suffices.

A good ordering for the elements should satisfy certain natural goals. First, the initial 169 elements should be distinct equivalence classes, so that every equivalence class is sampled at least once as soon as possible. Second, we should distribute samples such there is a balance of strong and weak hands. It would not do to process the hands in decreasing order of hand strength.

We constructed our sample by first sorting the 169 classes in order of hand strength, and then performing a single shuffle that interleaved strong and weak hands. The same sequence is used at the end of the 663 hands. The middle of the sequence was constructed by taking 1 instance of each pocket pair, and 4 instances of each non-suited non-pair, then sorting and interleaving as before. The result is that any short segment contains a balance of strong and weak hands, including pairs, suited and non-suited hands. Additionally, each equivalence class occurs within the first 169 trials.

Any time we run through the complete sequence we have an unbiased sample of the full distribution. If an experiment stops partway through a cycle, then some equivalence classes will have more than their share of trials, but in large runs we can ignore such biases. Or more simply, make the number of trials a multiple of 663.

In our implementation, only the hands of the first position are systematically sampled in this way. The hands of the second position are randomly sampled. It is possible to sample both hands systematically, but this would require cycling through a much larger and more complicated space of equivalence classes.

The standard deviation of this procedure is 0.047 small bets, versus Monte Carlo's 0.251. This is a spectacularly good result, the equivalent of a 28-fold increase in speed.

5.4 Focus Attention on Differences

Thus far we have focused on methods of controlling the variance. Now we shift attention to methods of emphasizing differences. The basic idea is to notice that two moves do not differ after certain continuations, but do differ after others. If you can characterize the continuations then you can emphasize those that matter.

In principle, you can make such discriminations experimentally. Taking the case of Hold'em poker again, we might notice that when the opponent holds Ace-Ace then he always wins. In such a case the difference between the Bet and Call options might be constant, or might have at any rate a low variance. In an analogous position, a suited 10-9 might have a huge variance because of straight and flush draws. Rather than continuing to sample the Ace-Ace continuation, it is better to sample the high-variance continuation.

In practice, you must sufficiently sample for patterns to emerge, so that the sampling does not become confused by a chance low-variance outcome. An epsilon-greedy policy may be able to avoid such problems. Normally, if a program emphasizes continuations then it risks biasing the simulation. To do this properly the program may have to adjust for non-uniformity in the sample. Below we provide some considerations why targeting high-variance samples in poker is to be advised.

5.4.1 Targeting High-Variance Samples in Poker

A different approach to reducing variance is to select the hole cards so as to target equivalence classes that contribute most to the variance of the measurement. The method is to simulate the equivalence class that leads to the greatest expected reduction in variance from one additional trial. A program can use various heuristics for deciding which equivalence class that is. The author suspects that a good heuristic is to differentiate the formula for variance with respect to the number of data points, and simulate the equivalence class that maximized the derivative.

In practice, this technique suffers from a "bootstrapping" problem: variance cannot be measured reliably using a small number of trials. In general, it is an easy matter to overcome this problem. For example, one can require that every equivalence class have at least $C \times N$ trials, where C is a small positive number.

In poker, the bootstrapping problem has a natural solution. One can systematically sample the hole cards for player 1 on even trials, and systematically sample the hole cards for player 2 on odd trials. On each trial, the hole cards for one player are systematically sampled, and the hole cards for the other player are randomly sampled. Random sampling of the hole cards for the other player guarantees a minimum level of coverage.

This technique has not been implemented and verified, so there are no results to report.

6 Time Control

At the start of a turn, a program can budget a number of iterations that it expects to complete. While the simulation is underway, the program will collect additional information, which may suggest altering the initial allocation. In this section we will consider three reasons for reducing search effort, viz. by obvious moves (in 6.1), by minimal differences (in 6.2), and by a search bound (in 6.3).

6.1 Obvious Moves

If one move stands out, then you can terminate the search. A 'difference of means' test will reveal when a move is forced. MAVEN employs a "six sigma" standard for permanently pruning moves, and if all moves are pruned then MAVEN declares that the decision is "obvious." The six-sigma standard was selected because numerical precision limitations prevent calculating one-tailed standard normal tests when sigma exceeds 6.

In practice, this terminates about 25 per cent of the simulations in a Scrabble game. As computers become faster, this fraction will rise. Each quadrupling of CPU speed is sufficient to halve the sigma of a simulation, so in a few years only genuinely close decisions will simulate for the full time limit.

6.2 Minimal Difference

It may also happen that two moves are so close together that it is impossible to distinguish them. This often happens in Scrabble. For example, if you can play BOARD or BROAD in the same spot and for the same score, then there might not be a hair's difference between them. Actually, in this case the variance of the difference will also be very small, so the search may terminate because of a difference of means test.

There may be other reasons for having minimal differences. It may be that the position transposes after two plies, so the future is the same, or it may be that both moves pay off to the same continuations.

6.3 Search Bound

Near the end of a simulation, the leading move may be uncatchable by any other move within the available number of iterations. That move may have a small advantage, but if the program has only a small number of continuations remaining to simulate, then the advantage may be insurmountable within the time limit.

7 Selecting a Sample Move

The inner loop of a simulation plays out a continuation after a move (see line 3c in Section 2). This gives us the opportunity to select both a move and a continuation. In general, the sequence of continuations that the simulation should pursue is fixed in advance. In this section we discuss a dynamic continuation selection (in 7.1), a dynamic move selection (in 7.2) and provide an example of dynamic move selection (in 7.3).

7.1 Dynamic Continuation Selection

We mention as a possibility that the continuations can be selected dynamically instead of following a fixed sequence. We do not know of any program that actually operates this way, and it is not immediately clear how to make this work. Many of our other ideas depend on maintaining comparability across continuations, which is difficult when the order of continuations varies across moves.

7.2 Dynamic Move Selection

Think of a simulation as having a fixed set of continuations 1, 2, 3, ... C. We also have a fixed list of moves 1, 2, 3, ... M. A simulation that runs to exhaustion needs M × C trials. But if we observe what happens in the early trials, then we can shift attention to moves that show the greatest chance of being better than the move that is currently ranked best. This heuristic focuses on moves that may have high potential of being best. In Scrabble, there are usually only a handful of good moves, and therefore the simulation runs in time that is linear in C + M.

Between each trial, the program should sort the moves according to the probability of being better than the move that is currently considered best. Let us introduce some notation to make this precise. Let the outcome of move M after continuation C be

E(M, C). Let the move ranked best be M = 0. Then one estimate of the difference between move M and move 0 is the mean of E(0, C) − E(M, C) over all continuations C for which we have outcomes for both move M and move 0. We can also compute the standard deviations of E(0, C) − E(M, C), and then rank the moves in increasing order of mean / sigma. Moves preferred by this metric are more likely to surpass the top-ranked move, so it makes sense to spend extra effort on them. MAVEN's simulations show that the vast majority of effort goes to only 2 to 3 moves, even when 100 moves are simulated.

Note that this trick works best when every continuation is feasible after every move. This is generally the case in games with hidden information, because the first move in such a game is to select hidden information for the opponent, and that selection cannot depend on our move.

Of course, we need some policies that guarantee that we make progress. Progress is guaranteed if search would ultimately force every move to be searched. For the tail-end moves, it is sufficient to use an epsilon-greedy strategy. The best move poses a different problem: the selection heuristic above never proposes that we spend any effort on the best move itself! Accordingly, MAVEN imposes a practical rule: the best move must always have at least as many trials as any move.

Thus, we should begin by simulating all of the moves for at least a few iterations apiece, so that the standard deviations of the differences E(0, C) − E(M, C) are defined.

Moreover, we need a special case to cover the possibility of two essentially identical moves, which could result in a standard deviation of zero, and then a division by zero. More generally, if two moves are sufficiently close, in the sense of having a very small standard deviation of E(0, C) − E(M, C) then it is possible for those two moves to freeze out the rest of the list (except for moves chosen by epsilon-greedy) if the two happen to be the first-ranked and second-ranked moves. MAVEN's approach is to regard such a move as "co-leaders." This means that (a) whenever MAVEN simulates one of these moves then it simulates the other, and (b) such moves are never considered to be second best.

7.3 Example of Dynamic Move Selection

Dynamic move selection is the most powerful method in this article, so an example is in order (see Table 2). The data was collected from a Scrabble game. MAVEN simulated 20 moves, using an evaluation function that estimated winning percentage.

At the end of the simulation, MAVEN had given the moves widely varying numbers of trials, as shown by the column labeled "Trials". In this position, three moves merited almost equal attention: GOOMBAH, HOMAGE, and GOMBO. Had all 20 moves been simulated for the maximum 2353 trials, there would have been 47060 trials. The actual simulation needed less than 1/5 as many: 9381 altogether. Of course, had all 20 moves been simulated for 2353 iterations then the simulation would have had a lower error rate. But only barely, imperceptibly lower, since all of the best moves received a full quota of trials.

The column labeled "Error" represents the amount of equity that was lost by terminating search. This is estimated from the standard normal scores at the point when simulation stopped. Note that the amounts are very small, totaling less than 1 game in

2000. That is, this simulation estimates that playing GOOMBAH results in a loss of equity of about 0.0005 compared to perfect play of this position. This loss arises because of statistical uncertainty concerning simulation results.

Moreover, we note that the values in the Error column are almost all approximately equal. The algorithm that distributes trials causes this. By selecting to simulate the move with the highest Error value, the algorithm drives down the highest member of this column. It keeps the simulation on a nearly optimal path to reduction in error.

There is one notable exception to the pattern. HOMAGE has an estimated Error that is 200 times greater than the other moves. How did that happen? At an earlier point in the simulation, HOMAGE appeared to be significantly worse than it does now, with an Error value in the same range as the other moves. At that point, HOMAGE went on a streak of good results that distinguished it from the other plays.[4] The large value of Error reflects those recent good results.

The simulation stopped with HOMAGE having 2248 trials compared to GOOMBAH's 2353. Given the good results that HOMAGE has achieved, it is probably the second-best move, if not the best. If simulation had continued, then HOMAGE would soon receive 2353 trials, at which point the simulation would continue by alternating trials to GOOMBAH and HOMAGE until the Error value of the lesser of the two dropped below 0.0000020, whereupon attention would go back to GOMBO and the other moves.

We note that every move has received at least 29 iterations. This simulation ran using 8 as the minimum number of trials. The fact that every move has received more than 8 trials suggests that the simulation should have admitted more moves. One possible dynamic strategy is to add new moves to the simula-

Table 2. Data for dynamic move selection

Word	P(Win)	Error	Trials
GOOMBAH	0.616	-	2353
HOMAGE	0.615	0.0004573	2248
GOMBO	0.611	0.0000020	2353
GABOON	0.609	0.0000020	713
BEHOOF	0.607	0.0000017	640
GENOA	0.594	0.0000018	185
BOGAN	0.592	0.0000019	29
BAH	0.590	0.0000012	131
BEGORAH	0.587	0.0000013	29
HOAGY	0.584	0.0000018	114
GOBAN	0.583	0.0000018	52
BOON	0.582	0.0000012	38
OHMAGE	0.581	0.0000012	38
GOBO	0.581	0.0000012	33
BEANO	0.581	0.0000019	33
GABOON	0.580	0.0000020	97
ABMHO	0.579	0.0000020	90
ABOON	0.577	0.0000013	58
BOOGY	0.577	0.0000019	71
OOH	0.57	0.0000018	76

---

[4] You can see the advantage of not pruning any moves. What if HOMAGE had been pruned before its hot streak? Dynamic move selection never permanently prunes a move. The six-sigma time control algorithm does prune moves, but only when the evidence is incontrovertible.

tion wherever every move in the simulation has, say, 20 iterations. Such a rule would dynamically scale the move list so as to use additional time to reduce the error rate at near optimal trajectories.

Finally, we note that the results of infrequently simulated moves are very uncertain. For example, OOH received only 76 trials. The standard deviation of the difference GOOMBAH − OOH is about 3.5 points. This is really poor accuracy. However, the mean of GOOMBAH − OOH is estimated as 12.1 points, which is 3.5 standard deviations. Accordingly, while we do not really know the value of OOH, we can be quite sure that GOOMBAH is a better play.

8 Evaluating the End Result

Simulated continuations eventually end, and then the end result should be evaluated. There are several approaches. Below we discuss the three most important ones, i.e., play to the end (in 8.1), heuristic evaluation (in 8.2), and compensation (in 8.3).

8.1 Play to the End

In the ideal case, a program can simulate continuations until it hits a terminal node. Then the evaluation is easy. For instance, in Scrabble, the program simply has to compare the points scored by both sides. But even here the program can have trouble. For example, in Go it can be difficult to know when the game is over. You can have seki, for example, in which the player that moves first will lose a group, so the proper play is to pass. Unfortunately, detecting seki is computationally expensive, and a better policy maybe to continue games until statically safe eyes are found. Playing to the end is the indicated method if (1) it is computationally feasible, and (2) if the game is guaranteed to end, and (3) a heuristic evaluation of middle-game positions is slower or inaccurate.

8.2 Heuristic Evaluation

Heuristic evaluation is often the best alternative, since playing to the end of the game is in most cases impractical. There is just one cautionary note. It is very important for heuristic evaluations to be comparable across all continuations considered by the program. This can be awkward in cases of early termination. For example, assume that a program has a heuristic evaluation function that measures progress towards a goal. The evaluation from one variation may be "need three more moves." Then, another branch may terminate early, resulting in the evaluation "the program wins." How do you average these outcomes? Below we provide an example of a heuristic evaluation for a non-deterministic strategy in Hold'em (see 8.2.1).

8.2.1 Accounting for Non-deterministic Strategy in Hold'em
The representation of the Hold'em strategy in terms of probability triples (Fold, Call, Bet) creates an attractive opportunity for the heuristic evaluation function. When a player decides to fold, the experimental trial is treated as though he folds 100 per cent of the time, whereas he actually folds only with some probability. In truth, the prob-

abilities selected are quite often 1.0, that is, the strategies are pure. Still, there is a percentage of mixed strategy, and it would be a shame not to exploit that.

When a mixed strategy includes a non-zero probability of folding, we can credit the opponent with the pot with that probability, and then carry on the rest of the game (which may include further non-determinism). This trick applies only to two-player games, and thus it appears that it cannot be used in a normal 10-player Hold'em game. However, generalizations of the same idea are applicable. For instance, the trick can be used when all but two players have folded. Many pots in Hold'em eventually come down to two players, so the trick would apply to some degree even in a 10-player game. Another example involves simulations where you only wish to measure the equity of a single player. Such a situation arises when simulations are used to select betting strategies, as only the equity of the bettor matters [2].

Another refinement of the same idea is to sample non-deterministically a final call. When a strategy gives both Call and Raise as an option, and the Call strategy would end the game, then we can settle the game under the Call strategy, and continue the game under the Raise strategy.

The author's experimental poker environment requires significant reorganization before this complex experiment can be carried out. It seems particularly interesting in conjunction with other methods. Consider two engines that differ only in some rare aspect of play. Simulations would not normally be able to detect such differences, because they do not occur often enough to stand out above statistical noise. But assume that the simulation used duplicate trials, with duplicate random number generators. In that case, two engines would always agree except in those cases where they differ, and the simulation would correctly find no differences in those cases. If the simulations also sampled cases that have a higher variance, then the focus would go to regions of the search space where differences were found. Finally, if the simulation accounted for non-determinism, then in case the two engines differed we would see the real effect of those differences, even if the only change were a small shift in probabilities.

It may be that two very similar programs can be tested more directly. For instance, you could create a test bed of situations where they disagree on the strategy, and then measure the disagreements. But that only works if you can characterize the differences. In many situations, particularly where automatic tuning is involved, all we know is that the engines are similar. Having a combination of techniques that *automatically* identified differences would be a huge boost to the productivity of research.

8.3 Compensation

If you can determine whether a continuation is better or worse than average, then a remarkable opportunity may be available to the program. Backgammon researchers, who have the benefit of excellent evaluation functions, developed this method. The idea is to correct for the "luck of the draw" by adjusting the outcomes of continuations for the effects of fortune. Below I will elaborate on this issue.

Compensation is important enough that I will give four examples, viz. Backgammon rollouts (in 8.3.1), Backgammon races (in 8.3.2), Scrabble simulations (in 8.3.3), and reducing variance of uncontrolled sequences (in 8.3.4). The first example is the original idea, which I believe is due to Fredrik Dahl (JELLYFISH), but does not appear to be described in the literature. Other examples are from the author's research.

8.3.1 Backgammon Rollouts

Backgammon engines use the term 'variance reduction' for the concept called compensation. In a Backgammon rollout "with variance reduction" the engine's evaluation function (which is presumed to be unbiased) is used to compensate for the effect of luck. This is a very effective compensation factor. Compensation is most beneficial in situations where the evaluation function is well tuned, but it is highly beneficial in every case. Variance-reduced rollouts are frequently two to four times as fast as Monte-Carlo rollouts.

When a Backgammon engine chooses moves in rollouts using at least a 2-ply search, then they can use compensation. Backgammon engines often count plies of search in peculiar ways, so I should describe what I mean. A 2-ply search consists of (1) generating the legal moves for the roll that a player has to play, and (2) evaluating each alternative by generating all 21 different rolls for the opponent and by averaging a static evaluation of the best opponent's reply for each roll. This accounting of search depth is consistent with conventions used in other games, though I have seen it called "1-ply" in Backgammon writing. Notice that when a Backgammon engine uses a 2-ply search, it generates an evaluation of each possible opponent's roll. This is very useful, because the next thing that happens in the rollout is that the opponent will randomly select a roll. It follows that a 2-ply (or deeper) search generates all of the information needed to estimate the extent to which the opponent's roll exceeded expectations. The sequence of amounts by which rolls exceed expectations can be summed and used to offset the outcome of the game.

8.3.2 Backgammon Races

Backgammon evaluation functions are very accurate, which is enormously helpful when trying to correct for the effects of luck. But it is not necessary to be tremendously accurate.

The basic idea is to offset the luck of the dice. That is, instead of computing the sum of the trials, we compute the sum of the trials and subtract a value that represents the quality of the rolls. We will call this term a *compensator*. A compensator will reduce variance if it satisfies the following three conditions.

1. It has mean zero when averaged over all possible continuations. This is required so that subtracting the compensator does not bias the mean.
2. It correlates well to the quality of the draw. To ensure that it can compensate for the luck of the draw.
3. It has roughly the same magnitude as wins and losses. This requirement ensures that variance will be cancelled, rather than be magnified.

A typical compensator will evaluate all possible outcomes at the point of a random choice, and determine a distribution of outcomes at that point. The actual outcome is offset by its evaluation within the list. The author believes that this is the first publication that states the sufficiency of these conditions. In this section I will present an example that shows that a program does not need to have a particularly good evaluation function in order to reduce variance. The domain is racing positions in Backgammon.

The occurrence of 6-6 or 5-5 at any time in a race is a huge swing. A program can control the frequency of such rolls early in the rollout, but beyond the first few ply there are not enough trials to fill a random distribution. Nevertheless, the program can compensate for the impact of the dice on races.

Here is how it works: a program has an evaluation function that roughly measures the effect of any roll on the race. The evaluation does not have to be sophisticated, as long as it is unbiased. That is, the program simply needs the expected sum of the evaluation to be zero and the evaluation to be positively correlated with the effect of the roll. The program might take the evaluation to be $C \times$ (pip-count $- 8.166666$), where C is a small positive parameter, and the magic number 8.166666 is chosen to make the evaluation unbiased. Every time a roll occurs, *subtract* this evaluation from the outcome of the simulated continuation. Big rolls will help a player to win the race, but they will also be offsets against the outcome.

The program can experimentally choose the parameter C to maximize the stability of rollouts. Alternatively, based on the expert rule of thumb that having a lead equal to 10 per cent of your pip count is equivalent to a 75 per-cent chance of victory, it is possible to choose C as a function of the lead in the game, which should achieve nearly the same degree of variance reduction as you would achieve using a neural-network evaluation function, but with much better speed.

8.3.3 Scrabble Simulations

In Scrabble, the only tiles that a program can force into a known distribution are the opponent's tiles. The tiles that the side-to-move will draw after its turn are a function of the tiles that the opponent receives, so they cannot be controlled independently. Even though those tiles cannot be forced to conform to a fixed distribution, they can be controlled by compensation.

MAVEN contains a table that represents the first-order impact of a tile on scoring. It contains facts such as "Holding an S increases the score by 7.75 points." Such first-order values were estimated by simulations, and the values are widely applicable. We can use these estimates to adjust the scores of variations, by offsetting the point differentials of continuations by subtracting the first-order value of the tiles drawn by both sides.

The effect of this adjustment is fairly minor for the opponent's tiles, since the tiles are controlled through selective sampling and duplicate continuations. But for the future racks of the side to move, or if simulations extend beyond two ply of lookahead, then such adjustments can reduce the variance by a substantial amount. My data is somewhat speculative, since I have not implemented this feature in MAVEN and I therefore have no direct measurements. I have indirect data from counting the tiles in a small sample of human expert games. That data suggests that the first-order quality of one's tiles explains at least 70 per cent of the variance in scoring. Accordingly, for long-range simulations that encompass many turns, I would expect a substantial reduction in variance by this adjustment. For the short-range, 2-ply simulations that MAVEN actually uses to select moves, I expect the effect to be small.

8.3.4 Reducing Variance of Uncontrolled Sequences

All of the techniques described thus far reduce variance by controlling the cards or randomness of the simulation. Such techniques are not always available. One important example is when measuring the quality of a program's play in games against Internet players. In such cases, the server distributes cards, and other players will not allow a person to control how it is done! Billings, Davidson, Schaeffer, and Szafron [4] remarked: "Since no variance reduction methods are available for online games, we generally test new algorithms for a minimum of 20,000 hands before interpreting the results." However, while it is true that a person or program cannot control the cards, there are still techniques for reducing variance.

First, one can apply class weighting. This simple technique attacks the variance caused by unequal distributions of hole cards. The drawback of class weighting (loss of information) is lessened when a lot of data is available.

But the real opportunity is compensation. In poker games, a suitable compensator is all-in equity. All-in equity is calculated by assuming that all current players will call for the rest of the game. It is easy to verify that all-in equity satisfies the conditions given above. Any expert computer program typically computes all-in equity, so there is no run-time implication of this technique.

In my implementation, each trial is compensated by the all-in equity of the hole cards only. Note that this technique is redundant when using duplicate trials, because the compensator would cancel across trials. No compensation is made for the community cards. Results should be better if compensation applies throughout the deal.

Applying this technique to a Monte-Carlo simulation produces a reduction in variance from 0.251 to 0.153. When this compensator is used in conjunction with duplicate continuations and an enforced uniform distribution of hole cards, the variance is reduced to 0.030, equivalent to a 70-fold speedup.

9 Summary

Selective search control is an effective tool in many domains of current interest. In suitable domains, selective simulation uses a large amount of CPU power to leverage a small amount of skill into a large amount of skill. The trick is to make the CPU burden manageable.

The methods described in this paper can go a long way towards reducing the CPU cost. The author suspects that some of these ideas are so obvious that they have been discovered many times, yet they do not appear to be described in the literature. It is our hope that by gathering these ideas into one source, we may eliminate the need to rediscover them in the future.

References

1. Anantharaman, T. (1997). Evaluation Tuning for Computer Chess: Linear Discriminant Methods. *ICCA Journal*, Vol. 20, No. 4, pp. 224-242.
2. Billings, D., Papp, D., Schaeffer, J., and Szafron D. (1998). Poker as an Experimental Testbed for Artificial Intelligence Research. *Proceedings of AI '98, (Canadian Society for Computational Studies in Intelligence)*.
3. Billings, D., Pena, L., Schaeffer, J., and Szafron, D. (1999). Using Probabilistic Knowledge and Simulation to Play Poker. *AAAI National Conference*, pp 697-703.

4. Billings, D., Davidson, A., Schaeffer, J., and Szafron, D. (2002). The Challenge of Poker. *Artificial Intelligence*, Vol. 134, Nos. 1-2, pp. 201-240. ISSN 0004-3702.
5. Bouzy, B. (2003). Associating domain-dependent knowledge and Monte Carlo approaches within a Go program. *Joint Conference on Information Sciences*, pp. 505-508. ISBN 0-9707890-2-5.
6. Bouzy, B. and Helmstetter, B. (2003). Monte-Carlo Go Developments. *Advances in Computer Games. Many Games, Many Challenges (ACG-10 Proceedings)* (eds. H.J. van den Herik, H. Iida, and E.A. Heinz), pp. 159-174. Kluwer Academic Publishers, Boston, MA. ISBN 1-4020-7709-2.
7. Brügmann, B. (1993). Monte Carlo Go. Available by FTP from ftp://ftp.cgl.ucsf.edu/pub/pett/go/ladder/mcgo.ps
8. Frank, I., Basin, D., and Matsubara, H., (1998). *Monte-Carlo Sampling in Games with Imperfect Information: Empirical Investigation and Analysis*. Research report of the Complex Games Lab, Ibaraki, Japan.
9. Galperin, G., and Viola, P. (1998). *Machine Learning for Prediction and Control*. Research Report of the Learning & Vision Group, Artificial Intelligence Laboratory, MIT, Cambridge, MA.
10. Ginsberg, M. (1999). GIB: Steps toward an Expert-Level Bridge-Playing Program. *Proc. IJCAI-99*, Stockholm, Sweden, pp. 584-589.
11. Gordon, S. (1994). A Comparison Between Probabilistic Search and Weighted Heuristics in a Game with Incomplete Information. AAAI Fall 1993 Symposium on Games: Playing and Learning, *AAAI Technical Press Report FS9302*, Menlo Park, CA.
12. Sheppard, B. (2002). *Towards Perfect Play of Scrabble*. Ph.D. Thesis, Universiteit Maastricht. Universitaire Pers Maastricht, Maastricht, The Netherlands. ISBN 90 5278 351 9.
13. Tesauro, G. (1995). Temporal Difference Learning and TD-Gammon. *Communications of the ACM,* March 1995, Vol. 38, No. 3. ISSN 0001-0782.

Game-Tree Search with Adaptation in Stochastic Imperfect-Information Games

Darse Billings, Aaron Davidson, Terence Schauenberg, Neil Burch,
Michael Bowling, Robert Holte, Jonathan Schaeffer, and Duane Szafron

Department of Computing Science,
University of Alberta, Edmonton, Alberta, Canada
{darse, davidson, terence, burch, bowling, holte,
jonathan, duane}@cs.ualberta.ca

Abstract. Building a high-performance poker-playing program is a challenging project. The best program to date, PSOPTI, uses game theory to solve a simplified version of the game. Although the program plays reasonably well, it is oblivious to the opponent's weaknesses and biases. Modeling the opponent to exploit predictability is critical to success at poker. This paper introduces VEXBOT, a program that uses a game-tree search algorithm to compute the expected value of each betting option, and does real-time opponent modeling to improve its evaluation function estimates. The result is a program that defeats PSOPTI convincingly, and poses a much tougher challenge for strong human players.

1 Introduction

Modeling the preferences and biases of users is an important topic in recent artificial intelligence (AI) research. For example, this type of information can be used to anticipate user program commands [1], predict web buying and access patterns [2], and automatically generate customized interfaces [3]. It is usually easy to gather a corpus of data on a user (e.g., web page accesses), but mining that data to predict future patterns (e.g., the next web page request) is challenging. Predicting human strategies in a competitive environment is even more challenging.

The game of poker has become a popular domain for exploring challenging AI problems. This has led to the development of programs that are competitive with strong human players. The current best program, PSOPTI [4], is based on approximating a game-theoretic Nash equilibrium solution. However, there remains a significant obstacle to overcome before programs can play at a world-class level: opponent modeling. As the world-class poker player used to test PSOPTI insightfully recognized [4]:

> "You have a very strong program. Once you add opponent modeling to it, it will kill everyone."

This issue has been studied in two-player perfect information games (e.g., [5,6,7]), but has not played a significant role in developing strong programs. In poker, however, opponent modeling is a critical facet of strong play. Since a player has imperfect information (does not know the opponent's cards), any information that a player can glean from the opponent's past history of play can be used to improve the quality of future decisions. Skillful opponent modeling is often the differentiating factor among world-class players.

Opponent modeling is a challenging learning problem, and there have been several attempts to apply standard machine learning techniques to poker. Recent efforts include: neural nets [8], reinforcement learning [9], and Bayesian nets [10], which have had only limited success. There are a number of key issues that make this problem difficult.

1. Learning must be rapid (within 100 games, preferably fewer). Matches with human players do not last for many thousands of hands, but the information presented over a short term can provide valuable insights into the opponent's strategy.
2. Strong players change their playing style during a session; a fixed style is predictable and exploitable.
3. There is only partial feedback on opponent decisions. When a player folds (a common scenario), their cards are not revealed. They might have had little choice with a very weak hand; or they might have made a very good play by folding a strong hand that was losing; or they might have made a mistake by folding the best hand. Moreover, understanding the betting decisions they made earlier in that game becomes a matter of speculation.

This paper presents a novel technique to automatically compute an exploitive counter-strategy in stochastic imperfect-information domains like poker. The program searches an imperfect-information game tree, consulting an opponent model at all opponent decision nodes and all leaf nodes. The most challenging aspects to the computation are determining: 1) the probability that each branch will be taken at an opponent decision node, and 2) the expected value (EV) of a leaf node. These difficulties are due to the hidden information and partial observability of the domain, and opponent models are used to estimate the unknown probabilities. As more hands are played, the opponent-modeling information used by the tree search generally becomes more accurate, thus improving the quality of the evaluations. Opponents can and will change their style during a playing session, so old data needs to be gradually phased out.

This paper makes the following contributions.

1. *Miximax* and *Miximix*, applications of the Expectimax search algorithm to stochastic imperfect information adversarial game domains.
2. Using opponent modeling to refine expected values in an imperfect-information game-tree search.
3. Abstractions for compressing the large set of observable poker situations into a small number of highly correlated classes.
4. The program VEXBOT, which convincingly defeats strong poker programs including PSOPTI, and is competitive with strong human players.

2 Texas Hold'em Poker

Texas Hold'em is generally regarded as the most strategically complex poker variant that is widely played in casinos and card clubs. It is the game used in the annual World Series of Poker to determine the world champion.

A good introduction to the rules of the game can be found in [8]. The salient points needed for this paper are that each player has two private cards (hidden from the opponent), some public cards are revealed that are shared by all players (community cards), and each player is asked to make numerous betting decisions: either *bet/raise* (increase the stakes), *check/call* (match the current wager and stay in the game), or *fold* (quit and lose all money invested thus far). A game ends when all but one player folds, or when all betting rounds are finished, in which case each player reveals their private cards and the best poker hand wins (the *showdown*). The work discussed in this paper concentrates on two-player Limit Texas Hold'em.

Computer Poker Programs

The history of computer poker programs goes back more than 30 years to the initial work by Findler [11]. Some of the mathematical foundations go back to the dawn of game theory [12,13]. Recently, most of the AI literature has been produced by the Computer Poker Research Group at the University of Alberta. Those programs—LOKI, POKI, and PSOPTI— illustrate an evolution of ideas that has taken place in the quest to build a program capable of world-class play.

- Rule-based [14]: Much of the program's knowledge is explicit in the form of expert-defined rules, formulas, and procedures.
- Simulations [15,8]: Betting decisions are determined by simulating the rest of the game. Likely card holdings are dealt to the opponents and the hand is played out. A large sample of hands is played, and the betting decision with the highest expected value is chosen.
- Game theory [4]: Two-player Texas Hold'em has a search space size of $O(10^{18})$. This was abstracted down to a structurally similar game of size $O(10^7)$. Linear programming was used to find a Nash equilibrium solution to that game (using techniques described in [16]), and the solution was then mapped back onto real poker. The resulting solution – in effect, a strategy lookup table – has the advantage of containing no expert-defined knowledge. The resulting *pseudo-optimal* poker program, PSOPTI, plays reasonably strong poker and is competitive with strong players. However, the technique is only a crude approximation of equilibrium play. Strong players can eventually find the seams in the abstraction and exploit the resulting flaws in strategy. Nevertheless, this represented a large leap forward in the abilities of poker-playing programs.

As has been seen in many other game-playing domains, progressively stronger programs have resulted from better algorithms and *less* explicit knowledge.

3 Optimal Versus Maximal Play

In the literature on game theory, a Nash equilibrium solution is often referred to as an *optimal strategy*. However, the adjective "optimal" is dangerously misleading when applied to a poker program, because there is an implication that an equilibrium strategy will perform better than any other possible solution. "Optimal" in the game theory sense has a specific technical meaning that is quite different.

A Nash equilibrium strategy is one in which no player has an incentive to deviate from the strategy, because the alternatives *could* lead to a worse result. This simply maximizes the minimum outcome (sometimes referred to as the *minimax solution* for two-player zero-sum games). This is essentially a defensive strategy that implicitly assumes the opponent is perfect in some sense (which is definitely not the case in real poker, where the opponents are highly fallible).

A Nash equilibrium player will not necessarily defeat a non-optimal opponent. For example, in the game of rock-paper-scissors, the equilibrium strategy is to select an action uniformly at random among the three choices. Using that strategy means that no one can defeat you in the long term, but it also means that you will not win, since you have an expected value of zero against *any* other strategy.

Unlike rock-paper-scissors, poker is a game in which some strategies are *dominated*, and could potentially lose to an equilibrium player. Nevertheless, even a relatively weak and simplistic strategy might break even against a Nash equilibrium opponent, or not lose by very much over the long term. There are many concrete examples of this principle, but one of the clearest demonstrations was seen in the game of Oshi-Zumo [17].

In contrast, a *maximal* player can make moves that are non-optimal (in the game-theoretic sense) when it believes that such a move has a higher expected value. The *best response* strategy is one example of a maximal player.

Consider the case of rock-paper-scissors where a opponent has played "rock" 100 times in a row. A Nash equilibrium program is completely oblivious to the other player's tendencies, and does not attempt to punish predictable play in any way. A maximal player, on the other hand, will attempt to exploit perceived patterns or biases. This always incurs some risk (the opponent *might* have been setting a trap with the intention of deviating on the 101st hand). A maximal player would normally accept this small risk, playing "paper" with a belief of positive expectation [18].

Similarly, a poker program can profitably deviate from an equilibrium strategy by observing the opponent's play and biasing its decision-making process to exploit the perceived weaknesses.

If PsOpti was based on a true Nash equilibrium solution, then no human or computer player could expect to defeat it in the long run. However, PsOpti is only an approximation of an equilibrium strategy, and it will not be feasible to compute a true Nash equilibrium solution for Texas Hold'em in the foreseeable future. There is also an important practical limitation to this approach. Since PsOpti uses a fixed strategy, and is oblivious to the opponent's strategy, a strong human player can systematically explore various options, probing for weaknesses without fear of being punished for using a highly predictable style. This kind of methodical exploration for the most effective counter-strategy is not possible against a rapidly adapting opponent.

Moreover, the key to defeating all human poker players is to exploit their highly non-optimal play. This requires a program that can observe an opponent's play and adapt to dynamically changing conditions.

4 Miximax and Miximix Search

EXPECTIMAX search is the counterpart of minimax search for domains with a stochastic element [19]. EXPECTIMAX combines the minimization and maximization nodes of minimax search with the addition of chance nodes, where a stochastic event happens (for example, a dice roll). The value of a chance node is the sum of the values of each of the children of that node, weighted by the probability of that event occurring (1/6 for each roll in the case of a die).

For perfect information stochastic games such as backgammon, an opponent decision node is treated as a normal max (or min) node. However, this cannot be done for imperfect information games like poker, because the nodes of the tree are not independent. Several opponent decision nodes belong to the same information set, and are therefore indistinguishable from each other. In poker, the information set is comprised of all the possible opponent hands, and our policy must be the same for all of those cases, since we do not know the opponent's cards. Furthermore, a player can, in general, use a *randomized mixed strategy* (a probability distribution over the possible actions), so taking the maximum (or minimum) value of the subtrees is not appropriate.

To extend EXPECTIMAX for poker, we handle all of the opponent decision nodes within a particular information set as a single node, implicitly maintaining a probability distribution over the range of possible hands they might hold. We cannot treat all possible combinations as having equal probability (for example, weak hands might be folded early and strong hands played through to the end). Imperfect information adds an extra challenge in evaluating leaf nodes in the search, since we can only estimate the relative probabilities for the opponent's possible holdings, rather than having exact values.

We have implemented two variants of EXPECTIMAX for search on poker trees, which we call MIXIMAX and MIXIMIX. These algorithms compute the expected value (EV) at decision nodes of an imperfect-information game tree by modeling them as chance nodes with probabilities based on the information known or estimated about the domain, and the specific opponent.

The algorithm performs a full-width depth-first search to the leaf nodes of the imperfect-information game tree. For two-player poker, the leaf nodes are terminal nodes that end the game — either at a showdown, or when one of the players folds. The search tree is the set of all possible betting sequences from the current state to the end of the game, over all possible outcomes of future chance nodes. At the showdown leaf nodes, the probability of winning is estimated with a heuristic evaluation function, and the resulting EV is backed-up the tree. This tree can be a fairly large (millions of nodes), but with efficient data structures and caching of intermediate calculations, it can be computed in real-time (about one second). In general, the search can be stopped at any depth, and the evaluation function used to estimate the EV of that subtree, as is done in traditional game-playing programs.

The EV calculation is used to decide which action the program should perform: bet/raise, check/call, or fold. Given the EV for each of our three possible actions, one could simply select the option with the maximum value. In that case, the tree will contain mixed nodes for the opponent's decisions and max nodes for our own decisions. Hence we call this algorithm MIXIMAX.

However, always taking the maximum EV could lead to predictable play that might be exploited by an observant opponent. Instead, we could choose to use a mixed strategy ourselves. Although we (presumably) know the randomized policy we will use, it can be viewed as both players having mixed nodes, and we call this more general algorithm MIXIMIX. (Thus MIXIMAX is a special case of MIXIMIX in which all of our own decision nodes use a *pure strategy*, choosing one action 100% of the time).

There are two unresolved issues.

1. How to determine the relative probabilities of the opponent's possible actions at each decision node. This is based on frequency counts of past actions at corresponding nodes (*i.e.*, given the same betting sequence so far).
2. How to determine the expected value of a leaf node. At fold nodes, computing the EV is easy — it is the net amount won or lost during the hand. At showdown nodes, a probability density function over the strength of the opponent's hand is used to estimate our probability of winning. This histogram is an empirical model of the opponent, based on the hands shown in corresponding (identical or similar) situations in the past.

In summary, the search tree consists of four types of nodes, each with different properties.

Chance nodes: For chance nodes in the game tree, the EV of the node is the weighted sum of the EVs of the subtrees associated with each possible outcome. In Texas Hold'em, chance outcomes correspond to the dealing of public board cards. To be perfectly precise, the probability of each possible chance outcome is dependent on the cards that each player will likely hold at that point in the game tree. However, since that is difficult or impossible to determine, we currently make the simplifying (but technically incorrect) assumption that the chance outcomes occur uniformly at random. Thus the EV of a chance node is simply the average EV over all of the expansions. Let $Pr(C_i)$ be the probability of each branch i of chance node C, and let n be the number of branches. The EV of node C is:

$$EV(C) = \sum_{1 \leq i \leq n} Pr(C_i) \times EV(C_i) \quad (1)$$

Opponent decision nodes: Let $Pr(O_i)$ be the estimated probability of each branch i (one of fold, call, or raise) at an opponent decision node O. The EV of node O is the weighted sum:

$$EV(O) = \sum_{i \epsilon \{f,c,r\}} Pr(O_i) \times EV(O_i) \qquad (2)$$

Program decision nodes: At decision node U, we can use a mixed policy as above (*Miximix*), or we can always take the maximum EV action for ourselves (*Miximax*), in which case:

$$EV(U) = \max(EV(U_f), EV(U_c), EV(U_r)) \qquad (3)$$

Leaf nodes: Let L be a leaf node, P_{win} be the probability of winning the pot, $L_{\$pot}$ be the size of the pot, and $L_{\$cost}$ be the cost of reaching the leaf node (normally half of $L_{\$pot}$). At terminal nodes resulting from a fold, P_{win} is either zero (if we folded) or one (if the opponent folded), so the EV is simply the amount won or lost during the hand. The net EV of a showdown leaf node is:

$$EV(L) = (P_{win} \times L_{\$pot}) - L_{\$cost} \qquad (4)$$

5 EV Calculation Example

For each showdown leaf node of the game tree, we store a histogram of the *hand rank* (HR, a percentile ranking between 0.0 and 1.0, broken into 20 cells with a range of 0.05 each) that the opponent has shown in previous games with that exact betting sequence. We will use 10-cell histograms in this section to simplify the explanation.

For example, suppose we are Player 1 (P1), the opponent is Player 2 (P2), and the pot contains four small bets (sb) on the final betting round. We bet (2 sb) and are faced with a raise from P2. We want to know what distribution of hands P2 would have in this particular situation. Suppose that a corresponding 10-cell histogram has relative weights of [1 1 0 0 0 0 0 4 4 0], like that shown in Figure 1. This means that based on our past experience, there is a 20% chance that P2's raise is a bluff (a hand in the HR range 0.0-0.2), and an 80% chance that P2 has a hand in the HR range 0.7-0.9 (but not higher).

The histogram for the showdown node after we re-raise and P2 calls (bRrC) will be related, probably having a shape like [0 0 0 0 0 0 0 5 5 0], because P2 will probably fold if he was bluffing, and call with all legitimate hands. The action frequency data we have on this opponent will be consistent, perhaps indicating that after we re-raise, P2 will fold 20% of the time, call 80%, and would not re-raise (because it is not profitable to do so). The probability triple of action frequencies is $Pr(F, C, R) = \{0.2, 0.8, 0.0\}$.

To decide what action to take in this situation, we compute the expected value for each choice: EV(fold), EV(call), and EV(raise). EV(fold) is easily determined from the betting history — the game will have cost us -4 small bets.

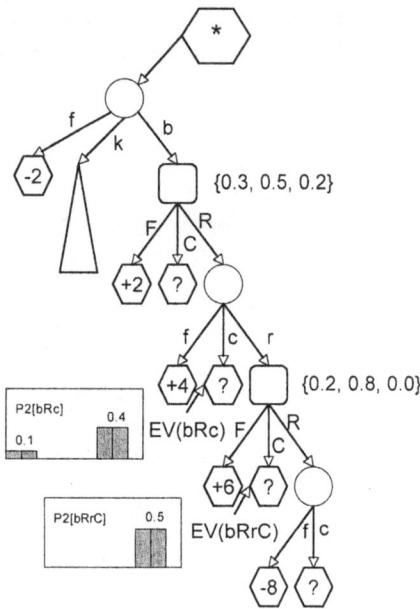

Fig. 1. Betting Tree for the EV Calculation Example

EV(call) depends on our probability of winning, which depends on the strength of our hand. If our hand rank is in the range 0.2-0.7, then we can only beat a bluff, and our chance of winning the showdown is $Pr(win|bRc) = 0.20$. Since the final pot will contain 12 sb, of which we have contributed 6 sb, the net EV(call) = -3.6 sb. Therefore, we would not fold a hand in the range 0.2-0.7, because we expect to lose less in the long run by calling (0.4 sb less).

If our hand rank is only HR = 0.1 (we were bluffing), then EV(call) = -4.8 sb, and we would be better off folding. If hand rank is HR = 0.8, then we can also beat half of P2's legitimate hands, yielding an expected profit of EV(call) = +1.2 sb. Table 1 gives the EV for calling with selected hand ranks in the range 0.7-0.9.

To calculate the expected value for re-raising, we must compute the weighted average of all cases in that subtree, namely: bRrF, bRrC, bRrRf, and bRrRc. Since the probability assigned to a P2 re-raise is zero, the two latter cases will not

Table 1. Expected Values for call or raise for selected hand ranks

HR	Pr(w|c)	EV(c)	Pr(w|rC)	EV(r)	Action
0.70	0.2	-3.6	0.0	-5.2	call
0.75	0.4	-1.2	0.25	-2.0	call
0.80	0.6	+1.2	0.5	+1.2	c or r
0.85	0.8	+3.6	0.75	+4.4	raise
0.90	1.0	+6.0	1.0	+7.6	raise

affect the overall EV and can be ignored. The share from bRrF is $0.2 \times 6 = +1.2$ sb, and is independent of our hand strength. The probability of winning after bRrC is determined by the histogram for that case, as before. Thus, in this example EV(raise) $= 1.2 + 0.8 \times (16 * Pr(win|rC) - 8)$, as shown in Table 1 for the same selected hand ranks.

As a consequence of this analysis, if we are playing a strictly maximizing strategy, we would decide to fold if our hand is weaker than HR $= 0.167$, call if it is in the range 0.167 to 0.80, and re-raise if it is stronger than HR $= 0.80$. Computing the viability of a possible bluff re-raise is done similarly.

6 Abstractions for Opponent Modeling

After each hand is played against a particular opponent, the observations made during that hand are used to update our opponent model. The action decisions made by the opponent are used to update the betting frequencies corresponding to the sequence of actions during the hand. When showdowns occur, the hand rank (HR) shown by the opponent is used to update a leaf node histogram, as illustrated in the previous section.

The *context tree* is an explicit representation of the imperfect-information game tree, having the same skeletal structure with respect to decision nodes. Chance nodes in the tree are represented implicitly (all possible chance outcomes are accounted for during the EV calculation).

A leaf node of the context tree corresponds to all of the leaves of the game tree with the same betting sequence (regardless of the preceding chance nodes). Associated with this is an efficient data structure for maintaining the empirically observed action frequencies and showdown histograms for the opponent. For this we use a trie, based on the natural prefix structure of related betting sequences. Hash tables are used for low-overhead indexing.

6.1 Motivation for Multiple Abstractions

The MIXIMAX and MIXIMIX search algorithms perform the type of mathematical computation that underlies a theoretically correct decision procedure for poker. For the game of two-player Limit Texas Hold'em, there are $9^4 = 6561$ showdown nodes for each player, or 13122 leaf-level histograms to be maintained and considered. This fine level of granularity is desirable for distinguishing different contexts and ensuring a high correlation within each class of observations.

However, having so many distinct contexts also means that most betting sequences occur relatively rarely. As a result, many thousands of games may be required before enough data is collected to ensure reliable conclusions and effective learning. Moreover, by the time a sufficient number of observations have been made, the information may no longer be current.

This formulation alone is not adequate for practical poker. It is common for top human players to radically change their style of play many times over the course of a match. A worthy adversary will constantly use deception to disguise the strength of their hand, mask their intentions, and try to confuse our model of their overall strategy. To be effective, we need to accumulate knowledge very quickly, and have a preference toward more recent observations. Ideally, we would like to begin applying our experience (to some degree) immediately, and be basing decisions primarily on what we have learned over a scope of dozens or hundreds of recent hands, rather than many thousands. This must be an ongoing process, since we may need to keep up with a rapidly changing opponent.

Theoretically, this is a more challenging learning task than most of the problems studied in the machine learning and artificial intelligence literature. Unlike most Markov decision process (MDP) problems, we are not trying to determine a static property of the domain, but rather the dynamic characteristics of an adversarial opponent, where historical perspective is essential.

In order to give a preference toward more recent data, we gradually "forget" old observations using exponential history decay functions. Each time an observation is made in a given context, the previously accumulated data is diminished by a history decay factor, h, and the new data point is then added. Thus for $h = 0.95$, the most recent event accounts for 5% of the total weight, the last $1/(1 - h) = 20$ observations account for $(1 - 1/e) = 0.63$ of the total, and so on.

6.2 Abstraction

In order to learn faster and base our inferences on more observations, we would like to combine contexts that we expect to have a high mutual correlation. This allows us to generalize the observations we have made, and apply that knowledge to other related situations. There are many possible ways of accomplishing these abstractions, and we will address only a few basic techniques.

An important consideration is how to handle the *zero frequency problem*, when there has yet to be any observations for a given context; and more generally, how to initialize the trie with good default data. Early versions of the system employed somewhat simplistic defaults, which resulted in rather unbalanced play early in a match. More recent implementations use default data based on rational play for both players, derived in a manner analogous to Nash equilibrium strategies.

The finest level of granularity is the context tree itself, where every possible betting sequence is distinct, and a different histogram is used for each. The opponent action frequencies are determined from the number of times each action was chosen at each decision node (again using a history decay factor to favour recent events). Unfortunately, having little data in each class will result in unreliable inferences.

One coarse-grained abstraction groups all betting sequences where the opponent made an equal number of bets and raises throughout the hand, ignoring what stage of the hand they were made. A finer-grained version of the same idea maintains an ordered pair for the number of bets and raises by each player.

However, testing reveals that an even courser-grained abstraction may be desirable. Summing the total number of raises by both players (no longer distinguishing which player initiated the action) yields only nine distinct classes. Despite the crudeness of this abstraction, the favorable effects of grouping the data is often more important than the lower expected correlations between those lines of play.

Another similar type of coarse-grained abstraction considers only the final size of the pot, adjusting the resolution (*i.e.,* the range of pot sizes) to provide whatever number of abstraction classes is desired.

An abstraction system can be hierarchical, in which case we also need to consider how much weight should be assigned to each tier of abstraction. This is based on the number of actual observations covered at each level, striving for an effective balance, which will vary depending on the opponent.

Our method of combining different abstraction classes is based on an exponential mixing parameter (say $m = 0.95$) as follows. Let the lowest-level context tree (no abstraction) be called A0, a fine-grained abstraction be called A1, a cruder amalgam of those classes be called A2, and the broadest classification level be called A3. Suppose the showdown situation in question has five data points that match the context exactly, in A0. This data is given a weight of $(1 - m^5) = 0.23$ of the total. If the next level of abstraction, A1, has 20 data points (including those from A0), it is assigned $(1 - m^{20}) = 0.64$ of the remaining weight, or about 50% of the total. The next abstraction level might cover 75 data points, and be given $(1 - m^{75}) = 0.98$ of the remainder, or 26% of the total. The small remaining weight is given to the crudest level of abstraction. Thus all levels contribute to the overall profile, depending on how relevant each is to the current situation.

7 Experiments

To evaluate the strength of VEXBOT, we conducted both computer *vs.* human experiments, and a round-robin tournament of computer *vs.* computer matches.

The field of computer opponents consisted of the following six programs.

1) SPARBOT, the publicly available version of PSOPTI-4 [1], which surpassed all previous programs for two-player Limit Hold'em by a large margin [4].

2) POKI, a formula-based program that incorporates opponent modeling to adjust its hand evaluation. Although POKI is the strongest known program for the ten-player game, it was not designed to play the two-player game, and thus does not play that variation very well [8].

3) HOBBYBOT, a slowly adapting program written by a hobbyist, specifically designed to exploit POKI's flaws in the two-player game.

4) JAGBOT, a simple static formula-based program that plays a rational, but unadaptive game.

5) ALWAYS CALL and 6) ALWAYS RAISE, extreme cases of weak exploitable players, included as a simple benchmark.

---

[1] Available at www.cs.ualberta.ca/ games/.

Table 2. Computer vs. computer matches (small bets per hand)

Program	Vexbot	Sparbot	Hobbot	Poki	Jagbot	A.Call	A.Raise
Vexbot		+0.052	+0.349	+0.601	+0.477	+1.042	+2.983
Sparbot	-0.052		+0.033	+0.093	+0.059	+0.474	+1.354
Hobbybot	-0.349	-0.033		+0.287	+0.099	+0.044	+0.463
Poki	-0.601	-0.093	-0.287		+0.149	+0.510	+2.139
Jagbot	-0.477	-0.059	-0.099	-0.149		+0.597	+1.599
Always Call	-1.042	-0.474	-0.044	-0.510	-0.597		=0.000
Always Raise	-2.983	-1.354	-0.463	-2.139	-1.599	=0.000	

Table 3. VEXBOT vs. Human matches

Num	Rating	sb/h	Hands
1	Expert	-0.022	3739
2	Intermediate	+0.136	1507
3	Intermediate	+0.440	821
4	Intermediate	+0.371	773

The results of the computer vs. computer matches are presented in Table 2. Each match consisted of at least 40,000 hands of poker. The outcomes are statistically significant, with a standard deviation of approximately ±0.03 sb/hand.

VEXBOT won every match it played, and had the largest margin of victory over each opponent. VEXBOT approaches the theoretical maximum exploitation against ALWAYS CALL and ALWAYS RAISE. No other programs came close to this level, despite those opponents being perfectly predictable.

Against SPARBOT, the strongest previous program for the two-player game, VEXBOT was able to find and exploit flaws in the pseudo-optimal strategy. The learning phase was much longer against SPARBOT than any other program, typically requiring several thousand hands. However, once an effective counter-strategy is discovered, VEXBOT will continue to win at that rate or higher, due to the oblivious nature of the game-theoretic player.

The testing against humans (Table 3) involves a smaller number of trials, and should therefore be taken only as anecdotal evidence (the outcomes being largely dominated by short-term swings in luck). Most humans players available for testing did not have the patience to play a statistically significant number of hands (especially when frustrated by losing). However, it is safe to say that VEXBOT easily exploited weaker players, and was compotitive against expert level players. The results also consistently showed a marked increase in its win rate after the first 200-400 hands of the match, presumably due to the opponent-specific modeling coming into effect.

A more recent implementation of the *Miximax* algorithm was able to improve considerably on the VEXBOT results against computer opponents. That version defeated SPARBOT by +0.145 small bets per hand — three times the win rate of the previous version. [2] Moreover, its win rate against SPARBOT is comparable to that of the most successful human player (the first author), and is more than three times the win rate achieved by the world-class player cited in [4].

Revised versions of both programs competed in the 2003 Computer Olympiad, with VEXBOT again dominating SPARBOT, winning the gold and silver medals respectively [20].

8 Conclusions and Future Work

Limit poker is primarily a game of mathematics and opponent modeling. We have built a program that "does the math" in order to make its decisions. As a result, many sophisticated poker strategies emerge without any explicit encoding of expert knowledge. The adaptive and exploitive nature of the program produces a much more dangerous opponent than is possible with a purely game-theoretic approach.

One limitation not yet adequately addressed is the need for effective defaults, to ensure that the program does not lose too much while learning about a new (unknown) opponent at the beginning of a match. If good default data is not easily derivable by direct means, there are several ways that existing programs can be combined to form hybrids that are less exploitable than any of the component programs in isolation.

With a smoothly adapting program and a good starting point, it may be possible to use self-play matches and automated machine learning to refine continuously the default data. In principle, that process could eventually approach a game-theoretic near-optimal default that is much closer to a true Nash equilibrium strategy than has been obtained to date.

The performance of VEXBOT can be improved further in numerous ways. While we believe that the modeling framework is theoretically sound, the parameter settings for the program could be improved considerably. Beyond that, there is much room for improving the context tree abstractions, to obtain higher correlations among grouped sequences.

Only the two-player variant has been studied so far. Generalization of these techniques to handle the multi-player game should be more straightforward than with other approaches, such as those using approximations for game-theoretic solutions.

Refinements to the architecture and algorithms described in this paper will undoubtedly produce increasingly strong computer players. It is our belief that these programs will have something to teach all human poker players, and that they will eventually surpass all human players in overall skill.

Acknowledgements. This research was supported by grants from the Natural Sciences and Engineering Research Council of Canada (NSERC), Alberta's

---

[2] The improved version has not been described in detail in this paper because it is still in the process of being tested against quality human opposition.

Informatics Circle of Research Excellence (iCORE), and by the Alberta Ingenuity Center for Machine Learning (AICML). The first author was supported in part by an Izaak Walton Killam Memorial Scholarship and by the Province of Alberta's Ralph Steinhauer Award of Distinction.

References

1. Horvitz, E., Breese, L., Heckerman, D., Hovel, D., Rommeke, K.: The Lumiere project: Bayesian user modeling for inferring the goals and needs of software users. In: UAI. (1998) 256–265
2. Brusilovsky, P., Corbett, A., de Rosis, F., eds.: User Modeling 2003. Springer-Verlag (2003)
3. Weld, D., Anderson, C., Domingos, P., Etzioni, O., Lau, T., Gajos, K., Wolfman, S.: Automatically personalizing user interfaces. In: IJCAI. (2003) 1613–1619
4. Billings, D., Burch, N., Davidson, A., Holte, R., Schaeffer, J., Schauenberg, T., Szafron, D.: Approximating game-theoretic optimal strategies for full-scale poker. In: IJCAI. (2003) 661–668
5. Jansen, P.: Using Knowledge about the Opponent in Game-Tree Search. PhD thesis, Computer Science, Carnegie-Mellon University (1992)
6. Carmel, D., Markovitch, S.: Opponent modeling in adversary search. In: AAAI. (1996) 120–125
7. Iida, H., Uiterwijk, J.W.H.M., van den Herik, H.J., Herschberg, I.S.: Potential applications of opponent-model search. ICCA Journal 16 (1993) 201–208
8. Billings, D., Davidson, A., Schaeffer, J., Szafron, D.: The challenge of poker. Artificial Intelligence 134 (2002) 201–240
9. Dahl, F.: A reinforcement learning algorithm to simplified two-player Texas Hold'em poker. In: ECML. (2001) 85–96
10. Korb, K., Nicholson, A., Jitnah, N.: Bayesian poker. In: UAI. (1999) 343–350
11. Findler, N.: Studies in machine cognition using the game of poker. CACM 20 (1977) 230–245
12. von Neumann, J., Morgenstern, O.: The Theory of Games and Economic Behavior. Princeton University Press (1944)
13. Kuhn, H.W.: A simplified two-person poker. Contributions to the Theory of Games 1 (1950) 97–103
14. Billings, D., Papp, D., Schaeffer, J., Szafron, D.: Opponent modeling in poker. In: AAAI. (1998) 493–499
15. Billings, D., Peña, L., Schaeffer, J., Szafron, D.: Using probabilistic knowledge and simulation to play poker. In: AAAI. (1999) 697–703
16. Koller, D., Pfeffer, A.: Representations and solutions for game-theoretic problems. Artificial Intelligence (1997) 167–215
17. Buro, M.: Solving the oshi-zumo game. In: Advances in Computer Games 10. (2004) 361–366
18. Billings, D.: The first international RoShamBo programming competition. International Computer Games Association Journal 23 (2000) 3–8, 42–50
19. Russell, S., Norvig, P.: Artificial Intelligence: A Modern Approach. Prentice Hall (2003)
20. Billings, D.: Vexbot wins poker tournament. International Computer Games Association Journal 26 (2003) 281

Rediscovering *-MINIMAX Search

Thomas Hauk, Michael Buro, and Jonathan Schaeffer

Department of Computing Science,
University of Alberta, Edmonton, Alberta, Canada
{hauk, mburo, jonathan}@cs.ualberta.ca

Abstract. The games research community has devoted little effort to investigating search techniques for stochastic domains. The predominant method used in these domains is based on statistical sampling. When full search is required, EXPECTIMAX is often the algorithm of choice. However, EXPECTIMAX is a full-width search algorithm. A class of algorithms were developed by Bruce Ballard to improve on EXPECTIMAX's runtime. They allow for cutoffs in trees with chance nodes similar to how ALPHA-BETA allows for cutoffs in MINIMAX trees. These algorithms were published in 1983—and then apparently forgotten. This paper "rediscovers" Ballard's *-MINIMAX algorithms (STAR1 and STAR2).

1 Introduction

Games can be classified as requiring skill, luck, or a combination of both. There are many games which involve both skill and chance, but often simple games of chance do not involve much strategy. For example, chess is clearly a game of skill, but luck only factors into the equation when we hope that our opponent makes a mistake. On the other hand, games of chance like roulette offer little opportunity to use skill, besides, perhaps, knowing when to quit. Games that involve chance usually involve dice or cards. Competitive card games combine skill and chance by requiring players to use strategic thinking, while they manage the uncertainty involved; since card decks are shuffled, the game's outcome is not certain. Since a player's cards tend to be hidden in card games, each player will have *imperfect information* about the game state. There are few *perfect information* games which blend skill and chance – games where nothing is hidden, yet nothing is certain. Backgammon is one such game.

For games where chance is a critical component of the game, statistical sampling has been often used to factor the random element out of the search. The method involves repeated trials where the outcome of the chance events are randomly decided before the search begins, and then search is run normally on the resulting tree. Since each chance event has one successor, they just become intermediary nodes in the tree. For example, for games that involve dice, the chance events can be determined in advance or on-demand when a chance node is met in the tree. To get a good statistical sample, the number of trials must be high enough to approximate the true distribution (for backgammon, this is often in the tens or hundreds of thousands of trials). Although the method is popular (it is easy to implement), it is not without its limitations (see, for example, [3]). The technique has been successfully used in backgammon (for post-mortem roll-outs [8]), bridge (for the play of the hand [3]), poker (computing expected values for betting decisions [2]), and Scrabble (anticipating opponent responses [7]).

Perfect information games without chance benefit from deep search. A natural question is whether there is a counterpart to the ALPHA-BETA algorithm for perfect information games with chance. Backgammon programs typically do almost no (or limited) search as part of their move decision process [4]. Would these programs benefit from an ALPHA-BETA-like algorithm?

Bruce Ballard enhanced ALPHA-BETA search to traverse trees consisting of min, max and chance nodes [1]. The ideas centered around a family of algorithms that he called *-MINIMAX, with STAR1 and STAR2 being the main variants. His work was published in 1983, but has received very little attention. Indeed, it appears that his work has been all but forgotten (there are few references to it, and the paper is not online).

The main contributions in this paper are as follows.

1. Re-introducing Ballard's ideas to the research community.
2. Negamax pseudo-code for STAR2.
3. Updating the algorithms to reflect non-uniform probabilities for chance nodes.
4. An understanding of the relative strengths of STAR1 and STAR2.
5. Updating the algorithms to take advantage of 20 years of ALPHA-BETA search enhancements.

This paper "re-discovers" Ballard's work. The algorithms are important enough that they need to be updated and re-introduced to the games research community. This work shows that STAR1 and STAR2 are worthy of consideration as a full-width search-based approach to dealing with stochastic domains. A companion work presents experimental results for these algorithms for the game of backgammon [5].

2 Search in Stochastic Domains

Most games extensively studied in the AI literature are non-stochastic games of perfect information. With the addition of a random element (roll of the dice; dealing of cards), chance events are introduced into the game tree. Hence, we need to add a new kind of node to our game tree: a chance node. A chance node will have successor states like minimization (min) or maximization (max) nodes, but each successor is associated with a probability of that state being reached. For example, a chance node in a game involving a single dice would have six successor nodes below it, each representing the state of the game after one of the possible rolls of the dice, and each reachable with the same probability of $\frac{1}{6}$.

The element of chance completely changes the landscape that search algorithms work on. In games of chance, we cannot say for certain what set of legal moves the opponent will have available on their turn, so we cannot be certain to avoid certain outcomes. The introduction of chance nodes means that we can no longer directly apply standard ALPHA-BETA to games of chance. Chance nodes act as intermediaries, by specifying the state the game will take before a choice of actions becomes available. Before we can search trees with chance nodes, we have to figure out how to handle them.

2.1 EXPECTIMAX

The baseline algorithm for trees with chance nodes (analogous to MINIMAX for games without chance nodes) is the EXPECTIMAX algorithm [6]. Just like MINIMAX, EXPECTIMAX is a full-width, brute-force algorithm. EXPECTIMAX behaves exactly like MINIMAX except it adds an extra component for dealing with chance nodes (in addition to min and max nodes). At chance nodes, the heuristic value of the node (or EXPECTIMAX value) is equal to the weighted sum of the heuristic values of its successors. For a state s, its EXPECTIMAX value is calculated with the function:

$$Expectimax(s) = \sum_i P(child_i) \times U(child_i)$$

where $child_i$ represents the ith child of s, $P(c)$ is the probability that state c will be reached, and $U(c)$ is the utility of reaching state c. Evaluating a chance node in this way is directly analogous to finding the utility of a state in a Markov Decision Process.

EXPECTIMAX is given in Figure 1, which makes use of the following functions:

1. `terminal()` that returns true if and only if a given state is terminal,
2. `evaluate()` that returns the heuristic evaluation of a state,
3. `numChanceEvents()` to specify how many different values the chance event can take,
4. `applyChanceEvent()` to apply the chance event to the state,
5. `eventProb()` to determine the probability of the chance event taking that value, and
6. `search()` calls the appropriate search function depending on the type of node that follows the chance node (min, max, or chance).

For games where chance nodes alternate with player turns, we can use MINIMAX for searching, with the modification that MINIMAX's recursive call uses

```
float Expectimax(Board board, int depth, int is_max_node) {
    if(terminal(board) || depth == 0) return (evaluate(board));
    N = numChanceEvents(board);
    for(sum = 0, i = 1; i <= N; i++) {
        succ = applyChanceEvent(board,i);
        sum += eventProb(board,i) * search(succ, depth-1, is_max_node);
    }
    return (sum);
}
```

Fig. 1. The EXPECTIMAX algorithm

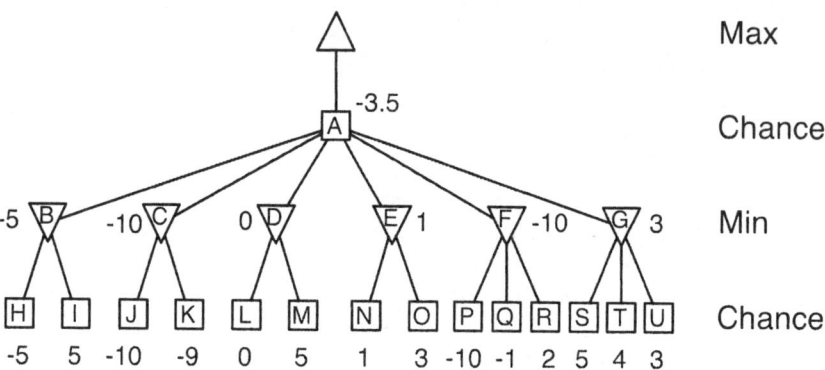

Fig. 2. An EXPECTIMAX tree

EXPECTIMAX instead of itself. We also use floating point numbers instead of integers for return values, since probabilities are real numbers and the sum may have a fractional component.

Figure 2 illustrates how EXPECTIMAX works. If we assume that each of the 6 branches at the chance node have the same probability (such as would be the case for a single dice), then each child contributes 1/6th of the value of the node: $value = -5 \times \frac{1}{6} + -10 \times \frac{1}{6} + 0 \times \frac{1}{6} + 1 \times \frac{1}{6} + -10 \times \frac{1}{6} + 3 \times \frac{1}{6} = -3.5$.

Assume that the search tree has a fixed branching factor B, and a search is being conducted to depth D (where a depth, or ply, consists of a min, max, or chance node). While the worst-case time complexity for MINIMAX is $O(B^D)$, the worst case for EXPECTIMAX (for trees with alternating levels of chance nodes) is $O(B \times B^{\frac{D-1}{2}} \times N^{\frac{D-1}{2}})$ (for D odd), where N is the branching factor at chance nodes (for example, in backgammon's case, $N = 21$ since there are twenty-one distinct rolls). As an example of the explosive effect of chance nodes even on shallow searches, there would be approximately 3.5 million nodes in a 3-ply search of an arbitrary backgammon position. If an evaluation function took 0.05 ms to complete (about the speed of GNU backgammon's neural network on a modern computer), then a 3-ply search would take about 3 minutes to complete, a 4-ply search would take about 21 hours, and a 5-ply search would be roughly a year.

2.2 *-MINIMAX

Bruce Ballard was the first to develop a technique, called *-MINIMAX, for enabling chance node cutoffs [1]. He proposed two versions of his algorithm, called STAR1 and STAR2. He also further refined the second algorithm to handle more general cases and have parameters to control functionality, and called the new version STAR2.5. All the experiments that Ballard performed were in a rather abstract domain. He did not use a real domain to validate his results.

The basic idea of EXPECTIMAX is sound, but slow. Just as we can derive a strategy for obtaining cutoffs in MINIMAX to obtain ALPHA-BETA, so too can we derive a strategy for obtaining cutoffs in EXPECTIMAX. Since there are three different types of nodes in a game tree for games with chance, there are three cases we need to consider for cutoffs. Since max and min nodes work the same way in trees with chance nodes as they do in trees without chance nodes, we get the cutoff strategies for those nodes for "free". All we need to concern ourselves with are chance nodes. If we pass alpha and beta values to chance nodes as we do min and max nodes, and we pass alpha and beta values from chance nodes to min and max nodes, all that is left to consider is exactly what values we can pass, and how they will be used.

In the first case, chance nodes can have a search window just like min and max nodes, using alpha and beta values to determine if further search below the node is relevant. However, these alpha and beta values cannot be used just like they are used in min or max nodes, because the child of a chance node cannot be chosen deterministically (unless there is only one child). We can obtain a cutoff, however, if the EXPECTIMAX value of a chance node falls outside the alpha-beta window. The problem is that we cannot know the exact EXPECTIMAX value of a chance node before we search all of its children. However, if we know bounds on the range of values leaf nodes can take (called L and U, respectively, using Ballard's notation), we can determine bounds on the value of a chance node based on the worst-case conditions for the alpha and beta values.

If we have reached the ith successor of a chance node, after having searched the first $i-1$ successors and obtained their backed-up values (which we will call $V_1 \ldots V_{i-1}$), then we can determine a bound for the value of the chance node. In the worst case, all the unsearched children will have a value of L, and in the best case, all the unsearched children will have a value of U. Therefore, the lower bound of a chance node's value, where V_i represents the true value of successor i and there are N different equally-likely chance events, is equal to

$$\frac{1}{N}((V_1 + \ldots + V_{i-1}) + V_i + L \times (N - i))$$

and the upper bound is equal to

$$\frac{1}{N}((V_1 + \ldots + V_{i-1}) + V_i + U \times (N - i))$$

These bounds determine the range in which the EXPECTIMAX value for a chance node must lie. We can use this range to generate cutoffs. Recall that the chance node itself was passed alpha and beta values. We can cut off our search if the lower bound of the EXPECTIMAX range for the chance node ever exceeds or equals beta,

$$\frac{1}{N}((V_1 + \ldots + V_{i-1}) + V_i + L \times (N - i)) \geq beta$$

or the upper bound is ever less than or equal to alpha,

$$\frac{1}{N}((V_1 + \ldots + V_{i-1}) + V_i + U \times (N - i)) \leq alpha \qquad (1)$$

where $(V_1 + \ldots + V_{i-1})$ are the accurate values for the first $i-1$ children of a node, V_i is the value for current node being searched, and $(U \times (N-i))$ and $(L \times (N-i))$ represent the best/worst-case assumptions for the values of the remaining nodes. In either equation, we can solve for V_i, and use the value as either an alpha or a beta value for the next child.

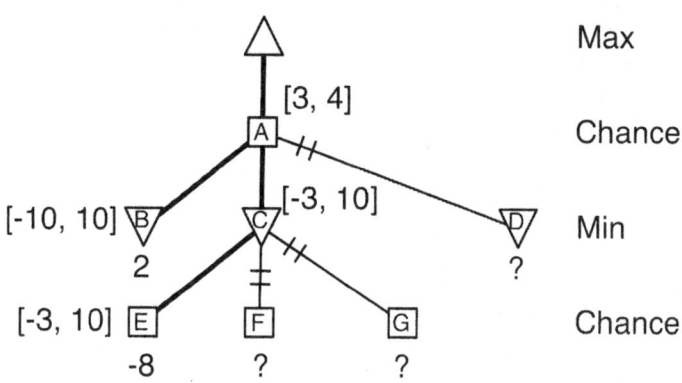

Fig. 3. Fragment of a *-MINIMAX tree

Take the following example shown in Figure 3, where heuristic values range from $L = -10$ to $U = 10$, inclusive. The top-most chance node, A, is entered with a window of alpha=3 and beta=4 (we will write this as [3,4]). Because we have not searched any of its children yet, we know its value lies in the range [-10,10], and the alpha and beta values for the first child are equal to $\frac{1}{3}(3 \times L) = L$ and $\frac{1}{3}(3 \times U) = U$, which is also [-10,10]. Assume that the first child (B) is searched and a value of 2 is returned. We now know the EXPECTIMAX range for the chance node is between $\frac{1}{3}(2+2 \times L) = \frac{1}{3}(-18) = -6$ and $\frac{1}{3}(2+2 \times U) = \frac{1}{3}(22) = 7\frac{1}{3}$. Since -6 is not greater than 4 and $7\frac{1}{3}$ is not less than 3, this child did not create a cutoff. Before we search the next child, we need to recalculate the alpha and beta values we want to pass down to it: $\frac{1}{3}(2 + V_i + (1) \times L) \geq beta \Rightarrow V_i \geq 20$, and $\frac{1}{3}(2 + V_i + (1) \times U) \leq alpha \Rightarrow V_i \leq -3$.

We will call the V_i value associated with alpha A_i, and the V_i value associated with beta B_i, at chance nodes, and so we will pass a window of $[A_i, B_i]$ to successor i when we search it.

Since the upper bound on a leaf node is 10, we will pass a window of [-3, 10] to the next child, C. Assume the next node searched at the bottom, E, has a value of -8. This will trigger a cutoff at C, because -8 lies outside the lower bound of the window (which is -3). The cutoff at C will also trigger a cutoff at the topmost chance node A. In fact, this could also trigger further cutoffs along this branch all the way up to the root; the possibility for two or more cutoffs to occur without intervening leaf searches is unique to trees with chance nodes, and not found in typical MINIMAX trees.

2.3 STAR1

When we translate the ability to obtain chance node cutoffs into a procedural representation, we end up with STAR1, Ballard's first version of the *-MINIMAX algorithm. By re-arranging equations (1) and (2), the alpha value for the ith successor, A_i, can be determined with

$$A_i = N \times alpha - (V_1 + \ldots + V_{i-1}) - U \times (N - i)$$

and the beta value for the ith successor, B_i, with

$$B_i = N \times beta - (V_1 + \ldots + V_{i-1}) - L \times (N - i)$$

where alpha and beta are the respective values passed to the chance node. These equations can be rewritten to be more efficient by initializing the two values as:

$$A_1 = N \times (alpha - U) + U; \quad B_1 = N \times (beta - L) + L$$

and updating them with

$$A_{i+1} = A_i + U - V_i; \quad B_{i+1} = B_i + L - V_i$$

where $i = 2 \ldots N$. When a chance node only has one successor ($N = 1$), the initial A and B values for the chance node take on the alpha and beta values initially passed to the node.

Figure 4 shows the resulting STAR1 algorithm. The algorithm makes use of the following additional functions:

1. numSuccessors() that returns the number of successors a state has,
2. successor() that returns a new state, and

```
float Star1(Board board, float alpha, float beta, int depth) {
    if(terminal(board) || depth == 0) return (evaluate(board));
    N = numSuccessors(board);
    A = N*(alpha-U) + U;
    B = N*(beta-L) + L;
    for(vsum = 0, i = 1; i <= N; i++) {
        AX = max(A, L);
        BX = min(B, U);
        v = search(successor(board,i), AX, BX, depth-1);
        vsum += v;
        if(v <= A) { vsum += U*(N-i); return (vsum/N); } // Fail soft
        if(v >= B) { vsum += L*(N-i); return (vsum/N); } // Fail soft
        A += U - v;
        B += L - v;
    }
    return (vsum/N);
}
```

Fig. 4. The STAR1 algorithm, adapted from [1]

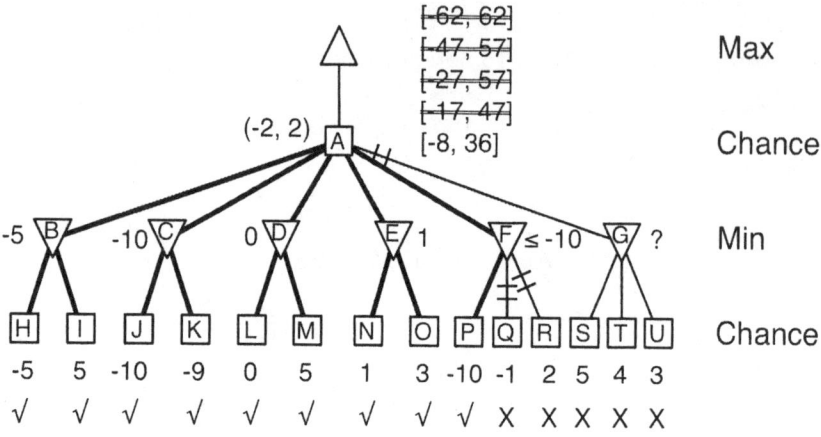

Fig. 5. A STAR1 tree

3. search() which calls the appropriate function, either STAR1 for a chance node or ALPHA-BETA for a min or max node.

This assumes that all values for the chance event have equal probability.

Note that our version of STAR1 extends the algorithm to include the fail-soft ALPHA-BETA enhancement. When further search at a node is unnecessary, rather than returning the window bound (alpha or beta) the code returns the lowest upper bound or highest lower bound that would be achievable if the remaining successors were searched.

An example of STAR1 cutoffs is shown in Figure 5. The uppermost chance node is initially passed bounds of [-2,2]. The initial value for A is equal to $N \times (alpha - U) + U = 6 \times (-2 - 10) + 10 = -62$ and B is equal to $N \times (beta - L) + L = 6 \times (2 + 10) - 10 = 62$. After searching the root's first successor, the A and B values are adjusted for the second successor (C), where A becomes $-62 + 10 + 5 = -47$ and B becomes $62 - 10 + 5 = 57$. As we continue to search the children of the root sequentially, we can see that the root node's $[A,B]$ window is equal to [-8,36] by the time it reaches its fifth child F, who gets an ALPHA-BETA window of [-8,10]. After searching P, which has a value of -10, F gets an immediate cutoff and returns this value to its parent A, the uppermost chance node, which triggers another cutoff because -10 falls outside its lower bound of -8. The other children of F, as well as the sixth successor G, do not need to be searched, as we can prove that the EXPECTIMAX value of A must be less than -2 (it is in fact $-3\frac{1}{2}$, which we can read from Figure 2).

2.4 STAR2

While STAR1 results in an algorithm which returns the same result as EXPECTIMAX, and uses fewer node expansions to obtain the same result, its results are

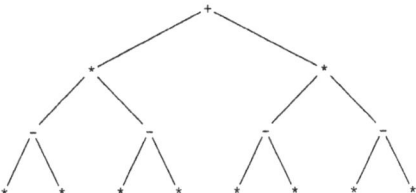

Fig. 6. A regular *-MINIMAX tree

generally not very impressive. One reason is that STAR1 is agnostic about its successors; it has no idea what kind of node (min, max or chance) will follow it, but even if it did, it would not be able to take advantage of that knowledge. However, game domains are fairly regular; for example, in a standard MINIMAX tree, min and max nodes are on levels that strictly alternate. Min always follows max, and max always follows min. In games like backgammon, where each player rolls the dice, then moves, we end up with a tree like a MINIMAX tree, except we insert a chance node immediately after any non-terminal min or max node. In other words, we add a layer of chance nodes between each layer of nodes in a standard MINIMAX tree. Ballard refers to trees with this structure as *regular* *-MINIMAX trees, an example of which is shown in Figure 6, where +, - and * refer to max, min and chance nodes, respectively. The regular structure assumed for STAR2 is not essential, as any tree can be transformed into such a form.

Another drawback to STAR1 is due to its pessimistic nature. We may potentially search nearly all the children of a chance node before a cutoff is obtained, because we assume that all unseen children have a worst-case evaluation. However, children of a successor of a chance node will tend to have values which are highly correlated. Instead of searching each child of a chance node fully and sequentially, and give a value of L to any children we have not seen yet, we can get a more accurate picture just by searching a single successor of each child. This value we get for the child then becomes a bound on the true value for the child (a lower bound if the child is a max node, and an upper bound if the child is a min node). It is likely that the bound will be much better than L, especially if we chose the child well. We will therefore introduce this phase of speculative search (which we will call the *probing phase*) before sequentially searching each child, in order to obtain a quicker cutoff.

We need to modify the equations used to generate A and B in STAR1 to reflect the new use of a probing phase in STAR2. For STAR2's probing phase, we derive the bounds for A and B just like we do in STAR1's search phase, except we do not have alpha cutoffs at chance nodes followed by min nodes (since we can only get an upper bound on those children), and we do not have beta cutoffs at chance nodes followed by max nodes (since we can only get a lower bound on those children).

We obtain a cutoff in STAR2's search phase if

$$\frac{(V_1 + \ldots + V_{i-1}) + V_i + (W_{i+1} + \ldots + W_N)}{N} \leq alpha$$

or

$$\frac{(V_1 + \ldots + V_{i-1}) + V_i + (W_{i+1} + \ldots + W_N)}{N} \geq beta$$

where $(W_1, \ldots, W_N)$ are the probed values for the N children of a node, obtained during the probing phase.

The alpha value for the ith successor, A_i is now obtained with

$$A_i = N \times alpha - (V_1 + \ldots + V_{i-1}) - (W_{i+1} + \ldots + W_N) \qquad (2)$$

and the beta value for the ith successor, B_i with

$$B_i = N \times beta - (V_1 + \ldots + V_{i-1}) - (W_{i+1} + \ldots + W_N) \qquad (3)$$

Like with STAR1, these equations can be rewritten:

$$A_1 = N \times alpha - (W_2 + \ldots + W_N); \quad B_1 = N \times beta - (W_2 + \ldots + W_N)$$

and updated by

$$A_{i+1} = A_i + W_{i+1} - V_i; \quad B_{i+1} = B_i + W_{i+1} - V_i$$

where $i = 2 \ldots N$.

Figure 7 shows the resulting STAR2 algorithm using a Negamax formulation (hence, a chance node is always followed by a max node). To get values for the probing phase, we need a procedure similar to ALPHA-BETA since successors are min or max nodes. Figure 8 shows the Probe algorithm. Figure 9 shows the PickSuccessor algorithm used by Probe, which is explained in more detail below.

Consider the tree in Figure 10, to see STAR2's strength. It is the same tree used in the previous example with STAR1. For the probing phase, the alpha value changes just like with STAR1 but the beta value does not. In this case, we only need to search five leaves: H, J, L, N and P, because by the time we reach child F, we give it a window of [-8,10]. Since P has a value of -10, this causes a cutoff at F. It also causes a cutoff at A since F returns a value of -10, which is less than or equal to A. In this example our Probe function did a good job and we always chose the best child for probing (fortuitously), so we obtained a cutoff after searching about half the nodes STAR1 searches.

As the branching factor increases, probing becomes even more effective, because sequential searching of children becomes more and more time-consuming. But even with small branching factors, probing can still be effective.

In his paper, Ballard did not specify how Probe should choose a successor besides to say it could be done "at random or by appeal to a static evaluation

```
float nStar2(Board board, float alpha, float beta, int depth) {
   if(terminal(board) || depth == 0) return (evaluate(board));
   N = numSuccessors(board);
   A = N*(alpha-U);
   B = N*(beta-L);
   AX = max(A, L);
   /* Probing phase */
   for(vsum = 0, i = 1; i <= N; i++) {
      B += L;
      BX = min(B, U);
      w[i] = nProbe(successor(board,i), AX, BX, depth-1);
      vsum += w[i];
      if(w[i] >= B) { vsum += L*(N-i); return (vsum/N); }
      B -= w[i];
   }
   /* Search phase */
   for(vsum = 0, i = 1; i <= N; i++) {
      A += U;
      B += w[i];
      AX = max(A, L);
      BX = min(B, U);
      v = nAlphaBeta_MM(successor(board,i), AX, BX, depth-1);
      vsum += v;
      if(v <= A) return { vsum += U*(N-i); return (vsum/N); }
      if(v >= B) return { vsum += L*(N-i); return (vsum/N); }
      A -= v;
      B -= v;
   }
   return (vsum/N);
}
```

Fig. 7. Negamax formulation of the STAR2 algorithm

function" [1]. Since the domain he used was limited to a depth=3 tree, all the probes done in his experiments were on leaf nodes. His domains also only had chance nodes at depth=1 (the nodes at depth=3 are technically chance nodes, but since they are leaves, they are just statically evaluated), so probing was always relatively inexpensive.

For STAR2 to be successful, Probe must search a "good" child. We can abstract the selection process away from Probe and create another function, which we will call PickSuccessor. PickSuccessor, shown in Figure 9, will take a set of nodes and return the node it thinks is the "best". We want this selection process to be relatively fast and not use much overhead, so PickSuccessor may not want to use the evaluation function used for leaf evaluations, but instead use domain-specific knowledge to heuristically select a child. For example, in backgammon we may first select moves that result in hitting the opponent's blots, moves that form primes, or moves that form points. As soon as we see a successor that meets the best quality, we can simply exit with that successor as

```
float Probe_Min(Board board, float alpha, float beta, int depth) {
   if(terminal(board) || depth == 0) return (evaluate(board));
   choice = PickSuccessor(board);
   return (AlphaBeta(successor(board,choice), alpha, beta, depth-1));
}
```

Fig. 8. The Probe algorithm

```
int PickSuccessor(Board board) {
   choice = 1;
   N = numSuccessors(board);
   if(N < 2) return (1)
   else {
      for(i = 1; i <= N; i++) {
         if(hasBestQuality(successor(board,i))) return (i);
         else if(hasGoodQuality(successor(board,i))) choice = i;
      }
   }
   return (choice);
}
```

Fig. 9. The PickSuccessor algorithm

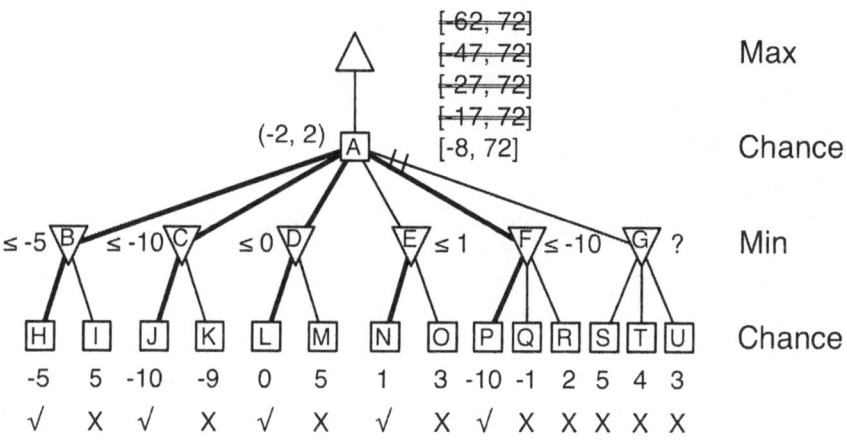

Fig. 10. A STAR2 tree, with good probing

the choice. Failing that, we can keep track of a successor that has the next best quality. If no successors have either quality, then the first can just be chosen.

Even if we do not obtain a quick cutoff during the probing phase, we will have a tighter window for the search phase, which in itself will lead to quicker cutoffs, because we have better estimates of the values of the children. Reconsider once

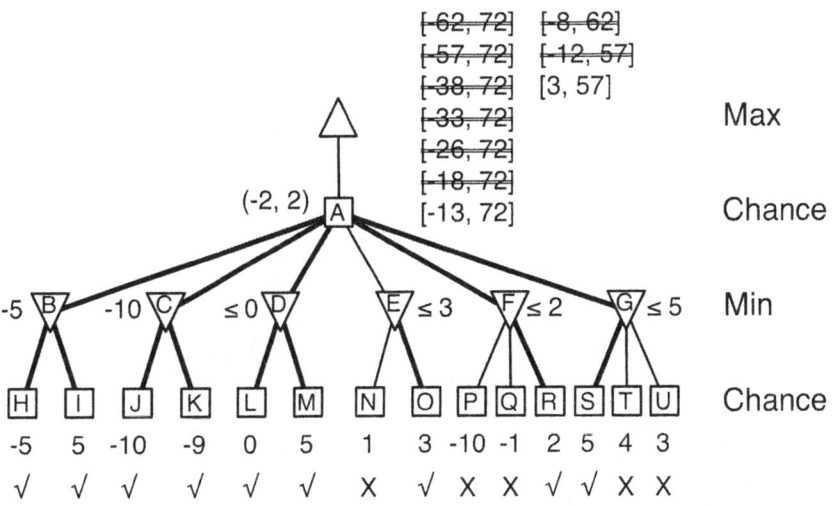

Fig. 11. A STAR2 tree, with bad probing

more the tree we have been using, but this time we will see what happens if Probe does a bad job. Figure 11 represents this situation. Assume that at the min nodes, we probe with the child that has the worst score for helping obtain a cutoff (the child with the maximum score). Now the probing phase will finish before we have obtained a cutoff, and so we will end up searching almost half of the leaves already. However, before the searching phase begins, notice that the window has been almost halved, because we have better upper bounds for the childrens' values. Instead of starting with a window of [-62,62] as STAR1 would, we start searching sequentially with a window of [-8, 62]. Now, by the time we start to search the third successor D, we have passed it a window of [3, 10]. If we assume that the leaf node L is searched first, then we get a cutoff at D (because 0 is less than 3) as well as at A. We end up searching six leaves in the probing phase, and an additional three leaves in the search phase, for a total of nine leaves. In this particular situation, even the worst-case probing resulted in the same number of leaves expanded as STAR1.

2.5 Non-uniform Chance Event Probabilities

For many applications (including backgammon), the probability of each chance event is not uniform. The formulas used to derive the equations for A and B need to be modified to accommodate this generalization. Ballard mentions the modifications needed but does not go into detail [1]. Note that this process does not affect EXPECTIMAX, just STAR1 and STAR2.

Recall equation 1 for obtaining A_i. The entire left-hand side of the inequality is divided by N because each of the N values has an equal chance of occurring. For non-uniform chance probabilities, this inequality changes to

$$(P_1 \times V_1 + \ldots + P_{i-1} \times V_{i-1}) + P_i \times V_i + U \times (1 - P_1 - \ldots - P_i) \leq alpha$$

or

$$A_i = \frac{alpha - U \times (1 - P_1 - \ldots - P_i) - (P_1 \times V_1 + \ldots + P_{i-1} \times V_{i-1})}{P_i}$$

where P_i is the probability that the ith chance occurs, for A. Similarly,

$$B_i = \frac{beta - L \times (1 - P_1 - \ldots - P_i) - (P_1 \times V_1 + \ldots + P_{i-1} \times V_{i-1})}{P_i}$$

We will make the substitution $Y = (1 - P_1 - \ldots - P_i)$, which can be computed incrementally, where $Y_0 = 1$ and updates are made with $Y_i = Y_{i-1} - P_i$. We will make another substitution $X = (P_1 \times V_1 + \ldots + P_{i-1} \times V_{i-1})$, which can also be computed incrementally where $X_1 = 0$ and updates are made with $X_{i+1} = X_i + P_i \times V_i$. We can then calculate A_i and B_i with

$$A_i = \frac{(alpha - U \times Y_i - X_i)}{P_i}; \quad B_i = \frac{(beta - L \times Y_i - X_i)}{P_i} \quad (4)$$

When there is only one successor, $A = alpha$ and $B = beta$, as desired.

These equations can be used for STAR1 and also for STAR2's probing phase. When calculating A and B values in STAR2's search phase, we can still use equation 4 to get A, but for B we need to modify equation 3:

$$B_i = \frac{(beta - W_i - X_i)}{P_i}$$

where $W_i = (W_{i+1} + \ldots + W_N)$, the sum of the probed values for nodes not yet searched.

2.6 Star2.5

Ballard proposed variations on the probing done by STAR2. For example, having probed one child of each node and not obtained a cutoff, additional probing effort could be invested. For example, each child could have a second probe performed. Ballard called the number of probes done at each node the *probing factor*. STAR1 can be viewed as having a probing factor of 0, while STAR2 has a probing factor of 1. Ballard proposed several algorithm variants with probing factors greater than 2, and called this family of algorithms STAR2.5.

2.7 Enhancements

ALPHA-BETA search has numerous enhancements that can greatly improve search efficiency. Here we briefly mention the enhancements that will have the most impact on *-MINIMAX algorithms.

- Transposition table. Besides the usual transpositions and move ordering benefits, transposition tables can help by re-using the results from STAR2's probing phase.

- Move ordering. Move ordering is always critical in any ALPHA-BETA-based search program. For STAR2, it is even more critical since it is needed to identify a "best" candidate for probing.
- Iterative deepening. Iterative deepening can be used to improve move ordering, both for the search and the probing.
- Fail soft. This is a simple enhancement that essentially comes for free (and has been added to the STAR1 and STAR2 pseudo-code). It helps narrow the search window bounds, resulting in earlier cutoffs.

These ideas have been implemented in a backgammon program and shown to be important enhancements to *-MINIMAX search [5].

3 Conclusions

Backgammon is the obvious domain for exploring performance issues of *-MINIMAX. The results are very encouraging, typically giving an extra ply or two of search. These results are reported in a companion article [5].

Besides games, the *-MINIMAX algorithms seem to also be applicable to Markov Decision Processes (MDPs), especially in the area of multi-agent MDPs. While solving MDPs usually involves an EXPECTIMAX-type evaluation of states one step away during value iteration, perhaps that component could be changed to a depth-N search of states, where the action at any given state would be determined by the current policy at that iteration. This may produce quicker convergence, or in the case of multi-agent MDPs, a better method for choosing actions that lead to higher rewards.

A general approach to solving games that combine elements of skill and chance will remain an open research problem for a while to come, but they provide some of the most interesting domains as they often have elements at which computers excel but humans do not (optimization, uncertainty calculation), and vice versa (long-term planning, opponent modeling). Games that combine skill, chance, imperfect information and opponent interaction are the most difficult domains for computers, so cross-disciplinary approaches involving combining elements of heuristic search, machine learning, agent theory, game theory, and even psychology may prove the most fruitful in the years to come.

Acknowledgements

This research was supported by grants from the Natural Sciences and Engineering Research Council of Canada (NSERC) and Alberta's Informatics Circle of Research Excellence (iCORE).

References

1. Bruce Ballard. The *-Minimax search procedure for trees containing chance nodes. *Artificial Intelligence*, 21(3):327–350, 1983.
2. Darse Billings, Aaron Davidson, Jonathan Schaeffer, and Duane Szafron. The challenge of poker. *Artificial Intelligence*, 134(1–2):201–240, 2002.

3. Matt Ginsberg. GIB: Steps toward an expert-level bridge-playing program. In *International Joint Conference on Artificial Intelligence*, pages 584–589, 1999.
4. Thomas Hauk. Search in trees with chance nodes. Master's thesis, Computing Science, University of Alberta, 2004.
5. Thomas Hauk, Michael Buro, and Jonathan Schaeffer. *-Minimax performance in backgammon. In H.J. van den Herik, Y. Björnsson, and N. Netanyahu, editors, *Computers and Games 2004*, volume 3864 of *LNCS*, pages 51–66, Ramat-Gan, Israel, 2005. Springer-Verlag.
6. Donald Michie. Game-playing and game-learning automata. In L. Fox, editor, *Advances in Programming and Non-Numerical Computation*, pages 183–200. Pergamon, New York, 1966.
7. Brian Sheppard. *Towards Perfect Play of Scrabble*. PhD thesis, Institute for Knowledge and Agent Technology (IKAT), Universiteit Maastricht, 2002.
8. Gerald Tesauro. Temporal difference learning and TD-Gammon. *Communications of the ACM*, 38(3):58–68, 1995.

*-MINIMAX Performance in Backgammon

Thomas Hauk, Michael Buro, and Jonathan Schaeffer

Department of Computing Science,
University of Alberta, Edmonton, Alberta, Canada
{hauk, mburo, jonathan}@cs.ualberta.ca

Abstract. This paper presents the first performance results for Ballard's *-MINIMAX algorithms applied to a real–world domain: backgammon. It is shown that with effective move ordering and probing the STAR2 algorithm considerably outperforms EXPECTIMAX. STAR2 allows strong backgammon programs to conduct depth-5 full-width searches (up from 3) under tournament conditions on regular hardware without using risky forward-pruning techniques. We also present empirical evidence that with today's sophisticated evaluation functions good checker play in backgammon does not require deep searches.

1 Introduction

*-MINIMAX is a generalization of ALPHA-BETA search for minimax trees with chance nodes [4,6,7]. Like ALPHA-BETA search, *-MINIMAX can safely prune subtrees which provably do not influence the move decision at the root node. Although introduced by Ballard as early as 1983, *-MINIMAX has not received much attention in the AI research community. In fact, it never found its way into strong backgammon programs. This is surprising in view of the importance of searching deeper in deterministic two-player perfect-information games like chess, checkers, and Othello. Curious about the reasons for this, we set out to investigate. Here we report the results we obtained.

The paper is organized as follows: we first briefly discuss search algorithms for trees with chance nodes. Then we survey the development of strong backgammon programs and describe the GNU backgammon program in some detail because we used its evaluation function in our *-MINIMAX experiments. Thereafter, we discuss implementation issues and our experimental setup, and present empirical results on the search performance and playing strength obtained by *-MINIMAX in backgammon. We conclude the paper by suggesting future research directions.

2 Heuristic Search in Trees with Chance Nodes

The baseline algorithm for trees with chance nodes analogous to MINIMAX search is the EXPECTIMAX algorithm [9]. Just like MINIMAX, EXPECTIMAX is a full-width search algorithm. It behaves exactly like MINIMAX except it adds an extra component for dealing with chance nodes (in addition to Min or Max nodes): at chance nodes, the heuristic value (or EXPECTIMAX value) is equal to the sum of the heuristic values of its successors weighted by their individual probabilities.

Just as cutoffs in MINIMAX search can be obtained by computing value bounds which are passed down to successor nodes (the ALPHA-BETA algorithm), so too can we derive a strategy for pruning subtrees in EXPECTIMAX based on value windows. Cutoffs at Min and Max nodes are found in the same way as in ALPHA-BETA search, i.e., whenever a value of a Max node successor is not smaller than the upper bound β passed down from the parent node, no further successors have to be considered. Similarly, cutoffs can be obtained at Min nodes by comparing successor values with the lower bound α. Ballard discovered that also at chance nodes cutoffs are possible [4]: if we know lower and upper bounds on the values leaf nodes can take (called L and U respectively), we can determine bounds on the value of a chance node based on the passed down search window (α, β), the values of successors we already determined $(V_1, \ldots, V_{i-1})$, the current successor value V_i, and pessimistic/optimistic assumptions about the successor values yet to be determined $(V_{i+1}, \ldots, V_n)$. Assuming probability P_j for event j we obtain the following sufficient conditions for the node value V to fall outside the current search window (α, β):

$$V_i \leq (\alpha - \sum_{j=1}^{i-1} P_j \cdot V_j - U \cdot \sum_{j=i+1}^{n} P_j)/P_i \quad \Rightarrow \quad V \leq \alpha$$

$$V_i \geq (\beta - \sum_{j=1}^{i-1} P_j \cdot V_j - L \cdot \sum_{j=i+1}^{n} P_j)/P_i \quad \Rightarrow \quad V \geq \beta$$

These value bound computations are at the core of Ballard's STAR1 algorithm which prunes move i when the resulting window for V_i is empty.

Ballard's family of STAR2 algorithms improves upon STAR1 by exploiting the regularity of trees in which successors of chance nodes have the same type (either Min or Max). This condition, however, is not a severe constraint because any tree with chance nodes can be transformed into the required form above by merging nodes and introducing artificial single action nodes. To illustrate the idea behind STAR2, we consider the case where a chance node is followed by Max nodes. In order to establish non-trivial lower bounds on the chance node value it is sufficient to *probe* just a subset of moves at the following Max nodes. I.e.,

$$\sum_{j=1}^{n} P_j V_j \geq \sum_{j=1}^{m} P_j V_j^*,$$

where $m \leq n$ and V_j^* is obtained by maximizing values over a subset of moves available at successor j. A straightforward and simple probing strategy — which we use in our backgammon experiments — is to consider only single good moves at each successor node and continuing the search like STAR1 if the probing phase does not produce a cutoff. Note that probing itself calls the STAR2 routine recursively.

For a more detailed description and analysis the reader is referred to Ballard's original work [4] and our companion paper [7] which also provides pseudo-code.

3 Backgammon Programs: Past and Present

With a large state space (estimated to be bigger than 10^{20} states [12]) and an imposing branching factor (there are 21 unique dice rolls, and about 20 moves per roll, on average), it is not surprising that most of the early computer backgammon programs were *knowledge-based*. These systems do not rely much on search, but rather attempt to choose moves based on *knowledge* about the domain, usually programmed into the system by a human expert.

The first real success in computer backgammon was BKG, developed by Hans Berliner. In 1979, BKG played the world champion at the time, Luigi Villa, and managed to defeat him 7-1 in a five point match [5]. While many people were shocked, even Berliner himself would concede weeks after the match that BKG had been lucky with rolls and made several technical blunders. However, Villa had not been able to capitalize on those mistakes – such is the life with dice.

The second milestone in computer backgammon was NEUROGAMMON [11], the work of IBM researcher Gerald Tesauro. NEUROGAMMON used an *artificial neural network* for evaluating backgammon positions. NEUROGAMMON was trained with *supervised learning*; it was fed examples labeled by a human expert, and told what the answer should be. The program quickly became the best in computer backgammon, but still only played at the level of a strong human amateur player.

Tesauro went back to the drawing board. One of the first things he changed was the data the program was training on. Instead of using hand-labeled positions, he decided he would rely solely on *self-play* to generate training data – the program would simply play against itself. This has advantages over the previous method since a human expert may label positions incorrectly, or tire quickly (NEUROGAMMON only used selected positions from about 3000 games [11] to train checker play, culled from games where Tesauro had played both sides), but self-play also may lead a program into a *local* area of play. For example, a program can learn how to play well against itself, but not against another opponent. This local minima problem in backgammon is partially overcome due to the fact that the environment is stochastic – dice insert a certain level of randomness – so a program is forced to explore different areas of the state space.

The other thing Tesauro changed was the training method itself. Instead of using a supervised learning approach that adjusted the network after each move (which he could do before because each training example was labeled), Tesauro decided on adapting *temporal-difference learning* for use with his neural network [10,12]. TD learning is based on the idea that an evaluation for a state should depend on the state that follows it. In a game sense, the computer keeps track of each position from start to finish, and then works backward. It trains itself on the last position, with the target score being the outcome of the game. Then it trains itself on the second last position, trying to predict more accurately the score it got for the last position (*not* the final score). The last position is the only position which is given a *reward signal*, or absolute value; all other positions are only trained to better predict the position that followed it. In games, the reward signal is related to the outcome of the game. If the program had lost, the reward signal would be low (to act like a punishment). If the program won, the reward signal would be high. Since backgammon cannot end in a draw, the reward signal could never be zero.

In this manner, Tesauro delayed the final reward signal for the neural network until the game was won or lost, at which point the network would begin adjusting itself. This new program was called TD-GAMMON in honour of its training method. Tesauro trained the first version of TD-GAMMON against itself for 300,000 games, at which point the program was able to play as well as NEUROGAMMON – quite surprising, considering the program had essentially "discovered" good play on its own, with no human intervention, and zero explicit knowledge. Later versions of TD-GAMMON increased the size of the *hidden units* in the network, added hand-crafted *features* to the input representation, trained for longer amounts of time, and included a selective search algorithm to extend the search process deeper than a single ply. TD-GAMMON is considered to safely be in the top-3 players of the world. One human expert even ventured to say it was probably better than any human, since it does not suffer from mental exhaustion or emotional play.

TD-GAMMON's use of temporal difference learning and a neural network evaluation function has lead to several copy-cat ventures, including the commercial programs JELLYFISH [2] and SNOWIE [3], as well as the open-source GNU Backgammon [1] (also known as GNUBG). Several versions of GNU Backgammon have sprung up on the Internet, and it has quickly become one of the most popular codebases for developers.

4 Search in Top Backgammon Programs

Much effort has been put into creating backgammon programs with increasingly stronger evaluation functions. Compared to other classic games like chess and checkers, however, not much research on improving search algorithms for games with chance nodes has been conducted in the past. Both TD-GAMMON and GNUBG use a *forward-pruning* approach to search, where some possible moves are eliminated before they are searched in order to reduce the branching factor of the game. Depending on the approach, using forward pruning can be a bit of a gamble, since the program is risking never seeing a good line of play, and therefore never having the chance to take it. Section 5.2 discusses GNUBG's search approach in some detail.

There are two important reasons why improvements in search have not been developed in backgammon. The first is that the current crop of neural network-based evaluation functions are very accurate, but take far too long in processing terms. For example, a complete 3-ply search of an arbitrary position in backgammon can take several minutes to complete. This is clearly undesirable from a performance perspective. The second reason has to do with the game itself. Since there are 21 distinct rolls in backgammon (with varying probability), and an average of 20 moves per roll, the effective branching factor becomes so large that, especially for a slow heuristic, searching anything deeper than a ply or two becomes impractical. It is clear due to these reasons why efforts have concentrated on developing an evaluation function that is as accurate as possible, instead of trying to grapple with the large branching factor inherent in the game.

But search *is* still important. Deeper search allows for the inaccuracies of a heuristic to be reduced – the deeper a program can search, the better that program can play. Backgammon is no exception, even with a trained neural network acting as a near-oracle. Still, it is interesting to note that improving search in backgammon programs has not been a priority, to the point where some of the GNU backgammon team are unfamiliar with the concept of ALPHA-BETA search. Tesauro thinks that improvements in search will come as a result of faster processors and Moore's Law [13]. In [14] he also considered on-line Monte Carlo sampling or so-called rollout analysis run on a parallel computer. The idea is to play many games starting in a given position using only a shallow search at each decision point and to pick the root move that on average yields the highest value. The results obtained by conducting on-line rollouts on a multi-processor machine are promising and indicate top performance when run on one of today's much faster single processor machines while using less sophisticated evaluation functions than the competition.

5 Overview of GNU Backgammon

GNU Backgammon is an open-source backgammon program developed through the GNU Project. Development began in 1997 by Gary Wong, and has continued up to this time with contributions from dozens of people. The other five primary members today are Joseph Heled, Øystein Johansen, David Montgomery, Jim Segrave, and Jørn Thyssen. The current version of GNUBG, 0.14, boasts an impressive list of features, including TD-trained neural network evaluation functions, detailed analysis of matches (including rollouts), a tutor mode, bearoff (endgame) databases, variable computer-skill levels, and a graphical user interface. GNUBG is also free, and since its exposure to the backgammon community was heightened, it is one of the most popular and strongest backgammon programs available. In fact, in September 2003 the results of a duel between GNUBG and JELLYFISH were posted to the rec.games.backgammon group on UseNet [8]. Both programs played 5,000 money games, each using their "optimal" settings. GNUBG came out the winner by an average of 0.12 points per game which is a statistically significant indication that GNUBG is stronger than the expensive "professional" backgammon program JELLYFISH [15].

5.1 The Evaluation Function

GNUBG has three different neural networks it uses for evaluating a backgammon position, depending on the classification of that position: either contact (at least one checker of a player is behind a checker of the other player), crashed (same as contact but with the added restriction that the player has six or less checkers left on the board, not including any checkers on the opponent's 1 or 2 points) or race (the opposite of a contact position). Since each of the three types of positions are quite different from the others, using three different neural networks improves the quality of the evaluation.

Each neural network is first trained using temporal difference learning, using self-play, similar to TD-GAMMON. The input and output representations of the neural networks are also similar to TD-GAMMON. The input neurons are comprised of both a raw board representation (with four neurons per point per player) as well as several hand-crafted features, such as the position of back anchors, mobility, as well as probabilities for hitting blots.

After self-play, the networks are trained against a position database (one each for the contact, crashed and race networks). The databases contain "interesting" positions, so-named because a network would return different moves depending on if they searched to either depth=1 or depth=5; and whenever a depth=5 search retains a better result than depth=1, *two* entries are made in the database for that position: the position after the depth=1 move, and the position after the depth=5 move. The positions are a mixture of randomly-generated positions as well as drawn from a large collection of human versus bot or bot self-play games, with the idea that the networks should gain more exposure to "real-life" playing situations than random situations. In total, over 110,000 positions form the position database collection used by the GNUBG team.

There is an entry for each position's cubeless evaluation in the database, along with five legal moves and their evaluations. An evaluation consists of the probabilities of normal win, gammon win, backgammon win, gammon loss and backgammon loss for the player to move (a normal loss is not explicitly evaluated, as it is just equal to $1 - P_{\text{normal win}}$). The moves in the database are chosen by first completing a depth=1 search using GNUBG, taking the top 20 moves from that search, and then searching those to depth=5; the best five moves from the depth=5 search are then kept. These moves are then "rolled out", meaning that the resulting position after the move is then played by GNUBG (doing the moves for both sides) until the game is over. Typically the number of rollouts is equal to a multiple of 36 (say, 1296) by using "quasi-random dice" in order to reduce the variance in the result, where each of the 36 possible rolls after the move is explored, with random dice thereafter. When a race condition is met in the game, the remaining rolls are played using a One-Sided Race (OSR) evaluator. The OSR is basically a table which gives the expected number of rolls needed to bear off all checkers, for a given position. It does not include any strategic elements. By using the OSR, the contact and crashed networks are judged on their own merits, and not based on the luck of the dice in the endgame. This is because race games are generally devoid of strategic play, because there is no interaction between the players anymore, not counting cube actions. Each rollout is performed in a 7-point money game setting, without cubeful evaluations.

A new network is trained against this database so its depth=1 evaluations more closely resemble a depth=5 search, and after the new network is fully trained, it then provides new entries for each position in the database. GNUBG was able to obtain a rating of about 1930 at a depth=1 setting on the First Internet Backgammon Server (FIBS), which put it roughly at an expert level on the server.

5.2 The Search Algorithm

GNUBG's search is based on heavy use of forward pruning to either completely eliminate or greatly reduce the branching factor at move nodes, and lower the branching factor at the root, in order to keep the search fast. Pruning is based on *move filters* that define how many moves are kept at the root node (and, depending on the depth of the search, at other move nodes lower in the tree). A move filter guarantees a fixed number of candidates that will be kept at a move node (if there are enough moves), plus the addition of n candidates which are added if they are within e equity of the best move. Search is performed using iterative deepening, and root move pruning is done after each iteration. At all other move nodes, the move filter will either limit the number of moves or only keep one move. Candidate moves are chosen by doing a static evaluation of all children of the move node and choosing the n moves with the best scores; in other words, a small depth=1 search is done at all move nodes.

The branching factor at chance nodes can also be optionally reduced by limiting the number of rolls to a smaller set than 21. All roll sets are hard-coded, so no attempt is made to order rolls nor bias roll selection when a reduced set is desired.

Unfortunately, GNUBG has an unusual definition of ply. In GNUBG, a depth=1 search is called "0-ply", a depth=3 search is considered "1-ply", and so on. While most users quickly adapt to this quirk, it makes working with the code potentially tricky, since one must always remember this to avoid bugs.

For depth=1 searches, GNUBG simply performs a static evaluation of all root move candidates (a candidate being a move that has not been pruned by the move filter), and the move with the highest score is chosen. At chance nodes in the search tree, all rolls in the roll set (the set is usually all 21 rolls but it can be reduced for speed) are investigated, and the best move for each roll (chosen by simple static evaluation) is applied and expanded, until the depth cutoff is reached. In homogeneous search trees, EXPECTIMAX visits $b(nb)^{(d-1)/2}$ leaves, where b is the branching factor at move nodes, n is the branching factor at chance nodes, and d is the odd search depth. By only doing a static evaluation of children at move nodes and then choosing only one for further expansion, the number of leaves of a GNUBG search tree for increasing d is $= b$ $(d = 1)$, $=bnb$ $(d = 3)$, $\leq bnnb + bnb$ $(d = 5)$, $\leq bnnnb + bnnb + bnb$ $(d = 7)$, etc. which can be bounded above by $b^2 \sum_{i=1}^{(d-1)/2} n^i \leq 2b^2 n^{(d-1)/2}$ for $d, n > 1$. Therefore, this pruning technique allows the search tree to be exponentially smaller than the full tree (with savings of at least $b^{(d-3)/2}/2$), but error is also introduced.

6 Implementation Issues and Experimental Setup

In this section we describe important implementation details and the environment in which the *-MINIMAX experiments were conducted.

6.1 Move Generation

Backgammon is not a trivial game to implement. While the board itself can be fairly easily represented by a two-dimensional array of integers, generating moves is rather complicated to not only do correctly, but also efficiently. Avoiding duplicating moves is also an important consideration because of the large branching factor for some situations (like a doubles roll for a player with checkers on several different points). The use of a transposition table can help reducing the complexity of the move-generation algorithm.

6.2 Evaluation Function

Instead of going out and designing a new evaluation function for our experiments, there was already one available for use: the GNUBG codebase, which is a very strong set of trained neural networks.

While many game programs are using integer-valued evaluation functions, the GNUBG evaluation function returns a floating point number (the value representing the equity of the player who just moved). Whenever search programs use floating point numbers, there is always the risk of floating point operations having rounding errors; even comparing two (seemingly) identical values may not result in the expected truth value.

To work around the uncertainty presented by floats and the continuous values they may have, we can *discretize* the values by putting them onto a one-dimensional *grid*. This involves taking the floating point number and multiplying it by a large number, and then rounding the value to the nearest integer number. That integer can then be divided by the same large number used for the multiplication. The *granularity* of the grid can be adjusted to meet the desired level of precision. A resolution of 262144 (2^{18}) was used to discretize the floating point numbers in our experiments, to ensure a fine enough granularity without being too fine for the floating point mantissa. Using floating point numbers instead of integers also meant a small performance hit.

6.3 Transposition Table

A transposition table (TT) was used to speed up the search. The TT was implemented as a simple hash table of 128 MB (more or less space could be used, depending on the amount of main memory available). Each entry was 20 bytes large, containing the value for the stored state, a flag to indicate if the entry was in use, an indicator for the depth searched, two flags to determine what kind of value for the state is stored (a lower bound, upper bound, or an exact value), the best move chosen at that state, and the hash key for that state. A Zobrist [16] hashing scheme was used.

6.4 Move Ordering and Probing

Move ordering and probe successor selection are both done with a different heuristic than the evaluation function. This is especially a concern with a heavy evaluation function such as GNUBG's. Probe successors were selected as follows: moves that hit opponent blots were taken first (best quality), moves that formed a point were taken second (good quality), and if no moves met either condition, the first move was chosen. Move ordering worked a little differently. Move sorting was done by scoring a move based on a number of criteria: the number of opponent checkers moved to the bar, the number of free blots it left open to hit, and the number of safe points (2 or more checkers) made. These criteria remained the same for all moves during the game.

6.5 Experimental Design

For obtaining quality results, all experiments were run on relatively new hardware. Two undergraduate labs (one of 22 machines and one of 34 machines) were made available for distributed processing. All machines were identical, each with an Athlon-XP 1.8 GHz processor and 512 MB of RAM, as well as 27 GB of local disk space (to bypass using NFS). Each machine used Slackware Linux with kernel version 2.4.23 and had gcc version 3.2.2 installed. All search software was coded in C.

While all experiments were performed when the labs were largely idle, all experiments were nevertheless subjected to possible skewing if students logged into a host to use it. However, less than a dozen students logged into any one of the machines during the entire experimental phase, so fluctuations in results due to lost CPU cycles are negligible. Since each machine only had a modest amount of free RAM, the transposition table was kept to a relatively small size of 128 MB. GNUBG's codebase was used for the evaluation function and all executables were compiled under gcc with -O3 optimization.

7 Performance

Using randomly-seeded positions does not make sense for backgammon, since it is difficult to generate random positions which look "reasonable" in backgammon terms. Instead of randomly generating positions, a position database was used. The database came from the GNUBG team, used for training the neural network. It contains several thousands of positions classified into different categories. The contact position database was made available for experiments. The results of searching these positions are therefore more applicable to real-world performance compared to random positions.

Five hundred randomly selected contact positions were used for testing. Each was searched to depths of 1, 3 and 5 by EXPECTIMAX, STAR1 and STAR2. There is a direct relation between time and node expansions, as the GNUBG evaluation function is very heavy in terms of CPU usage (over 90%).

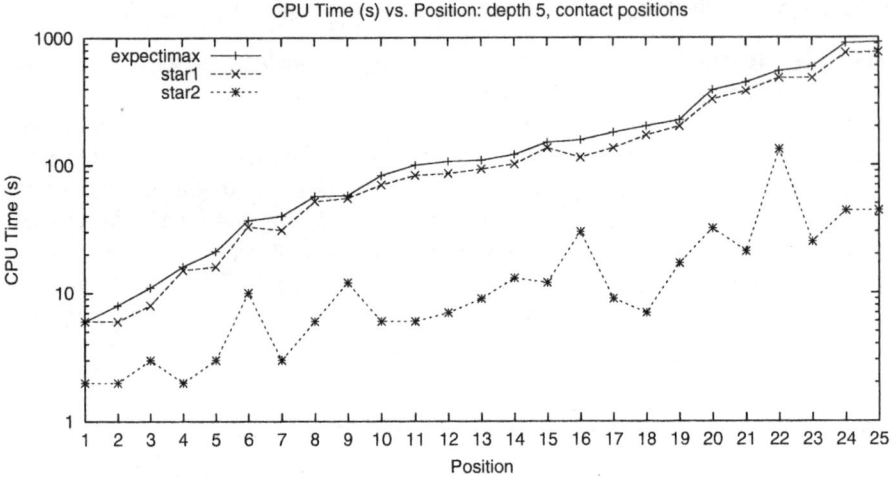

Fig. 1. Time used (s) at d=5 for 25 contact positions

In Figures 1 (CPU time) and 2 (node expansions) graphed on a logarithmic scale, we can see some variation in the amount of effort STAR2 requires to complete a search at depth=5. Each of the 25 positions shown were selected at random from the GNUBG contact position database, searched by all three algorithms, and then sorted in order of EXPECTIMAX time. The variation in savings for STAR2 for the 25 positions goes from about 75% to about 95%. EXPECTIMAX and STAR1 closely follow each other, where STAR1 has only a slight decrease in overall costs.

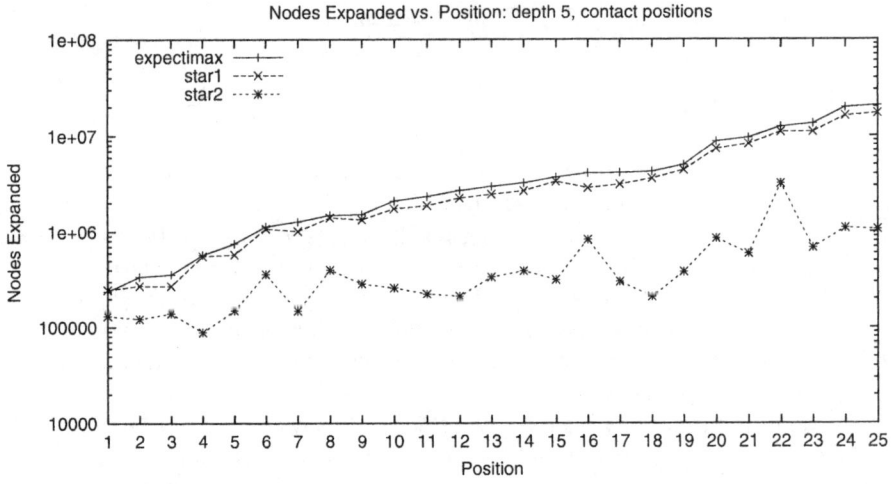

Fig. 2. Node expansions at d=5 for 25 contact positions

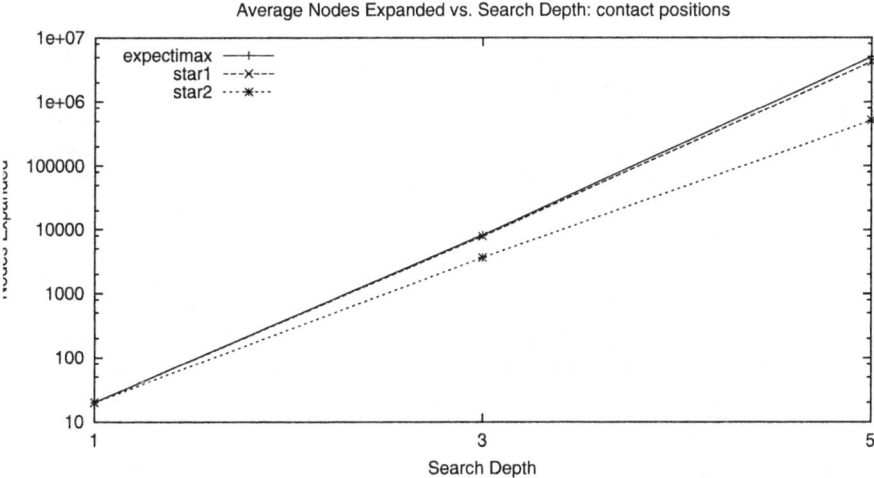

Fig. 3. Average number of node expansions over 500 contact positions (interpolation is used to indicate the trend)

Table 1. Average time (s) over 500 contact positions

	EXPECTIMAX μ	STAR1 μ	STAR1 %	STAR2 μ	STAR2 %
$d=3$	1.1	1.1	100	1.0	91
$d=5$	315.0	258.6	82	21.0	7

Table 1 summarizes the time usage over 500 positions. STAR2 is clearly the most efficient of the algorithms by over a factor of 10, but even at 21 seconds per search, this would probably still be too slow for tournament play. Figure 3 shows the average number of node expansions over 500 positions, graphed on a logarithmic scale.

7.1 Probe Efficiency

Table 2 shows the resulting probe efficiency for using STAR2. The "quick" successor selection scheme for backgammon is relatively weak because backgammon is a complicated game. Still, these results are better than Ballard's, whose probing was never successful more than about 45% of the time. The improvement here is probably due to better move ordering.

Table 2. Probe efficiency for backgammon

$d=3$	$d=5$
68.9%	64.2%

7.2 Odd-Even Effect

Many 2-player game-playing computer programs suffer from what is called the *odd-even* effect, where alternating levels of move nodes will give scores that are either optimistic (for odd-ply searches) or pessimistic (for even plies). For example, a depth=1 search in any game will tend to be optimistic, since we are only investigating the moves currently available to us. The odd-even effect comes from the way in which an evaluation function is created, which generally tries to score the position from the point of view of the player to move.

Table 3 shows the results of two different trials of 3200 backgammon positions. The positions were generated as a continuous sequence of cubeless money games, with the computer playing both sides at different search depths. This generated a decent set of "real-world" moves for backgammon. The table shows the average difference, absolute average difference, and absolute standard deviation in the root node value when comparing searches of the same positions to different depths.

Table 3. Root value difference statistics for two trials (3200 moves each)

	Average	Abs. Average	Abs. Std. Dev.
d=1 vs. d=3	0.0280	0.0336	0.0397
d=1 vs. d=5	0.0018	0.0134	0.0184

	Average	Abs. Average	Abs. Std. Dev.
d=1 vs. d=3	0.0267	0.0328	0.0389
d=1 vs. d=5	0.0014	0.0126	0.0172

Numbers in both tables are very similar. The results show that the evaluation of the root node for a depth=1 search is very close to the evaluation for a depth=5 search, on average. When absolute differences are used instead, depth=1 is not as good as a predictor for a depth=5 search, but the difference is reasonably small (only about 0.01 points).

The differences between depth=1 and depth=3 are much more striking. Both the average and the absolute average difference between them is nearly the same. In fact, the average difference is positive, which means that the depth=3 search value is usually significantly less than the value from a depth=1 search.

These results show a tangible odd-even effect with the GNUBG evaluation function. Even if searches to different depths produce different values for the root, the move chosen at the root usually is the same across searches of different depths. This means the evaluation function itself is very consistent between depths. These results also show that a depth=1 search value is a reasonable predictor for a depth=5 search value for the same position.

7.3 Tournaments

Another way to measure an algorithm's performance is to pit it against itself in a tournament, where each player is searching to a different depth. The question to answer, then, is if deeper search increases real performance in the game. Tournaments were therefore run between combinations of players searching to depths of 1, 3, and 5. Each tournament used a file containing a sequence of seed values, such that they all would then have the same sequence of dice rolls across each tournament. The starting roll for each game was pre-set by a testing script, and went through all combinations sequentially, in order to reduce variance. Furthermore, for each opening roll and sequence of dice rolls, both players were given an opportunity to be the starting player. A file containing 9,000 seeds was used, and therefore a total of 18,000 games per tournament were played. We chose cubeless money games as tournament mode to study *-MINIMAX checker play performance. Gammons and backgammons only counted as one-point wins. Tournaments were set up between the GNUBG search function and itself, and STAR2 against itself. Since GNUBG also has a facility for adding deterministic noise to an evaluation, different noise settings were also investigated.

Table 4. Tournament results for GNUBG with no noise versus GNUBG with no noise, 18,000 games per matchup. Reported are winning percentages and average points per game in view of the player whose search depth is named in the top row (1, 3, and 5).

	1	3	5
1	*	50.92 +0.018	51.79 +0.036
3	49.08 −0.018	*	51.63 +0.037
5	48.21 −0.036	48.37 −0.037	*

While we expected that deep search was beneficial for tournament performance just like in chess and Othello, this was not evident in backgammon. Table 4 shows the results of GNUBG playing against itself at different depth settings. We can see from the table that a depth=5 search barely shows any significant improvement over shallower searches. In fact, the three depth settings are nearly identical. This suggests that deeper searches are only finding better moves a small fraction of time, which suggests that the three searches are choosing the same move just about every time. That means the GNUBG evaluation function must be extremely consistent between depth levels. Table 5 A) shows the STAR2 performance when playing against itself, in the same manner as Table 4. Deep search is still pretty much irrelevant using the GNUBG evaluation function as-is. Since the evaluation function is so consistent, results were also desired for a less consistent setting. Instead of developing a new evaluation function, noise can just be added to the evaluation function. GNUBG has a built-in noise generator already, which can add either deterministic or non-deterministic noise to each evaluation. Since it is highly desirable that the evaluation for a state be always deterministic, especially when transpositions are possible, another tournament using deterministic noise was added. Only modest amounts of noise were added, consistent with an "intermediate" and "advanced" level of play for GNUBG (noise settings $n = 0.03$ and $n = 0.015$). Tables 5 B) and C) show tournament results in an identical manner to the previous two tables. Now, deeper search is paying

Table 5. Tournament results for STAR2 vs. STAR2 (4000 games per matchup). Reported are winning percentages and average points per game for various evaluation function noise levels and depth pairs.

A) noise level $n = 0.000$

	1	3	5
1	*	50.60 +0.012	51.60 +0.032
3	49.40 −0.012	*	52.52 +0.050
5	48.40 −0.032	47.48 −0.050	*

B) noise level $n = 0.015$

	1	3	5
1	*	55.53 +0.111	54.67 +0.093
3	44.47 −0.111	*	51.57 +0.031
5	45.33 −0.093	48.43 −0.031	*

C) noise level $n = 0.030$

	1	3	5
1	*	59.00 +0.180	64.20 +0.284
3	41.00 −0.180	*	53.40 +0.068
5	35.80 −0.284	46.60 −0.068	*

off to a significant degree. For $n = 0.03$ a depth=1 search now loses to a depth=5 search 64% of the time. Depth=1 fares slightly better against depth=3. Depth=5 wins slightly less than 55% of the time against depth=3, but it is still a tangible amount. Deep search helps to mitigate evaluation function errors by adding more foresight to the move decision process. Adding deterministic noise to the GNUBG evaluation function shows that deep search becomes important again in backgammon.

8 Conclusions and Future Work

STAR2 and STAR1 both outperform EXPECTIMAX on single position searches. STAR2 has a significant savings in costs at depth=5, mostly due to the large branching factor inherent in backgammon. GNUBG's evaluation function is time consuming which means that performance is strongly linked to eliminating as many leaves as possible.

To our surprise strong cubeless money game tournament performance is not much improved by deeper search. The GNUBG evaluation function is sufficiently well-trained and consistent that searches to increasing depths almost always choose the same move at the root. When the searches do not agree on the best move it is usually because they are searching a tactical position. But even the occurrence of tactical positions is relatively infrequent, and the benefits of deep search in these situations is usually washed away by the randomness of the dice rolls. However, when small amounts of deterministic noise are introduced into the search, deep search once again becomes important as the evaluation function becomes less consistent and less accurate. This suggests future *-MINIMAX experiments in domains less affected by chance events or in which good evaluation functions are unknown. Comparing *-MINIMAX with Monte-Carlo search with respect to playing strength versus search time in various domains would also be interesting.

GNUBG's forward-pruning search method works very well for its evaluation function, since the best move candidate at the root is unlikely to change much from one iteration to the next. Deeper search catches some tactical errors in some situations, but because tactical situations can be thrown completely askew by a single lucky roll, deep search does not pay huge dividends.

With an excellent evaluation function such as GNUBG's set of neural networks, checker play is virtually perfect, even with shallow search. However, since backgammon matches are generally played with a doubling cube, and cube decisions are usually the most important part of the game, this work should be extended to cubeful games and cube decisions. Being able to see farther ahead in these situations and to estimate the winning probability more accurately can make or break a player's chances of winning. The odd-even fluctuations we recognized indicate that although neural networks may be almost optimal in ordering moves, there is still room for improvement with respect to absolute accuracy. Looking deeper by using *-MINIMAX and considering doubling actions in the search is a promising approach.

In view of the close tournament outcomes reported in this paper, future experiments should also consider gammons and backgammons to produce results more relevant to actual tournament play. Moreover, choosing starting positions unrelated to the GNUBG evaluation function tuning could generate results more favorable to deeper search.

Since STAR2 is so reliant on successful probing, a more powerful probing function would also increase performance. Right now successors are picked according to some *ad hoc* rules about good backgammon play for quickly choosing a child, but there are perhaps better techniques for making this decision — e.g., probing functions amenable to fast incremental updating similar to piece–square tables in chess.

There are other stochastic perfect-information games which could benefit greatly from the use of *-MINIMAX search. One excellent domain would be the German tile-laying game Carcassonne. Being able to see a line of play from even five or six tiles out could result in expert play. Because computers can also keep track of which tiles have been played better than humans, a computer player could also avoid many of the pitfalls which plague humans. However, since the branching factor at chance nodes after the root starts at 40 (when using the most common expansion tileset, Inns & Cathedrals), some form of statistical sampling may be required to jumpstart the computer player.

Acknowledgements

This research was supported by grants from the Natural Sciences and Engineering Research Council of Canada (NSERC) and Alberta's Informatics Circle of Research Excellence (iCORE).

References

1. GNU Backgammon (backgammon software). http://www.gnubg.org/.
2. Jellyfish (backgammon software). http://jelly.effect.no/.
3. Snowie (backgammon software). http://www.bgsnowie.com/.
4. B.W. Ballard. The *-Minimax Search Procedure for Trees Containing Chance Nodes. *Artificial Intelligence*, 21(3):327–350, 1983.
5. H. Berliner. Backgammon Computer Program Beats World Champion. *Artificial Intelligence*, 14:205–220, 1980.
6. T. Hauk. Search in Trees with Chance Nodes. *M.Sc. Thesis, Computing Science Department, University of Alberta*, 2004.
7. T. Hauk, M. Buro, and J. Schaeffer. Rediscovering *-Minimax. In H.J. van den Herik, Y. Björnsson, and N. Netanyahu, editors, *Computers and Games 2004*, volume 3864 of *LNCS*, pages 35–50, Ramat-Gan, Israel, 2005. Springer-Verlag.
8. M. Howard. The Duel, August 5, 2003. [Online] rec.games.backgammon.
9. D. Michie. Game-playing and game-learning automata. In L. Fox, editor, *Advances in Programming and Non-Numerical Computation*, pages 183–200. Pergamon, New York, 1966.
10. R.S. Sutton and A.G. Barto. *Reinforcement Learning: An Introduction*. MIT Press, Cambridge, MA, 1998.
11. G. Tesauro. Neurogammon: A neural-network backgammon learning program. In D.N. Levy and D.F. Beal, editors, *Heuristic Programming in Artificial Intelligence*, pages 78–80. Ellis Horwood, 1989.
12. G. Tesauro. Temporal Difference Learning and TD-Gammon. *Communications of the ACM*, 38(3), March 1995.
13. G. Tesauro. Programming Backgammon Using Self-Teaching Neural Nets. *Artificial Intelligence*, 134(1-2):181–199, 2002.
14. G. Tesauro and G.R. Galperin. On-line policy improvement using Monte Carlo search. *Proc. of NIPS*, pages 1068–1074, 1996.
15. J. Thyssen, 2003. Personal Communication.
16. A.L. Zobrist. A New Hashing Method with Applications for Game Playing. *ICCA Journal*, 13(2):69–73, 1990.

Associating Shallow and Selective Global Tree Search with Monte Carlo for 9×9 Go

Bruno Bouzy

UFR de mathématiques et d'informatique,
Université Paris 5, Paris, France
bouzy@math-info.univ-paris5.fr

Abstract. This paper explores the association of shallow and selective global tree search with Monte Carlo in 9×9 Go. This exploration is based on OLGA and INDIGO, two experimental Monte-Carlo programs. We provide a min-max algorithm that iteratively deepens the tree until one move at the root is proved to be superior to the other ones. At each iteration, random games are started at leaf nodes to compute mean values. The progressive pruning rule and the min-max rule are applied to non terminal nodes. We set up experiments demonstrating the relevance of this approach. INDIGO used this algorithm at the 8th Computer Olympiad held in Graz.

1 Introduction

Knowledge and tree search are the two main approaches to computer Go [8]. However, other approaches are worth considering. The Monte-Carlo approach has been developed by Brügmann [10], and recently by Bouzy and Helmstetter [9]. While using very little Go knowledge, Monte-Carlo Go programs have performed well on 9×9 boards. Furthermore, associating domain-dependent knowledge with Monte Carlo has been very effective too [6]. Therefore, the remaining question is to study the association of tree search with Monte Carlo. Because a strength of Monte Carlo consists in avoiding the breaking down of the whole game into sub-games, the tree search considered in this study is global. Moreover, because the association of knowledge and Monte Carlo partly relies on selectivity, the tree search is selective. Finally, because of the combinatorial explosion, the tree search is shallow and applied on 9×9 boards only. Thus, this paper aims to explore the association of shallow and selective global tree search with Monte Carlo in 9×9 Go. This exploration is based on our current research with the Go-playing programs INDIGO and OLGA.

Section 2 describes the related work about Monte Carlo and tree search. Section 3 provides an example and the algorithm that associates tree search and Monte Carlo. Section 4 gathers the results of experiments proving the relevance of this approach. Section 5 discusses some questions raised by this approach. Finally, Section 6 provides perspectives and arrives at a conclusion.

2 Related Work

This section first relates works on Monte Carlo and games, and then discusses tree-search works that have inspired our present work.

2.1 Monte Carlo and Games

The term Monte Carlo has a very broad meaning, using the random function of the computer and averaging outcomes [14]. Simulated annealing is a refinement that includes a temperature which decreases during the simulation process [17]. Monte-Carlo simulations have already been used in other games than Go. Abramson has proposed the expected-outcome model, in which the proper evaluation of a game-tree node is the expected value of the game's outcome given random play from that node on. He showed that the expected outcome is a powerful heuristic and concluded that the expected-outcome model of two-player games is "precise, accurate, easily estimable, efficiently calculable, and domain-independent" [1]. In games containing either randomness or hidden information, the use of simulations has nothing surprising. POKI uses simulations at Poker [4], and MAVEN at Scrabble [22]. Tesauro and Galperin have tried "truncated rollouts" in Backgammon by using a parallel approach [23]. In Go, the information is not hidden and randomness is absent, apparently yielding very little interest for simulations. However, ten years ago, Brügmann showed the adequacy of simulated annealing in Go, with his program GOBBLE [10]. Recently, Kaminski has performed Brügmann's experiment again with his program VEGOS [16]. Last year, Bouzy and Helmstetter studied Monte-Carlo Go programs, experimentally demonstrating their effectiveness on 9×9 boards [9]. Since then, Bouzy has successfully associated Monte Carlo and knowledge in his program [6], yielding INDIGO2003.

2.2 Tree-Search Works Relative to Our Study

The works about tree search and games are numerous. This subsection only mentions the works relative to our aim of integrating Monte Carlo and tree search. Because Monte Carlo averages samples of terminal position evaluations, we are most interested in studies assuming that the position evaluation is not a value but a set of values, such as a probability distribution or a sample of values. Palay suggested the use of a back-up rule when the evaluation is a probability distribution [19]. It has been used in the work by Baum and Smith about the Bayesian player [2], and applied to Othello. Moreover, Berliner proposed the B* algorithm [3], and Korf and Chickering described a general best-first min-max algorithm [18] that also inspired our work.

Buro's PROBCUT algorithm uses the results of shallow tree searches to prune moves [12]. Junghanns surveyed all the ALPHA-BETA works [15]. Rivest studied a back-up rule using a complex formula [20], using an exponent p. When $p = 0$, the formula yields the classical min-max back-up rule and when $p = \infty$, it gives the average back-up, that is a feature of Monte Carlo. Sadikov, Kononenko, and

Bratko have shown that evaluations containing errors introduce a bias in the min-max values of the tree. The bias varies in the search depth, but remains constant for two sibling nodes [21]. This would explain the success of tree search in practice, although, in theory, pathologies exist in the game tree. Chen has experimentally shown the effect of selectivity during tree search in Go [13].

3 Our Work

This section describes our work based on Go-playing programs. First, it defines the names of the programs mentioned along the paper. Second, it gives an intuitive view of the requirements. Third, it uncovers the algorithm that we used for 9×9 Go. Fourth, it shows the algorithm performing on an example. Finally, it highlights two enhancements to control the width of the tree.

3.1 The Programs' Names

Our work is based on experiments about Go-playing programs. Let us start by clearly defining the programs' names used in this paper. First, INDIGO is the generic name of the program we have been developing over the past years [5]. It regularly attends computer-Go competitions. Each year, we set up a new release of this program, and INDIGO2002 corresponds to INDIGO's release at the end of 2002. INDIGO2002 was mainly based on knowledge and tree search [7], and not on Monte Carlo. Second, OLGA is the name of the working release of INDIGO. Each year, we work on OLGA, and if our work turns out to be effective, then OLGA's effective features are integrated into INDIGO. In 2003, we processed our work in three stages. The first improvement tested in OLGA was the Mont-Carlo approach, thus, at the beginning of 2003, OLGA was a little-knowledge Monte-Carlo Go program described in [9]. Since Monte Carlo worked well, we also integrated knowledge into OLGA in a second stage, which was satisfactory and described in [6]. Furthermore, we added selective global tree search in OLGA in a final stage. Thus, in this paper, OLGA refers to a program containing Monte Carlo, knowledge, and selective global tree search. As this paper aims at describing the selective global tree-search aspect with different tree-search parameters, say X and Y, it might be useful to write OLGA($X = x$, $Y = y$) to refer to OLGA using the particular values x and y. By the end of 2003, OLGA was better than INDIGO2002, so we copied OLGA into INDIGO2003 which is consequently a knowledge-based tree-search and Monte-Carlo Go program too. INDIGO2003 attended the computer-Go olympiad held in Graz. Finally, we also mention OLEG in this paper because it is the Monte-Carlo Go program containing little knowledge which was developed by Bernard Helmstetter [9].

3.2 The Requirement

On the one hand, OLGA and OLEG with little knowledge, described in [9], were performing a depth-one global search without any selectivity. The attempt of performing a depth-two search failed because the branching factor of 9×9 game (about 80 in the beginning of a game) was too high. On the other hand, OLGA with knowledge, described in [6], was performing a depth-one global search with high selectivity adapted for 19x19 board. With such a selectivity, OLGA at depth one plays very quickly on 9×9 boards. Thus, the idea of the current work consists in improving 9×9 OLGA with a depth-n selective global tree search.

However, we should not overlook the idea of progressive pruning [9]. This is the back-up of average values on the evaluations of the random games that must drive the process. When many moves are equal, we want the process to use as much CPU time as needed to discriminate between them, while when one move is clearly superior to the other, we want the process to find it out quickly.

By pruning the bad moves quickly, the progressive pruning algorithm is rather efficient at the beginning of the process. However, the longer the algorithm performs, the fewer moves it prunes. At the end of the process, when few moves remain, many random games may be necessary to separate those moves to keep the best one. When the separation requires too much CPU time at depth one, would it be relevant to expand these moves to depth-two in order to speed up the process? Very often, looking one move ahead enables the player to observe a difference between the effect of moves. Thus, we need an algorithm that prefers expanding nodes one depth further to perform random games starting at this depth, rather than passively awaiting the end of the current depth process. This requirement remains valid at any depth. If some moves are nearly equal at a given depth, expanding them to the next depth is worthwhile.

The statistical evaluation of a node consists in the expectation of sampled evaluations, given that the sequence of moves from the root until this node has been played out. When considering two sibling leaf nodes, the two players have played the same number of moves, and their expected evaluation can be compared. When considering a parent node and its children, the parent node updates an expected value assuming that a given sequence of moves has been played out while the children have expectation values assuming one additional move has been played after the given sequence. Thus, the statistical evaluation of a parent node cannot be compared to its children's one. In addition, in the min-max context, the parent node must compute its min-max value with the statistical evaluations of its children. In this context, it is inappropriate to compare the min-max value of the parent node with its own statistical evaluation. Now, considering two sibling nodes, the first one being a leaf node with a statistical evaluation, and the second one being a parent node of leaf nodes with statistical evaluations, it is inappropriate to compare the statistical evaluation of the first node with the min-max value of the second one. This remark can be extended to the comparison of min max values of sibling nodes whose sub-trees have different depths. In conclusion, we need an algorithm that compares sibling nodes whose sub-trees have the same depth.

Moreover, Sadikov et al. [21] have shown that min-max back-ups on evaluations containing errors introduce a bias in the min-max values of the tree. As the bias depends on the search depth, we need a fixed-depth searching algorithm. Besides, the greater the error in the evaluations, the greater the bias in the min-max value. Consequently we need enough random games to lower the evaluation errors. When the algorithm reaches the maximal depth, it has to perform a sufficient number of random games.

To sum up, we need an algorithm that prunes some moves at depth one, expand the remaining ones to depth-two, run random games starting at depth-two, prune some depth-two moves, and so on, increasing the search depth iteratively either until one move remains at the root or until the maximal depth is reached.

3.3 The Algorithm

This subsection yields the algorithm answering the requirements. Before showing the algorithm, we define the two classes of interest: Node and J_stat (representing the statistical player). We do not mention the properties of a node not related to progressive pruning.

```
class Node {
  float mean;
  float mean_sd;
  int n_children;
  bool all_are_equal;
}
```

While the node is terminal, the slot mean contains the mean value of the sample of evaluations. When the node becomes interior to the tree, this slot contains the min-max value of the node. mean_sd is the standard deviation of the mean. Its value is used by the progressive pruning rule. The slot n_children contains the number of remaining moves in the progressive pruning meaning [9]. The slot all_are_equal indicates whether all remaining moves are "equal" or not in the progressive pruning meaning. This slot is used to check the end of the algorithm's internal loop.

```
class J_Stat {
  int    r_games_p_depth;
  int    depth;
  int    depth_max;
  int    width_p;
  int    width_m;
  Node   root;
}
```

r_games_p_depth is the current number of random games performed while processing a given depth. The slot depth is the current depth of the algorithm at which random games are started. The slot depth_max is the maximal depth at which the process performs the random games. root is the root node of the tree. width_p is the number of moves selected by the knowledge-based move generator. In other words, it is the maximal branching factor of the tree which is developed by the algorithm. While it does not appear explicitly in the pseudo-code below, it is a parameter of importance in the experiments. width_m enables the algorithm to control the tree growing. Its use will be explained in subsection 3.5. Other slots are useful but have been omitted for clarity.

```
int J_Stat.choose_move() {
  depth = 0;
  width_p = WIDTH_PLUS;   // 5, 7, 9 or 11
  width_m = WIDTH_MINUS;  // 2, 3, 4 or 5
  int n = N_R_GAMES;      // 2500
  do {
    depth = depth + 1;
    generate_nodes(depth);
    n *= 1+root.n_children;
    process(depth, n);
  } while ((root.n_children>1) && (depth<depth_max));
  return root.best_move();
}
```

The function choose_move() is the solution we offer. It returns the move chosen by the algorithm. The function best_move() returns the best move of a node. The function generate_nodes(int d) generates the nodes at depth d. The function process(int depth, int max) is defined below.

```
void J_Stat.process(int depth, int
max) {
  r_games_p_depth = 0;
  int max_r = width_p ;
  while (    (root.n_children>1)
          && (!root.all_are_equal)
          && (r_games_p_depth<max)
          && ((max_r>width_m)||(depth==depth_max))
        ) {
    perform_random_games(depth);
    for (int d=depth-1; d>=0; d--) {
      int r = update_remaining_moves(d);
      if (d==depth-1) max_r = r;              // (a)
      update_min_max_values(d);
    }
  }
  cut_nodes(depth, width_m);                  // (b)
}
```

Earlier in this subsection, we promised to explain the basic version of the algorithm only. The two lines containing a comment are not part of the basic version. They refer to enhancements for controlling the width of the tree, explained in subsection 3.5. The other lines explained below correspond to the basic version.

The function perform_random_games(int d) performs random games starting at depth d. After each random game, it updates mean and mean_sd of the current node. The function update_remaining_moves(int d) updates the remaining moves of nodes situated at depth d with the progressive pruning rule.

It updates n_children and all_are_equal of the current node. The function update_min_max_values(int d) applies the min-max back-up rule to nodes situated at depth d. After each min-max back-up, it updates mean and mean_sd.

To sum up, the algorithm is similar to iterative deepening. It stops when there is only one remaining move left at the root, or when the maximal depth is reached. Each depth has its own specificity. At the root, the goal is to find out the best move. At maximal depth, the random games are started. At non maximal depth, the progressive pruning rule and the min-max rule are applied. In order to perform all these updates, the whole tree is, of course, stored in the computer's memory.

3.4 An Example

This subsection shows how the algorithm works on a simple example. At the beginning, the moves are generated from the root node, resulting in the tree on the left of Figure 1. The leaf nodes are drawn with a white circle, and other nodes are gray. For clarity, we use width_p = 4. Thus, the root is expanded

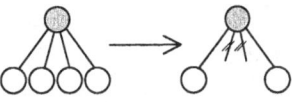

Fig. 1. Performing depth 1

with four children. Then, the function process() runs at depth one, and during this process two moves are pruned leading to the tree situated on the right of Figure 1. Let us assume that the ending conditions are true for depth one.

Thus, the leaf nodes are expanded into the tree drawn on the left of Figure 2. Again, the random games are performed starting on depth-two nodes, and some moves are pruned leading to the tree located on the right of Figure 2. Assuming that the ending conditions are true for this depth, the leaf nodes are expanded once more to bring about the tree drawn on the left of Figure 3.

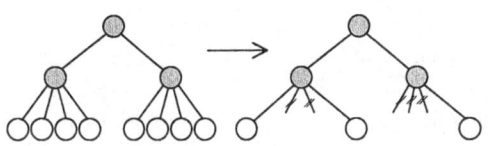

Fig. 2. Performing depth 2

At depth three, the algorithm prunes other nodes, leaf nodes or interior nodes. For example, as shown by the tree drawn in the middle of Figure 3, it prunes a node situated at depth two, pointing out the possibility of pruning an interior node. Finally, a move situated at the root is pruned. This corresponds to the tree situated on the right of figure 3. Because one move is left at the root, the algorithm stops and returns this move.

3.5 Controlling the Width of the Tree

This subsection highlights the way in which the algorithm drastically controls the width of the tree. The instructions enabling the algorithm to control the width of the tree were indicated with a comment in subsection 3.3. The comment marked

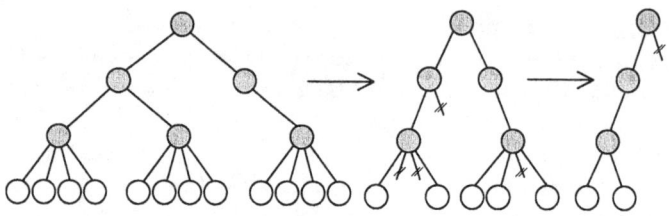

Fig. 3. Performing depth 3

up with (a) apply to expanding nodes difficult to discriminate at the current depth. The comments marked up with (b) mention the node number limitation after processing a given depth. This last enhancement is useful when the loop ends up with either r_games_p_depth reaching max or root.all_are_equal being true. max_r is the maximum of n_children over all the nodes situated at depth - 1. width_m is a threshold that determines the maximal width of the tree from depth 0 to depth - 2. In the example shown in subsection 3.4, it is set up with the value 2. With the enhancement (a), the algorithm does not perform random games if max_r reaches the threshold width_m and if the depth is not depth_max. Instead, the algorithm goes to the next depth. With the enhancement (b), the processing of a given depth always terminates with less than width_m nodes at depth-1. Of course, the lines marked up with comments can be removed and the algorithm may work without them.

4 Experiments

In this section, we provide the results of the experiments carried out on 9×9 boards with the algorithm described above. An experiment is a set of confrontations. One confrontation consists in a match of 100 games between 2 programs, each program playing 50 games with Black. The result of such a confrontation is the mean score and a winning percentage. Given that the standard deviation of games played on 9×9 boards is roughly 15 points, 100 games enable our experiments to lower σ down to 1.5 point and to obtain a 95% confidence interval of which the radius equals 2σ, i.e., 3 points. We have used 2.4 GHz computers, and we mention the response time of each program. The variety of games is guaranteed by the random seed values that are different from one game to another. The result of an experiment is generally a set of relative scores provided by a table assuming that the program of the column is the max player.

Subsection 4.1 highlights the relative strengths of programs using different depths. Subsection 4.2 underlines the relative strengths of programs using different widths. Then, subsection 4.3 shows the relative strengths of OLGA against GNUGO 3.2 [11]. Finally, subsection 4.4 mentions the result of INDIGO at the 9×9 Go competition held during the 8th Computer Olympiad.

4.1 Making the Depth Vary

This subsection contains the results of the experiment making depth_max vary. We set up different instances of OLGA, each of them using their own value of depth_max. In the following, d and *Depth* refer to depth_max for short. For the same reason, in the following sections, $W+$ refers to width_p, and $W-$ to width_m.

In the first experiment, each instance of OLGA uses $W+ = 7$, and $W- = 3$. Table 1 summarizes the results of the confrontations of OLGA(*Depth* = d) versus OLGA(*Depth* = $d - 1$) and OLGA(*Depth* = $d - 2$) for d ranging from 2 up to 5. The table mentions the mean score and the winning percentage assuming that the program of the column is the max player. The difference between OLGA(*Depth* = 2) and OLGA(*Depth* = 1) is clear as well as the difference between OLGA(*Depth* = 3) and OLGA(*Depth* = 2). These two confrontations experimentally prove the relevance of global tree search with Monte Carlo. For higher values of *Depth*, the upside is less significant, which confirms the fact that the returns diminish when the search depth increases. The results of depth five against depth four, and depth four against depth three are below the 3 points' threshold giving 95% confidence in the superiority of one program over the other one. More games should be performed to obtain a statistically significant conclusion. The winning percentages of OLGA($d = 5$) over OLGA($d = 4$) and over OLGA($d = 3$) are the same. This is not a mistake, but it results from a too low number of games. A similar remark can be made about the winning percentages of OLGA($d = 4$) over OLGA($d = 3$) and over OLGA($d = 2$).

In addition, Table 2 summarizes the results of OLGA(*Depth* = d) versus OLGA(*Depth* = $d - 1$) and versus OLGA(*Depth* = $d - 2$) with $W+ = 9$ and $W- = 4$. The results of this experiment confirms the results of the first one. Going up to depth two from depth one, returns about 8 points and, going up to depth three from depth two returns about 7 points. But the returns diminish at depth three and even deeper. As in the previous table, many more results should be collected to conclude on the superiority of one program over the other one.

Table 1. OLGA(*Depth* = d) versus OLGA(*Depth* = $d - 1$) and OLGA(*Depth* = $d - 2$), $W+ = 7, W- = 3$

d	2	3	4	5
d-1	+7.7 61%	+7.6 63%	+1.8 54%	+1.0 52%
d-2		+12.8 70%	+5.9 54%	+4.2 52%

Table 2. OLGA(*Depth* = d) versus OLGA(*Depth* = $d - 1$) and OLGA(*Depth* = $d - 2$), $W+ = 9, W- = 4$

d	2	3	4	5
d-1	+8.4 67%	+7.0 65%	+1.2 54%	+2.1 53%
d-2		+11.5 72%	+3.2 60%	+4.2 60%

4.2 Making the Width Vary

This subsection contains the results of the experiment making $W-$ vary. In this experiment, each instance of OLGA uses $Depth = 3$ and $W+ = 1 + 2W-$. Table 3 summarizes the results of the confrontations of OLGA($W- = w$) versus OLGA($W- = w - 1$) and OLGA($W- = w - 2$) for w ranging from 3 up to 5. The table mentions the mean score and the winning percentage assuming that the program of the column is the max player.

Table 3. OLGA($Width = d$) versus OLGA($W- = w - 1$) and OLGA($W- = w - 2$), with $Depth = 3$

w	3	4	5
w-1	+1.5 57%	+3.8 61%	+0.1 51%
w-2		+5.9 62%	+0.6 51%

The table shows that going from $W- = 3$ up to $W- = 4$ is worth considering. About four points are gained on average. But, going up to $W- = 5$ does not seem satisfactory. We can explain this fact by the good move ordering of the knowledge move generator for the first ranked moves and the bad ordering for the moves ranked after the fifth position.

Table 4 yields the CPU time in minutes used by OLGA($Depth$, $W-$) for playing one 9×9 game, for $Depth$ ranging from 1 up to 5 and $W-$ ranging from 2 up to 5.

Table 4. CPU time used by OLGA($Depth = d$, $W- = w$) on 2.4 Ghz computers

	d=1	d=2	d=3	d=4	d=5	d=6
w=3	2	5	10	16	24	35
w=4	3	7	12	27	42	55
w=5	3	10	22	44	65	100
w=6	3	10	30	60	100	160

This table shows that time increases with both $Depth$ and $W-$. Considering our Computer-Olympiad program, OLGA($Depth = 3$, $W- = 3$), is it better to increase $W-$ or $Depth$? Choosing between the two possibilities is not obvious. Increasing $Depth$ seems better than increasing $W-$ because no programming effort has to be made to increase $Depth$, while the knowledge-based move generation has to be improved if we want to increase $W-$.

4.3 Playing Against GNUGO 3.2

To measure the effect of the variation in the depth and the width of the tree, we have set up confrontations between OLGA($Depth$, $W-$) and GNUGO 3.2.

Table 5. Mean score obtained by OLGA($Depth = d$, $W- = w$) against GNUGO 3.2

	d=1	d=2	d=3	d=4	d=5	d=6	total
w=3	-5.4	-3.2	-5.7	-6.3	-6.5	-1.5	-4.8
w=4	-6.5	-8.7	-3.0	-5.0	-4.6	-3.7	-5.2
w=5	-8.1	-6.8	-4.1	-2.7	-2.9	-1.0	-4.3
w=6	-6.6	-5.1	-1.8	-0.8	-3.1		-3.5
total	-6.6	-6.0	-3.7	-3.7	-4.3	-2.1	-4.4

Table 6. Winning percentage obtained by OLGA($Depth = d$, $W- = w$) against GNUGO 3.2

	d=1	d=2	d=3	d=4	d=5	d=6	total
w=3	33	40	33	37	34	44	37
w=4	33	32	40	39	44	41	38
w=5	40	36	38	37	47	45	40
w=6	36	38	48	49	45		42
total	35	36	40	41	42	43	39

Table 5 shows the mean scores, and Table 6 shows the winning percentages. The results are given from OLGA's viewpoint. For each table, the last column (line respectively) indicates the mean of the previous columns (lines respectively).

First, the total line of Table 6 shows a correlation between $Depth$ and the winning percentage. Second, the total column of Table 6 also shows a correlation between $W-$ and the winning percentage. On average, OLGA wins 39% of the games. Third, in terms of mean score obtained by OLGA, the correlation still appears, but less clearly. On average, the mean score equals -4.4. We believe that the correlation clearly exists within self-play, because other elements than tree search remain constant. The correlation observed within self-play diminishes against differently designed opponents such as GNUGO. Unlike GNUGO, OLGA still lacks important elements such as a good life-and-death move generator or a good territory move generator. Consequently, to improve OLGA, some effort should be made on such elements that are different from global tree search and Monte Carlo. We did not compute the data for $w > 6$ and $d > 6$ because of the memory constraints: each internal node of the tree contains a whole board with its knowledge, and the whole tree is kept in the computer's memory. Hence, it is impossible to yield the results for $(w, d) = (6, 6)$.

4.4 9×9 Competition at the 2003 Computer Olympiad

INDIGO, a copy of OLGA(d=3, w=3), has participated in the 9×9 Go competition during the 8th Computer Olympiad in Graz, in November 2003. INDIGO ranked 4th upon 10 programs with 11 wins and 7 losses, which was a reasonable result, demonstrating the relevance of this approach against other differently designed programs.

5 Discussion

In this section we mention the advantages of the approach in term of complexity. Then, we discuss the relevance of adding classical tree-search enhancements within our statistical search. Finally, we mention the possibility of scaling the results up to 19×19.

5.1 Complexity of the Approach

W being the width of the tree and $Depth$ the search depth, full-width and full-depth tree search algorithms have a time complexity in W^{Depth}. Assume N to be the number of random games per candidate move, then depth-one Monte Carlo has a time complexity in $N \times W \times Depth$. The Monte-Carlo and tree-search approach developed in this work, has a time complexity in $N \times W_+ \times Depth \times W_-^{Depth_{max}-1}$ because it starts a depth-one Monte-Carlo tree search at leaf nodes of a tree whose width equals W_-, and whose depth equals $Depth_{max} - 1$. Thus, on 9×9 boards, the time complexity can fit the computing power by adjusting $Depth_{max}$, W_+, and W_- appropriately, and the program using this hybrid approach is endowed with some global tactical ability. Besides, the space complexity should also be underlined. The computer's memory is mostly occupied by internal nodes containing a board with its domain-dependent knowledge: the matched patterns, and the global evaluation. The size of leaf nodes is not taken into account because they only contain statistical information. Furthermore, the memory size occupied by the running random game is in $Depth$, and it is not taken into account either. At a depth inferior to $Depth_{max} - 1$, the tree branching factor being equal or inferior to W_-, the space complexity of the algorithm is in $W_-^{Depth_{max}-1}$.

5.2 Classical Tree-Search Enhancements

Our algorithm uses the min-max back-up rule and approximately follows the idea of iterative deepening. So far, it has not used the transposition-table principle. How could the transposition principle be integrated into our algorithm? When two leaf nodes refer to the same position, it would be interesting to merge the two samples. However, this enhancement is not urgent because the depths remain rather shallow at the moment, not greater than five.

Since PROBCUT includes the idea of correlation of position evaluations situated at different depths [12], how to establish the link between our algorithm and PROBCUT? Before performing a deep search, PROBCUT performs a shallow tree search to obtain rough alpha-beta values and prune moves with some confidence level. In our work, before performing the next depth search, the algorithm prunes moves by using the results of the current depth search. In this respect, at root node, our algorithm corresponds to a simple version of PROBCUT.

5.3 Scaling Up to 19×19

On 2.4 GHz computers, the algorithm performs well on 9×9 with $Depth_{max} = 3$ and $W+ = 7$ in about 10 minutes. The same algorithm plays on 19×19 with

$Depth_{max} = 1$ and $W+ = 7$ in about 50 minutes. Knowing the time used by the algorithm to play a 19×19 game with $Depth_{max} = 3$ and $W+ = 7$, and knowing the number of points corresponding to the self-play improvement are worth considering. Actually, a ten-game test shows that a 19×19 game lasts about 2 hours for $Depth_{max} = 2$, and 6 hours for $Depth_{max} = 3$. A ten-point improvement is observed with $Depth_{max} = 2$, and a fifteen-point improvement with $Depth_{max} = 3$, which has nothing exceptional. Of course, this assessment is not statistically significant, and it must be carried out again with more games in a few years' time. In conclusion, on 19×19 boards, as described in [6], instead of increasing $Depth_{max}$, our current strategy consists in increasing $W+$.

6 Perspectives and Conclusion

Various perspectives can be considered. First, to make it more general, we wish to apply our algorithm to another mind game. The game of Amazons may be a relevant choice. Besides, since most of the thinking time of the program is spent at the beginning of the game, we want to develop an opening book. Finally, as 5×5 Go was solved by [24], we also plan to apply this algorithm on small boards ranging from 5×5 up to 9×9.

To sum up, we have issued the algorithm that INDIGO used during the 9×9 Go competition at the last Computer Olympiad held in Graz in November 2003. This algorithm combines a shallow and selective global tree search with Monte Carlo. It illustrates the model developed by Abramson [1]. To our knowledge, this algorithm is new within the computer-Go community. It results from a work about Monte Carlo alone [9], and a work associating Monte Carlo and knowledge [6]. Following this line, the current work shows the association between Monte Carlo and tree search. As could be expected, we have observed an improvement when increasing the depth of the search, or when increasing the width of the tree. This improvement is clearer in the self-play context than against GNUGO. On today's computers, a tree search with $depth = 3$, $W+ = 7$, and $W- = 3$ offers the satisfactory compromise between time and level on 9×9 boards. However, depth-four and depth-five tree searches are possible, and furthermore they reach a better level. We believe that combining global tree search and Monte Carlo will strengthen Go programs in the future.

References

1. B. Abramson. Expected-outcome : a general model of static evaluation. *IEEE Transactions on PAMI*, 12:182–193, 1990.
2. E. Baum and W. Smith. A bayesian approach to relevance in game-playing. *Artificial Intelligence*, 97:195–242, 1997.
3. H. Berliner. The B* tree search algorithm: a best-first proof procedure. *Artificial Intelligence*, 12:23–40, 1979.
4. D. Billings, A. Davidson, J. Schaeffer, and D. Szafron. The challenge of poker. *Artificial Intelligence*, 134:201–240, 2002.

5. B. Bouzy. Indigo home page. www.math-info.univ-paris5.fr/~bouzy/INDIGO.html, 2002.
6. B. Bouzy. Associating knowledge and Monte Carlo approaches within a go program. In *7th Joint Conference on Information Sciences*, pages 505–508, Raleigh, 2003.
7. B. Bouzy. The move decision process of Indigo. *International Computer Game Association Journal*, 26(1):14–27, March 2003.
8. B. Bouzy and T. Cazenave. Computer Go: an AI oriented survey. *Artificial Intelligence*, 132:39–103, 2001.
9. B. Bouzy and B. Helmstetter. Monte-Carlo Go developments. In H. Jaap van den Herik, Hiroyuki Iida, and Ernst A. Heinz, editors, *10th Advances in Computer Games*, pages 159–174, Graz, 2003. Kluwer Academic Publishers.
10. B. Brügmann. Monte Carlo Go. www.joy.ne.jp/welcome/igs/Go/computer/mcgo.tex.Z, 1993.
11. D. Bump. GNUGO home page. www.gnu.org/software/gnugo/devel.html, 2003.
12. M. Buro. Probcut: an effective selective extension of the alpha-beta algorithm. *ICCA Journal*, 18(2):71–76, 1995.
13. K. Chen. A study of decision error in selective game tree search. *Information Sciences*, 135:177–186, 2001.
14. G. Fishman. *Monte-Carlo: Concepts, Algorithms, and Applications*. Springer, 1996.
15. A. Junghanns. Are there practical alternatives to alpha-beta? *ICCA Journal*, 21(1):14–32, March 1998.
16. P. Kaminski. Vegos home page. www.ideanest.com/vegos/, 2003.
17. S. Kirkpatrick, C.D. Gelatt, and M.P. Vecchi. Optimization by simulated annealing. *Science*, May 1983.
18. R. Korf and D. Chickering. Best-first search. *Artificial Intelligence*, 84:299–337, 1994.
19. A.J. Palay. *Searching with probabilities*. Morgan Kaufman, 1985.
20. R. Rivest. Game-tree searching by min-max approximation. *Artificial Intelligence*, 34(1):77–96, 1988.
21. A. Sadikov, I. Bratko, and I. Kononenko. Search versus knowledge: an empirical study of minimax on KRK. In H. Jaap van den Herik, Hiroyuki Iida, and Ernst A. Heinz, editors, *10th Advances in Computer Games*, pages 33–44, Graz, 2003. Kluwer Academic Publishers.
22. B. Sheppard. World-championship-caliber scrabble. *Artificial Intelligence*, 134:241–275, 2002.
23. G. Tesauro and G. Galperin. On-line policy improvement using Monte Carlo search. In *Advances in Neural Information Processing Systems*, pages 1068–1074, Cambridge MA, 1996. MIT Press.
24. E. van der Werf, H.J. van den Herik, and J.W.H.M. Uiterwijk. Solving Go on small boards. *International Computer Game Association Journal*, 26(2):92–107, June 2003.

Learning to Estimate Potential Territory in the Game of Go

Erik C.D. van der Werf, H. Jaap van den Herik, and Jos W.H.M. Uiterwijk

Department of Computer Science, Institute for Knowlegde and Agent Technology,
Universiteit Maastricht, Maastricht, The Netherlands
{e.vanderwerf, herik, uiterwijk}@cs.unimaas.nl

Abstract. This paper investigates methods for estimating potential territory in the game of Go. We have tested the performance of direct methods known from the literature, which do not require a notion of life and death. Several enhancements are introduced which can improve the performance of the direct methods. New trainable methods are presented for learning to estimate potential territory from examples. The trainable methods can be used in combination with our previously developed method for predicting life and death [25]. Experiments show that all methods are greatly improved by adding knowledge of life and death.

1 Introduction

Evaluating Go positions is a difficult task [7,17]. In the last decade the game of Go[1] has received significant attention from AI research [5,16]. Yet, despite all efforts, the best Go programs are still weak. An important reason lies in the lack of an adequate full-board evaluation function. Building such a function requires a method for estimating potential territory. At the end of the game territory is defined as the intersections that are controlled by one colour. Together with the captured or remaining stones, territory determines who wins the game. For final positions (where both sides have completely sealed off the territory by stones of their colour) territory is determined by detecting and removing dead stones and assigning the empty intersections to their surrounding colour. Recently we have developed a system that learns to detect dead stones and that scores final positions at the level of at least a 7-kyu player [23]. Even more recently we have extended this system so that it is now able to predict life and death in non-final positions too [25].

In this paper we focus on evaluating non-final positions. In particular we deal with the task of estimating potential territory in non-final positions, which is much more difficult than determining territory in final positions. We believe that for both tasks predictions of life and death are a valuable component. We investigate several possible methods to estimate potential territory based on the

---
[1] For general information about the game including an introduction to the rules readers are referred to gobase.org [22].

predictions of life and death and compare them to other approaches, known from the literature, which do not require an explicit notion of life and death.

The remainder of this paper is organised as follows. First, in section 2 we define potential territory. Then, in section 3 we discuss five direct methods for estimating (potential) territory as well as two enhancements for supplying them with information about life and death. In section 4 we describe trainable methods for learning to estimate potential territory from examples. Section 5 presents our experimental setup. Then, in section 6 we present our experimental results. Finally, section 7 provides our conclusion and suggestions for future research.

2 Defining Potential Territory

The game of Go is played by two players, Black and White, who consecutively place a stone of their colour on an empty intersection of a square grid. At the start of the game the board is empty. During the game the moves gradually divide the intersections between Black and White. In the end the player who controls most intersections wins the game. Intersections that are controlled by one colour at the end of the game are called territory.

During the game human players typically try to estimate the territory that they will control at the end of the game. Moreover, they often distinguish between *secure territory*, which is assumed to be safe from attack, and *regions of influence*, which are unsafe. An important reason why human players like to distinguish secure territory from regions of influence is that, since the secure territory is assumed to be safe, they do not have to consider moves inside secure territory, which reduces the number of candidate moves to choose from.

In principle, secure territory can be recognised by extending Benson's method for recognising unconditional life [2], such as described in [15] or [24]. In practice, however, these methods are not sufficient to predict accurately the outcome of the game until the late end-game because they aim at 100 per cent certainty, which is assured by assumptions like losing all ko-fights, allowing the opponent to place several moves without the defender answering, and requiring completely enclosed regions. Therefore, such methods usually leave too many points undecided.

An alternative (probably more realistic) model of the human notion of secure territory may be obtained by identifying regions with a high confidence level. However, finding a good threshold for distinguishing regions with a high confidence level from regions with a low confidence level is a non-trivial task and admittedly always a bit arbitrary. As a consequence it may be debatable to compare heuristic methods to methods with a 100 per cent confidence level. Subsequently the debate continues when comparing among heuristic methods, e.g., a 77 per cent versus a 93 per cent confidence level (cf. Figure 1).

In this paper, our main interest is in evaluating positions with the purpose of estimating the score. For this purpose the distinction between secure territory and regions of influence is relatively unimportant. Therefore we combine the two notions into one definition of *potential territory*.

Definition 1. *In a position, available from a game record, an intersection is defined as potential territory of a certain colour if the game record shows that the intersection is controlled by that colour at the end of the game.*

Although it is not our main interest, it is possible to use our estimates of potential territory to provide a heuristic estimate of secure territory. This can be done by focusing on regions with a high confidence level, by setting an arbitrarily high threshold. In subsection 6.3 we will present results at various levels of confidence so that our methods can be compared more extensively to methods that are designed for regions with a high confidence level only.

3 Direct Methods for Estimating Territory

In this section we present five direct methods for estimating territory (subsections 3.1 to 3.5). They are known or derived from the literature and are easy to implement in a Go program. All methods assign a scalar value to each (empty) intersection. In general, positive values are used for intersections controlled by Black, and negative values for intersections controlled by White. In subsection 3.6 we mention two immediate enhancements for adding knowledge about life and death to the direct methods.

3.1 Explicit Control

The explicit-control function is obtained from the 'concrete evaluation function' as described by Bouzy and Cazenave [5]. It is probably the simplest possible evaluation function and is included here as a baseline reference of performance. The explicit-control function assigns +1 to empty intersections which are completely surrounded by black stones and −1 to empty intersections which are completely surrounded by white stones, all other empty intersections are assigned 0.

3.2 Direct Control

Since the explicit-control function only detects completely enclosed intersections (single-point eyes) as territory it performs quite weak. Therefore we propose a slight modification of the explicit-control function, called direct control. The direct-control function assigns +1 to empty intersections which are adjacent to a black stone and not adjacent to a white stone, −1 to empty intersections which are adjacent to a white stone and not adjacent to a black stone, and 0 to all other empty intersections.

3.3 Distance-Based Control

Both the explicit-control and the direct-control functions are not able to recognise larger regions surrounded by (loosely) connected stones. A possible alternative is the distance-based control (DBC) function. Distance-based control uses the Manhattan distance to assign +1 to each empty intersection which is closer to a black stone, −1 to each empty intersection which is closer to a white stone, and 0 to all other empty intersections.

3.4 Influence-Based Control

Although distance-based control is able to recognise larger territories a weakness is that it does not take into account the strength of stones in any way, i.e., a single stone is weighted equally important as a strong large block at the same distance. A way to overcome this weakness is by the use of influence functions, which were already described by the early researchers in computer Go, Zobrist [26] and Ryder [20], and are still in use in several of today's Go programs [8,9].

In this paper we adopt Zobrist's method to recognise influence; it works as follows. First, all intersections are initialised by one of three values: $+50$ if they are occupied by a black stone, -50 if they are occupied by a white stone, and 0 otherwise. (It should be noted that the value of 50 has no specific meaning and any other large value can be used in practice.) Then the following process is performed four times. For each intersection, add to the absolute value of the intersection the number of neighbouring intersections of the same sign minus the number of neighbouring intersections of the opposite sign.

3.5 Bouzy's Method

It is important to note that the repeating process used to radiate the influence of stones in the Zobrist method is quite similar to the dilation operator known from mathematical morphology. This was remarked by Bouzy [3] who proposed a numerical refinement of the classical dilation operator which is similar (but not identical) to Zobrist's dilation.

Bouzy's dilation operator D_z works as follows. For each non-zero intersection which is not adjacent to an intersection of the opposite sign, take the number of neighbouring intersections of the same sign and add it to the absolute value of the intersection. For each zero intersection with positive adjacent intersections only, add the number of positive adjacent intersections. For each zero intersection with negative adjacent intersections only, subtract the number of negative adjacent intersections.

Bouzy argued that dilations alone are not the best way to recognise territory. Therefore he suggested that the dilations should be followed by a number of erosions. This combined form is similar to the classical closing operator known from mathematical morphology.

To do this numerically Bouzy proposed the following refinement of the classical erosion operator E_z. For each non-zero intersection subtract from its absolute value the number of adjacent intersections which are zero or have the opposite sign. If this causes the value of the intersection to change its sign the value becomes zero.

The operators E_z and D_z are then combined by first performing d times D_z followed by e times E_z. Bouzy suggested the relation $e = d(d-1) + 1$ because this becomes the unity operator for a single stone in the centre of a sufficiently large board. He further recommended to use the values 4 or 5 for d.

The reader may be curious why the number of erosions is larger than the number of dilations. The main reason is that (unlike in the classical binary case) Bouzy's dilation operator propagates faster than his erosion operator. Furthermore, Bouzy's method seems to be more aimed at recognising secure territory with a high confidence level than Zobrist's method (the intersections with a lower confidence level are removed by the erosions). Since Bouzy's method leaves many intersections undecided it is expected to perform sub-optimal at estimating potential territory, which also includes regions with lower confidence levels (cf. subsection 6.3). To improve the estimations of potential territory it is therefore interesting to consider an extension of Bouzy's method for dividing the remaining empty intersections. A natural choice to extend Bouzy's method is to divide the undecided empty intersections using distance-based control. The reason why we expect this combination to be better than only performing distance-based control directly from the raw board is that radiating influence from a (relatively) safe base, as provided by Bouzy's method, implicitly introduces some understanding of life and death. (It should be noted that extending Bouzy's method with distance-based control is not the only possible choice, and extending with for example influence-based control provides nearly identical results.)

3.6 Enhanced Direct Methods

The direct methods all share one important weakness: the lack of understanding life and death. As a consequence, dead stones (which are removed at the end of the game) can give the misleading impression of providing territory or reducing the opponent's territory. Recognising dead stones is a difficult task, but many Go programs have available some kind of (usually heuristic) information about the life-and-death status of stones. We have this information, provided by our recently developed system which has been trained to predict life and death for non-final positions [25].

Here we mention two immediate enhancements for the direct methods. (1) The simplest approach to use information about life and death for the estimation of territory is to remove dead stones before applying one of the direct methods. (2) An alternative sometimes used is to reverse the colour of dead stones [4].

4 Trainable Methods

Although the direct methods can be improved by (1) removing dead stones, or (2) reversing their colour, neither approach seems optimal, especially because both lack the ability to exploit the more subtle differences in the strength of stones, which would be expressed by human concepts such as 'aji' or 'thickness'. However, since it is not well understood how such concepts should be modelled, it is tempting to try a machine-learning approach to train a general function approximator to provide an estimation of the potential territory. For this task we have selected the Multi-Layer Perceptron (MLP). The MLP has been used on similar tasks by several other researchers [10,11,12,21], so we believe it is a reasonable choice. Nevertheless it should be clear that any other general function approximator can be used for the task.

Our MLP has a feed-forward architecture which estimates potential territory on a per intersection basis. The estimates are based on a local representation which includes features that are relevant for predicting the status of the intersection under investigation. In this paper we test two representations, first a simple one which only looks at the raw configuration of stones, and second an enhanced representation that exploits additional information about life and death.

For our experiments we exploit the fact that the game is played on a square board with eight symmetries. Furthermore, positions with Black to move are equal to positions with White to move provided that all stones reverse colour. To simplify the learning task we remove the symmetries in our representation by rotating the view on the intersection under investigation to one canonical region in the corner, and reversing the colours if the player to move is White.

4.1 The Simple Representation

The simple representation is characterised by the configuration of all stones in the region of interest (ROI) which is defined by all intersections within a predefined Manhattan distance of the intersection under investigation. For each intersection in the ROI we include the following feature.

– *Colour:* $+1$ if the intersection contains a black stone, -1 if the intersection contains a white stone, and 0 otherwise.

The simple representation will be compared to the direct methods because it does not use any explicit information of life and death (although some knowledge of life and death may of course be learned from examples) and only looks at the local configuration of stones. Since both Zobrist's and Bouzy's method (see above) are diameter limited by the number of times the dilation operator is used, our simple representation should be able to provide results which are at least comparable. However, we actually expect it to do better because the MLP might learn some additional shape-dependent properties.

4.2 The Enhanced Representation

We enhanced the simple representation with features that incorporate explicit information about the life-and-death status of stones. Of course, we used our recently developed system for predicting life and death [25]. The most straightforward way to include the predictions of life and death would be to add these predictions as an additional feature for each intersection in the ROI. However, preliminary experiments showed that this was not the best way to add knowledge of life and death. (The reason is that adding features reduces performance due to peaking phenomena caused by the curse of dimensionality [1,14].) As an alternative which avoids increasing the dimensionality we decided to multiply the value of the colour feature in the simple representation with the estimated probability that the stones are alive. (This means that the sign of the value of an intersection indicates the colour, and the absolute value indicates some kind of strength.) Consequently, the following three features were added.

– *Edge:* encoded by a binary representation (board=0, edge=1) using a 9-bit string vector along the horizontal and vertical line from the intersection under investigation to the nearest edges.

- *Nearest colour:* the classification for the intersection using the distance-based control method on the raw board (black=1, empty=0, white=−1).
- *Nearest alive:* the classification for the intersection using the distance-based control method after removing dead stones (black=1, empty=0, white=−1).

5 Experimental Setup

In this section we discuss the data set used for training and evaluation (5.1) and the performance measures used to evaluate the various methods (5.2).

5.1 The Data Set

In the experiments we used our collection of 9×9 game records which were originally obtained from NNGS [18]. The games, played between 1995 and 2002, were all played to the end and then scored. Since the original NNGS game records only contained a single numeric value for the score, the fate of all intersections was labelled by a threefold combination of GNUGO [13], our own learning system, and some manual labelling. Details about the data set and the way we labelled the games can be found in [23].

In all experiments, test examples were extracted from games played in 1995, training examples were extracted from games played between 1996 and 2002. In total the test set contained 906 games, 46,616 positions, and 2,538,152 empty intersections. It is remarked that the 1995 games were also left out of the training set used for learning to predict life and death [25].

5.2 The Performance Measures

Now that we have introduced a series of methods (combinations of methods are possible too) to estimate (potential) territory, an important question is: how good are they? We attempt to answer this question (in section 6) using several measures of performance which can be calculated from labelled game records (see section 5.1). Although game records are not ideal as an absolute measure of performance (because the people who played those games surely have made mistakes) we believe that the performance averaged over large numbers of unseen game records is a reasonable indication of strength.

Probably the most important question in assessing the quality of an evaluation function is how well it can predict the winner at the end of the game. By combining the estimated territory with the (alive) stones we obtain the so-called area score, which is the number of intersections controlled by Black minus the number of intersections controlled by White. Together with a possible komi (which compensates the advantage of the first player) the sign of this score determines the winner. Therefore, our first performance measure P_{winner} is the percentage of positions in which the sign of the score is predicted correctly.

Our second performance measure P_{score} uses the same score to calculate the average absolute difference between the predicted score and the actual score at the end of the game.

Both P_{winner} and P_{score} combine predictions of stones and territory in one measure of performance. As a consequence these measures are not sufficiently informative to evaluate the task of estimating potential territory alone. To provide more detailed information about the errors that are made by the various methods we also calculate the confusion matrices (see section 6.1) for the estimates of potential territory alone.

Since some methods leave more intersections undecided (by assigning empty) than others, for example because they may have been designed originally for estimating secure territory only, it may seem unfair to compare them directly using only P_{winner} and P_{score}. As an alternative the fraction of intersections which are left undecided can be considered together with the performance on intersections which are decided. This typically leads to a trade-off curve where performance can be improved by rejecting intersections with a low confidence. The fraction of intersections that are left undecided, as well as the performance on the decided intersections is directly available from the confusion matrices of the various methods.

6 Experimental Results

We tested the performance of the various methods and in this section we present our experimental results. They are subdivided as follows: performance of direct methods in 6.1; performance of trainable methods in 6.2; comparing different levels of confidence in 6.3; and performance over the game in 6.4.

6.1 Performance of Direct Methods

The performance of the direct methods was tested on all positions from the labelled test games. The results for P_{winner} and P_{score} are shown in Table 1. In this table the columns 'remain' represent results without using knowledge of life and death, the columns 'remove' and 'reverse' represent results with predictions of life and death used to remove or reverse the colour of dead stones.

To compare the results of P_{winner} and P_{score} it is useful to have a confidence interval. Unfortunately, since positions of the test set are not all independent, it is non-trivial to provide exact results. Nevertheless it is easy to calculate lower and upper bounds, based on an estimate of the number of independent positions. If we pessimistically assume only one independent position per game an upper bound (for a 95% confidence interval) is roughly 3% for P_{winner} and 1.2 points for P_{score}. If we optimistically assume all positions to be independent a lower bound is roughly 0.4% for P_{winner} and 0.2 points for P_{score}. Of course this is only a crude approximation which ignores the underlying distribution and the fact that the accuracy increases drastically towards the end of the game. However, given the fact that the average game length is around 50 moves it seems safe to assume that the true confidence interval will be somewhere in the order of 1% for P_{winner} and 0.4 points for P_{score}.

More detailed results about the estimations (in percentages) for the empty intersections alone are presented in the confusion matrices shown in Table 2. The

Table 1. Average performance of direct methods

Predicted dead stones	P_{winner} (%)			P_{score} (points)		
	remain	remove	reverse	remain	remove	reverse
Explicit control	52.4	60.3	61.8	16.0	14.8	14.0
Direct control	54.7	66.5	66.9	15.9	12.9	12.7
Distance-based control	60.2	73.8	73.8	18.5	13.8	13.9
Influence-based control	61.0	73.6	73.6	17.3	12.8	12.9
Bouzy(4,13)	52.6	66.9	67.5	17.3	12.8	12.8
Bouzy(5,21)	55.5	70.2	70.4	17.0	12.3	12.4
Bouzy(5,21) + DBC	63.4	73.9	73.9	18.7	14.5	14.6

fraction of undecided intersections and the performance on the decided intersections, which can be calculated from the confusion matrices, will be discussed in subsection 6.3. (The rows of the confusion matrices contain the possible predictions which are either black (PB), white (PW), or empty (PE). The columns contain the actual labelling at the end of the game which are either black (B), white (W), or empty (E). Therefore, correct predictions are found on the trace, and errors are found in the upper right and lower left corners of the matrices.)

The difference in performance between (1) when stones remain on the board and (2) when dead stones are removed or reversed colour underlines the importance of understanding life and death. For the weakest direct methods reversing the colour of dead stones seems to improve performance compared to only removing them. For the stronger methods, however, it has no significant effect.

The best method for predicting the winner without understanding life and death is Bouzy's method extended with distance-based control to divide the remaining undecided intersections. It is interesting to see that this method also has a high P_{score} which would actually indicate a bad performance. The reason for this is instability of distance-based control in the opening, e.g., with only one stone on the board it assigns the whole board to the colour of that stone. We can filter out the instability near the opening by only looking at positions that occur after a certain minimal number of moves. When we do this for all positions with at least 20 moves made, as shown in Table 3, it becomes clear that Bouzy's method extended with distance-based control also achieves the best P_{score}. Our experiments indicate that radiating influence from a (relatively) safe base, as provided by Bouzy's method, outperforms other direct methods probably because it implicitly introduces some understanding of life and death. This conclusion is supported by the observation that the combination does not perform significantly better than for example influence-based control when knowledge about life and death is used.

At first glance the results presented in this subsection could lead to the tentative conclusion that for a method which only performs N dilations to estimate potential territory the performance keeps increasing with N; so the largest possible N might have the best performance. However, this is not the case and N should not be chosen too large. Especially in the beginning of the game a large N tends to perform significantly worse than a restricted setting with 4 or 5 di-

Table 2. Confusion matrices of direct methods

	B	W	E
PB	0.78	0.16	0
PW	0.1	0.88	0
PE	48.6	49.3	0.13

Explicit control

	B	W	E
PB	0.74	0.04	0
PW	0.04	0.82	0
PE	48.7	49.5	0.14

dead stones removed

	B	W	E
PB	0.95	0.09	0.01
PW	0.07	1.2	0.01
PE	48.4	49.0	0.12

dead colour reversed

	B	W	E
PB	15.4	4.33	0.02
PW	3.16	14.3	0.01
PE	30.9	31.6	0.1

Direct control

	B	W	E
PB	16.0	3.32	0.02
PW	2.56	15.2	0.02
PE	30.9	31.7	0.09

dead stones removed

	B	W	E
PB	16.6	3.44	0.02
PW	2.75	16.3	0.03
PE	30.0	30.6	0.08

dead colour reversed

	B	W	E
PB	36.0	11.6	0.05
PW	6.63	31.0	0.03
PE	6.86	7.71	0.06

Distance-based control

	B	W	E
PB	38.2	10.4	0.06
PW	6.21	34.3	0.06
PE	5.09	5.55	0.02

dead stones removed

	B	W	E
PB	38.0	10.3	0.05
PW	6.21	34.1	0.05
PE	5.29	5.87	0.04

dead colour reversed

	B	W	E
PB	37.4	12.2	0.07
PW	7.76	33.3	0.04
PE	4.25	4.79	0.03

Influence-based control

	B	W	E
PB	38.4	10.5	0.06
PW	7.02	35.3	0.06
PE	4	4.47	0.02

dead stones removed

	B	W	E
PB	38.4	10.5	0.06
PW	7.11	35.4	0.06
PE	3.98	4.46	0.02

dead colour reversed

	B	W	E
PB	17.7	1.82	0.02
PW	0.83	15.4	0.01
PE	30.9	33.1	0.11

Bouzy(4,13)

	B	W	E
PB	21.2	1.86	0.03
PW	1.06	19.9	0.03
PE	27.2	28.5	0.07

dead stones removed

	B	W	E
PB	21.2	1.9	0.03
PW	1.16	20.1	0.03
PE	27.0	28.3	0.08

dead colour reversed

	B	W	E
PB	19.0	1.87	0.02
PW	0.81	15.8	0.01
PE	29.6	32.6	0.1

Bouzy(5,21)

	B	W	E
PB	23	1.98	0.03
PW	1.13	20.8	0.04
PE	25.3	27.5	0.07

dead stones removed

	B	W	E
PB	22.9	1.99	0.03
PW	1.17	20.7	0.03
PE	25.3	27.6	0.08

dead colour reversed

	B	W	E
PB	37.9	12.1	0.05
PW	6.51	32.4	0.03
PE	5	5.73	0.05

Bouzy(5,21) + DBC

	B	W	E
PB	39.0	11.0	0.06
PW	6.3	34.8	0.06
PE	4.12	4.5	0.02

dead stones removed

	B	W	E
PB	38.9	10.9	0.05
PW	6.32	34.6	0.05
PE	4.28	4.74	0.03

dead colour reversed

lations such as used by Zobrist's method. Moreover, a too large N is a waste of time under tournament conditions.

6.2 Performance of Trainable Methods

Below we present the results of the trainable methods. All architectures were trained with the resilient propagation algorithm (RPROP) developed by Ried-

Table 3. Average performance of direct methods after 20 moves

Predicted dead stones	P_{winner} (%)			P_{score} (points)		
	remain	remove	reverse	remain	remove	reverse
Explicit control	55.0	66.2	68.2	16.4	14.5	13.3
Direct control	57.7	74.9	75.3	16.1	11.6	11.3
Distance-based control	61.9	82.1	82.1	16.4	9.6	9.7
Influence-based control	63.6	82.1	82.2	16.0	9.5	9.6
Bouzy(4,13)	56.7	77.9	78.3	17.2	10.4	10.5
Bouzy(5,21)	58.6	80.3	80.5	16.9	9.9	10
Bouzy(5,21) + DBC	66.7	82.2	82.3	15.5	9.6	9.7

miller and Braun [19]. The non-linear architectures all had one hidden layer with 25 units using the hyperbolic tangent sigmoid transfer function. (Preliminary experiments showed this to be a reasonable setting, though large networks may still provide a slightly better performance when more training examples are used.) For training, 200,000 examples were used. A validation set of 25,000 examples was used to stop training. For each architecture the weights were trained three times with different random initialisations, after which the best result was selected according to the performance on the validation set. (Note that the validation examples were taken, too, from games played between 1996 and 2002.)

We tested the various linear and non-linear architectures on all positions from the labelled test games. Results for P_{winner} and P_{score} are presented in Table 4, and the confusion matrices are shown in Table 5. The enhanced representation, which uses predictions of life and death, clearly performs much better than the simple representation. We further see that the performance tends to improve with increasing size of the ROI. (A ROI of size 24, 40, and 60 corresponds to the number of intersections within a Manhattan distance of 3, 4, and 5 respectively, excluding the centre point which is always empty.)

It is interesting to see that the non-linear architectures are not much better than the linear architectures. This seems to indicate that, once life and death has been established, influence spreads mostly linearly.

Table 4. Performance of the trainable methods

Architecture	Representation	ROI	P_{winner} (%)	P_{score} (points)
linear	simple	24	64.0	17.9
linear	simple	40	64.5	18.4
linear	simple	60	64.6	19.0
non-linear	simple	24	63.1	18.2
non-linear	simple	40	64.5	18.3
non-linear	simple	60	65.1	18.3
linear	enhanced	24	75.0	13.4
linear	enhanced	40	75.2	13.3
linear	enhanced	60	75.1	13.4
non-linear	enhanced	24	75.2	13.2
non-linear	enhanced	40	75.5	12.9
non-linear	enhanced	60	75.5	12.5

Table 5. Confusion matrices of trainable methods

	B	W	E
PB	40.5	13.6	0.08
PW	6.7	33.8	0.05
PE	2.3	2.9	0.01

Simple, linear, roi=24

	B	W	E
PB	41.5	14.2	0.08
PW	5.8	33.2	0.05
PE	2.1	2.9	0.01

Simple, linear, roi=40

	B	W	E
PB	42.0	14.6	0.08
PW	5.3	32.6	0.04
PE	2.2	3.1	0.01

Simple, linear, roi=60

	B	W	E
PB	40.5	13.6	0.07
PW	6.4	33.3	0.05
PE	2.6	3.5	0.02

Simple, non-linear, roi=24

	B	W	E
PB	41.4	14.0	0.08
PW	5.8	33.0	0.04
PE	2.4	3.3	0.02

Simple, non-linear, roi=40

	B	W	E
PB	41.8	14.2	0.0759
PW	5.5	33.0	0.04
PE	2.2	3.2	0.01

Simple, non-linear, roi=60

	B	W	E
PB	40.5	11.7	0.06
PW	7.0	36.4	0.07
PE	2.0	2.3	0.01

Enhanced, linear, roi=24

	B	W	E
PB	40.6	11.6	0.06
PW	6.8	36.3	0.07
PE	2.1	2.5	0.01

Enhanced, linear, roi=40

	B	W	E
PB	40.6	11.6	0.06
PW	6.6	36.2	0.07
PE	2.2	2.6	0.01

Enhanced, linear, roi=60

	B	W	E
PB	40.3	11.4	0.06
PW	6.8	36.2	0.06
PE	2.4	2.7	0.01

Enhanced, non-linear, roi=24

	B	W	E
PB	40.4	11.3	0.06
PW	6.8	36.4	0.06
PE	2.4	2.7	0.01

Enhanced, non-linear, roi=40

	B	W	E
PB	40.2	10.9	0.06
PW	6.8	36.6	0.07
PE	2.5	2.9	0.01

Enhanced, non-linear, roi=60

6.3 Comparing Different Levels of Confidence

The MLPs are trained to predict positive values for black territory and negative values for white territory. Small values close to zero indicate that intersections are undecided and by adjusting the size of the window around zero, in which we predict empty, we can modify the confidence level of the non-empty classifications. If we do this we can plot a trade-off curve which shows how the performance increases at the cost of rejecting undecided intersections.

In Figure 1 two such trade-off curves are shown for the simple MLP and the enhanced MLP, both non-linear with a ROI of size 60. For comparison, results for the various direct methods are also plotted. It is shown that the MLPs perform well at all levels of confidence. Moreover, it is interesting to see that at high confidence levels Bouzy(5,21) performs nearly as good as the MLPs.

Although Bouzy's methods and the influence methods provide numerical results, which could be used to plot trade-off curves, too, we did not do this because they would make the plot less readable. Moreover, for Bouzy's methods the lines would be quite short and uninteresting because they already start high.

6.4 Performance over the Game

In the previous subsections we looked at the average performance over complete games. Although this is interesting, it does not tell us how the performance

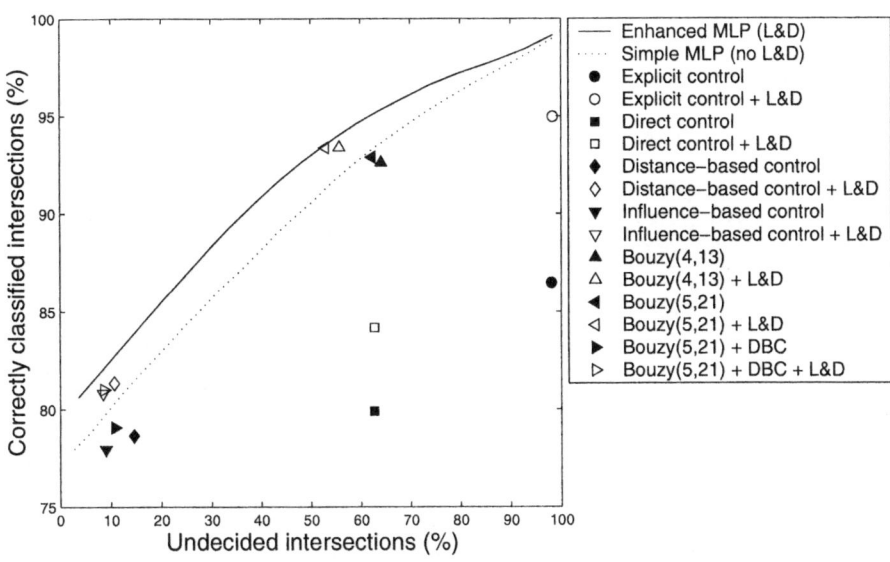

Fig. 1. Performance at different levels of confidence

changes as the game develops. Below we consider the performance changes and the adequacy of the MLP performance.

Since all games do not have equal length, there are two principal ways of looking at the performance. First, we can look forward from the start, and second, we can look backward from the end. The results for P_{winner} are shown in Figure 2a looking forward from the start and in Figure 2b looking backward from the end. We remark that the plotted points are between moves and their associated performance is the average obtained for the two directly adjacent positions (where one position has Black to move and the other has White to move). This was done to filter out some distracting odd-even effects caused by the alternation of the player to move. It is shown that the MLP using the enhanced representation performs best. However, close to the end Bouzy's method extended with distance-based control and predictions of life and death performs nearly as good. The results for P_{score} are shown in Figure 2c looking forward from the start and in Figure 2d looking backward from the end. Also here we see that the MLP using the enhanced representation performs best.

For clarity of presentation we did not plot the performance of DBC, which is rather similar to Influence-based control (IBC) (but over-all slightly worse). For the same reason we did not plot the results for DBC and IBC with knowledge of life and death, which perform quite similar to Bouzy(5,21)+DBC+L&D.

It is interesting to observe how good the simple MLP performs. It outperforms all direct methods without using life and death. Here it should be noted that the adequate performance of the simple MLP could still be improved considerably, if it would be allowed to make predictions for occupied intersections too, i.e., remove dead stones. (This was not done for a fair comparison with the direct methods.)

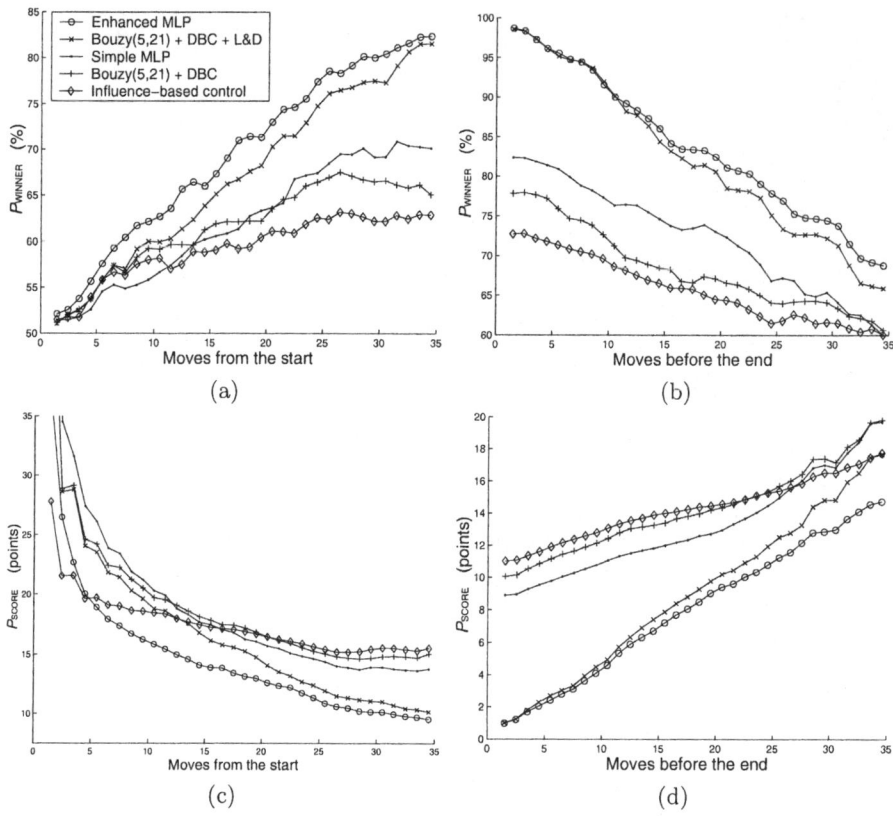

Fig. 2. Performance over the game

7 Conclusion and Future Research

We have investigated several direct and trainable methods for estimating potential territory in the game of Go. We tested the performance of the direct methods, known from the literature, which do not require an explicit notion of life and death. Additionally, two enhancements for adding knowledge of life and death and an extension of Bouzy's method were presented. From the experiments we may conclude that without explicit knowledge of life and death the best direct method is Bouzy's method extended with distance-based control to divide the remaining empty intersections. If information about life and death is used to remove dead stones this method also performs well. However, the difference with distance-based and influence-based control becomes small.

Moreover, we presented new trainable methods for estimating potential territory. They can be used as an extension of our system for predicting life and death. Using only the simple representation our trainable methods can estimate potential territory at a level outperforming the best direct methods. Experiments showed that all methods are greatly improved by adding knowledge of life and death, which leads us to conclude that good predictions of life and death are the most important ingredients for an adequate full-board evaluation function.

Future Research

Although our system for predicting life and death already performs quite well, we believe that it can still be improved significantly. The most important reason is that we only use static features, which do not require search. Incorporating features from specialised life-and-death searches should improve predictions of life and death as well as estimations of potential territory.

Previous work on learning to score final positions [23] indicated that our system for predicting life and death scales up well to the 19×19 board. Although we expect similar results for estimating potential territory, additional experiments should be performed to validate this claim.

In this paper we estimated potential territory based on knowledge extracted from game records. An interesting alternative for acquiring such knowledge may be obtaining it by simulation using, e.g., Monte Carlo methods [6].

Acknowledgements

This work was funded by the Netherlands Organisation for Scientific Research (NWO), dossier number 612.052.003. We would like to thank the anonymous referees for their helpful comments that significantly improved this paper.

References

1. R. E. Bellman. *Adaptive Control Processes: A Guided Tour.* Princeton University Press, Princeton, New Jersey, 1961.
2. D. B. Benson. Life in the game of Go. *Information Sciences*, 10:17–29, 1976. Reprinted in Computer Games, Levy, D.N.L (editor), Vol. II, pp. 203-213, Springer Verlag, New York, 1988. ISBN 0-387-96609-9.
3. B. Bouzy. Mathematical morphology applied to computer Go. *International Journal of Pattern Recognition and Artificial Intelligence*, 17(2):257–268, March 2003.
4. B. Bouzy, 2004. Personal communication.
5. B. Bouzy and T. Cazenave. Computer Go: An AI oriented survey. *Artificial Intelligence*, 132(1):39–102, October 2001.
6. B. Brügmann. Monte Carlo Go, March 1993. Available at ftp://ftp.cse.cuhk.edu.hk/pub/neuro/GO/mcgo.tex.
7. K. Chen. Some practical techniques for global search in Go. *ICGA Journal*, 23(2):67–74, 2000.
8. K. Chen. Computer Go: Knowledge, search, and move decision. *ICGA Journal*, 24(4):203–215, 2001.
9. Z. Chen. Semi-empirical quantitative theory of Go part i: Estimation of the influence of a wall. *ICGA Journal*, 25(4):211–218, 2002.
10. F. A. Dahl. Honte, a Go-playing program using neural nets. In J. Fürnkranz and M. Kubat, editors, *Machines that Learn to Play Games*, chapter 10, pages 205–223. Nova Science Publishers, Huntington, NY, 2001.

11. M. Enzenberger. The integration of a priori knowledge into a Go playing neural network, September 1996. Available at http://www.markus-enzenberger.de/neurogo1996.html.
12. M. Enzenberger. Evaluation in Go by a neural network using soft segmentation. In H. J. van den Herik, H. Iida, and E. A. Heinz, editors, *Advances in Computer Games: Many Games, Many Challenges*, pages 97–108, Boston, 2003. Kluwer Academic Publishers.
13. GNUGO, 2003.
14. A. Jain and B. Chandrasekaran. Dimensionality and sample size considerations in pattern recognition practice. In P. R. Krishnaiah and L. N. Kanal, editors, *Handbook of Statistics*, volume 2, pages 835–855. North-Holland, Amsterdam, 1982.
15. M. Müller. Playing it safe: Recognizing secure territories in computer Go by using static rules and search. In H. Matsubara, editor, *Proceedings of the Game Programming Workshop in Japan '97*, pages 80–86. Computer Shogi Association, Tokyo, Japan, 1997.
16. M. Müller. Computer Go. *Artificial Intelligence*, 134(1-2):145–179, 2002.
17. M. Müller. Position evaluation in computer Go. *ICGA Journal*, 25(4):219–228, 2002.
18. NNGS. The no name Go server game archive, 2002.
19. M. Riedmiller and H. Braun. A direct adaptive method for faster backpropagation: the RPROP algorithm. In H. Rusini, editor, *Proceedings of the IEEE Int. Conf. on Neural Networks (ICNN)*, pages 586–591, 1993.
20. J. L. Ryder. *Heuristic analysis of large trees as generated in the game of Go*. PhD thesis, Stanford University, 1971.
21. N. N. Schraudolph, P. Dayan, and T. J. Sejnowski. Temporal difference learning of position evaluation in the game of Go. In J. D. Cowan, G. Tesauro, and J. Alspector, editors, *Advances in Neural Information Processing 6*, pages 817–824. Morgan Kaufmann, San Francisco, 1994.
22. J. van der Steen. Gobase.org - Go games, Go information and Go study tools, 2003.
23. E. C. D. van der Werf, H. J. van den Herik, and J. W. H. M. Uiterwijk. Learning to score final positions in the game of Go. In H. J. van den Herik, H. Iida, and E. A. Heinz, editors, *Advances in Computer Games: Many Games, Many Challenges*, pages 143–158, Boston, 2003. Kluwer Academic Publishers.
24. E. C. D. van der Werf, H. J. van den Herik, and J. W. H. M. Uiterwijk. Solving Go on small boards. *ICGA Journal*, 26(2):92–107, 2003.
25. E. C. D. van der Werf, M. H. M. Winands, H. J. van den Herik, and J. W. H. M. Uiterwijk. Learning to predict life and death from Go game records. In Ken Chen et al., editor, *Proceedings of JCIS 2003 7th Joint Conference on Information Sciences*, pages 501–504. JCIS/Association for Intelligent Machinery, Inc., 2003.
26. A. L. Zobrist. A model of visual organization for the game Go. In *Proceedings of AFIPS 1969 Spring Joint Computer Conference*, volume 34, pages 103–112. AFIPS Press, 1969.

An Improved Safety Solver for Computer Go

Xiaozhen Niu and Martin Müller

Department of Computing Science,
University of Alberta, Edmonton, Canada
{xiaozhen, mmueller}@cs.ualberta.ca

Abstract. Most Go-playing programs use a combination of search and heuristics based on an influence function to determine whether territories are safe. However, to assure the correct evaluation of Go positions, the safety of stones and territories must be proved by an exact method.

The first exact algorithm, due to Benson [1], determines the unconditional safety of stones and completely surrounded territories. Müller [3] develops static rules for detecting safety by alternating play, and introduces search-based methods.

This paper describes new, stronger search-based techniques including region-merging and a new method for efficiently solving weakly dependent regions. In a typical final position, more than half the points on the board can be proved safe by our current solver. This almost doubles the number of proven points compared to the 26.4% reported in [3].

1 Introduction

This paper describes recent progress in building a search-based solver for proving the safety of stones and territories. The main application is late in the game, when much of the board has been partitioned into relatively small areas that are completely surrounded by stones of one player. In previous work [3], the analysis of such positions was done by a strict divide and conquer approach, analyzing one region at a time. The current solver implements several techniques that relax this strict separation approach. One technique merges several *strongly related* regions into a single one for the purpose of search. Another technique deals with how to search separately a set of *weakly dependent* regions in order to prove the safety of the union of all these regions.

This paper is organized as follows: the remainder of the introduction describes the terminology and previous work. Section 2 provides a formal framework for the safety prover, and Section 3 describes the solver: it gives an overview of the steps in the proving algorithm, describes the new technique of region merging and introduces the concept of weakly dependent regions. Section 4 deals with search enhancements, and Section 5 describes the experimental setup and results, followed by conclusions and future work in Section 6.

1.1 Preliminaries: Terminology and Go Rules

Our terminology is similar to [1, 3], with some additional definitions. Differences are indicated below. A *block* is a connected set of stones on the Go board. Each block has a number of adjacent empty points called *liberties*. A block that loses its last liberty is *captured*, i.e., removed from the board. A block that has only one liberty is said to be *in atari*. Figure 1 shows two black blocks and one white block. The small black block ▲ contains two stones, and has five liberties (two marked A and three marked B).

Given a color c, let $A_{\neg c}$ be the set of all points on the Go board which are *not* of color c. Then a *basic region* of color c (called a region in [1, 3]) is a maximal connected subset of $A_{\neg c}$. Each basic region is surrounded by blocks of color c. In this paper, we also use the concept of a *merged region*, which is the union of two or more basic regions of the same color. We will use the term region to refer to either a basic or a merged region. In Figure 1 A and B are basic regions and $A \cup B$ is a merged region.

Fig. 1. Blocks, basic regions and merged regions

We call a block b *adjacent* to a region r if at least one point of b is adjacent to one point in r. A block b is called *interior block* of a region r if it is adjacent to r but no other region. Otherwise, if b is adjacent to r and at least one more region it is called a *boundary block* of r. We denote the set of all boundary blocks of a region r by $Bd(r)$. In Figure 1, the black block ▲ is a boundary block of the basic region A but an interior block of the merged region $A \cup B$. The *defender* is the player playing the color of boundary blocks of a region. The other player is called the *attacker*.

Our results are mostly independent of the specific rule set used. As in previous work [1,3], suicide is forbidden. Our algorithm is incomplete in the sense that it can only find stones that are safe by two sure liberties [3]. This excludes cases such as conditional safety that depends on winning a ko, and also less frequent cases of safety due to double ko or snapback. The solver does not yet handle coexistence in *seki*.

1.2 Previous Work on Safety of Blocks and Territories

Benson's algorithm for *unconditionally alive blocks* [1] identifies sets of blocks and basic regions that are safe, even if the attacker can play an unlimited number of moves in a row, and the defender never plays. Müller [3] defines static rules for detecting safety by *alternating play*, where the defender is allowed to reply to each attacker move. [3] also introduces local search methods for identifying regions that provide one or two *sure liberties* for an adjacent block. Experimental results for a preliminary implementation in the program EXPLORER were presented for Benson's algorithm, static rules, and a 6-ply search.

Van der Werf implemented an extended version of Müller's static rules to provide input for his program that learns to score Go positions [6]. Vilà and Cazenave developed static classification rules for many classes of regions up to a size of 7 points [7]. These methods have not been implemented in our solver yet.

1.3 Contributions

The main contributions of this paper are as follows.

- A new method to merge regions and a search technique to prove that a merged region provides two sure liberties.
- A new divide-and-conquer analysis of weakly dependent regions.
- A greatly improved $\alpha\beta$ search routine for finding 2-vital regions (defined in Subsection 2.1), with improvements in move ordering, evaluation function, and pruning by recognizing forced moves.
- An improvement that takes external eyes of boundary blocks into account during the search.

2 Establishing the Safety of Blocks and Territories

Below we describe three definitions (in 2.1) and discuss the recognition of safe regions (in 2.2).

2.1 Definitions

The following definitions, adapted from [3], are the basis for our work. They are used to characterize blocks and territories that can be made safe under alternating play, by creating two sure liberties for blocks, and at the same time preventing the opponent from living inside the territories. During play, the liberty count of blocks may decrease to 1 (they can be in atari), but they are never captured and ultimately achieve two sure liberties.

Regions can be used to provide either one or two liberties for a boundary block. We call this number the *Liberty Target* $LT(b, r)$ of a block b in a region r. A search is used to decide whether all blocks can reach their liberty target in a region, under the condition of alternating play, with the attacker moving first and winning all ko fights.

Definition: Let r be a region, and let $Bd(r) = \{b_1, \ldots, b_n\}$ be the set of boundary blocks of r. Let $k_i = LT(b_i, r)$, $k_i \in \{1, 2\}$, be the liberty target of b_i in r. A defender strategy S is said to *achieve all liberty targets* in r if each b_i has at least k_i liberties in r initially, as well as after each defender move.

Each attacker move in r can reduce the liberties of a boundary block by at most one. The definition implies that the defender can always regain k_i liberties for each b_i with his next move in r. The following definition of life under alternating play is analogous to Benson's.

Definition: A set of blocks B is *alive under alternating play* in a set of regions R if there exist liberty targets $LT(b,r)$ and a strategy S that achieves all these liberty targets in each $r \in R$ and

$$\forall b \in B \quad \sum_{r \in R} LT(b,r) \geq 2$$

Note that this construction ensures that blocks will never be captured. Initially each block has two or more liberties. Each attacker move in a region r reduces only liberties of blocks adjacent to r, and by at most 1 liberty. By the invariant, the defender has a move in r that restores the previous liberty count. Each block in B has at least one liberty overall after any attacker move and two liberties after the defender's local reply.

It is easy to adapt this definition to the case where blocks have sure external liberties outside of R. The sum of liberty targets for such blocks can be reduced to 1, if the block has one sure external liberty, or 0, if the block is already safe.

Definition: We call a region r *one-vital* for a block b if b can achieve a liberty target of one in r, and *two-vital* if b can achieve a target of two.

2.2 Recognition of Safe Regions

The attacker *cannot live inside* a region surrounded by safe blocks if there are no two non-adjacent potential attacker eye points, or if the attacker eye area forms a *nakade* shape. Our current solver uses a simple static test for this condition as described in [3].

3 Methods for Processing Regions

We describe the structure of the safety solver (3.1), region merging (3.2), weakly dependent regions (3.3), and other improvements to the solver (3.4).

3.1 The Structure of the Safety Solver

Our safety solver includes five sub-solvers.

Benson solver – implements Benson's classic algorithm to analyze unconditional life.
Static solver – uses static rules to recognize safe blocks and regions under alternating play, as described in [3]. No search is used.
1-Vital solver – uses search to recognize regions that are 1-vital for one or more boundary blocks. As in [3] there is also a combined search for 1-vitality and connections in the same region, that is used to build chains of safely connected blocks.

Generalized 2-Vital solver – uses searches to prove that each boundary block can reach a predefined liberty target. For *safe blocks*, the target is 0, since their safety has already been established using other regions. Blocks that have one external eye outside of this region are defined as *external eye blocks*. For these blocks the target is 1. For all other non-safe blocks the target is 2 liberties in this region. All the search enhancements described in the next section were developed for this solver.

The 2-Vital solver in [3] could not handle external eye blocks, it would try to prove 2-vitality for all non-safe boundary blocks.

Expand-vital solver – uses searches to prove the safety of partially surrounded areas, as in [3]. This sub-solver can also be used to prove that non-safe stones can connect to safe stones in a region.

The basic algorithm of the safety solver is as follows.

1. The static solver is called first. It is very fast and resolves the simple cases.
2. The 2-Vital solver is called for each region. As a simple heuristic to avoid computations that most likely will not succeed, searches are performed only for regions up to size 30.
3. The Expand-vital solver is called for regions that have some safe boundary blocks. The safety of those blocks has been established by using other regions. Our previous solver in [3] only used the steps so far.
4. (New) Region merging. After the previous steps, all the easy-to-prove safe basic regions have been found. In this step the remaining unproven related regions are merged. For each small-enough merged region (up to size 14 in the current implementation) the generalized 2-Vital solver is called. The mechanism is described in detail in Subsection 3.2.
5. (New) Weakly dependent regions. A new algorithm deals with weakly dependent regions. In this step both the 1-Vital solver and the 2-Vital solver are used. A detailed description is given in Subsection 3.3.
6. (New) As in step 3, the Expand-vital solver is called for those regions for which one or more new safe boundary blocks have been found.

3.2 Region Merging

One of the major drawbacks of our previous solver is that it processes basic regions one by one and ignores the possible relationship between them. Figure 2 shows an example of two related regions. The previous solver treats regions A and B separately, and neither region can be solved. However the merged region $A \cup B$ can be solved easily.

The first algorithm step scans all regions and merges all related regions.

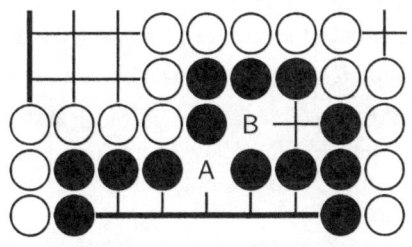

Fig. 2. Two related regions

Two regions are defined as *related* if they have a common boundary block. After the merging step, the 2-Vital solver is used to recognize safe merged regions.

This method can solve simple cases such as the one in Figure 2. However, since merging all related regions usually creates a very large merged region, the search space often becomes too large.

To improve the locality of search, we distinguish between *strongly dependent* regions, which share more than one common boundary block, and *weakly dependent* regions with exactly one common boundary block.

Our current solver uses a two-step merging process. In the first step, strongly dependent basic regions are merged. In the second step *groups of weakly dependent regions* are formed. A *group* can contain both basic regions and merged regions computed in the first step. Figure 3 shows an example.

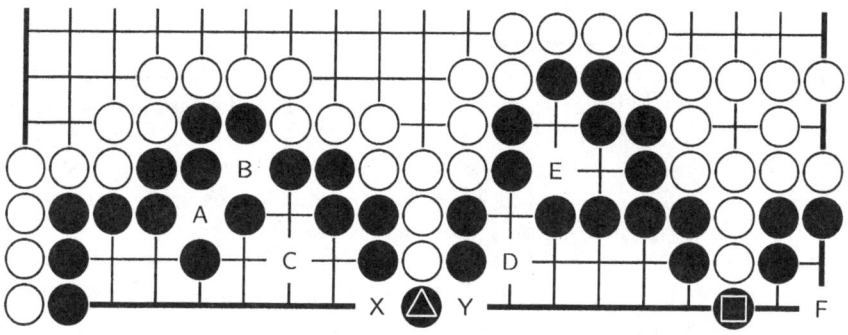

Fig. 3. Strongly and weakly dependent regions

In this figure, there are total of 6 related black regions A, B, C, D, E, and F. Since the huge outside region contains surrounding white stones that are already safe, we do not need to consider the huge outside region.

A complete merge of all six regions yields a combined new region with size 32, which is too large to be fully searched. Two-step merging creates the following result: The first step identifies connected components of strongly dependent regions and merges them. A, B and C are strongly dependent and are merged into a new region $R_1 = A \cup B \cup C$. Next D and E are merged into $R_2 = D \cup E$. Region F is not strongly dependent on any other region and is not merged. The second step identifies weak dependencies between R_1, R_2 and F and builds the group. R_1 and R_2 are weakly dependent through block ▲ , and R_2 and F are weakly dependent through block ■ . The result is a group of weakly dependent regions $\{R_1, R_2, F\}$ with region sizes of 15, 14 and 3 respectively. The regions within a group are not merged but searched separately, as explained in the next section.

The common boundary block between two weakly dependent regions has both *internal* and *external* liberties relative to each region. For example, for block ▲ and $R_2 = D \cup E$, the liberty Y is internal and the liberty X is external.

3.3 Weakly Dependent Regions

We distinguish between two types of weak dependencies. In type 1, the common boundary block has more than one liberty in each region. For example, in Figure 4 the shared boundary block of regions A and B has more than 1 liberty in each region. In type-1 dependencies, our search in one region does not consider the external liberties of the common block.

In type-2 weak dependencies, the common boundary block has only one liberty in at least one of the regions. In Figure 3 black block ▲ has only 1 liberty in both regions R_1 and R_2. We need to consider the external liberties for the common block because moves in one region might affect the result of the other. However, we do not want to merge these two regions because of the resulting increase in problem size.

The pseudo code in Figure 5 describes the method for processing groups of weakly dependent regions.

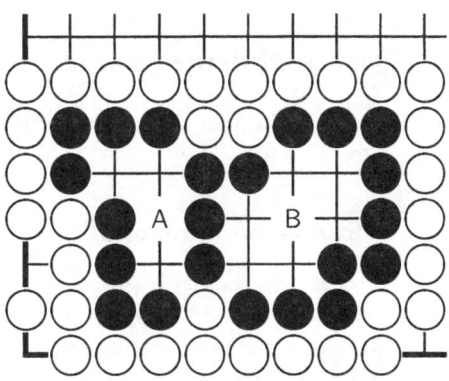

Fig. 4. First type of weakly dependent regions

```
for each weakly dependent group G
    if ( total size of all regions in G < 14) // 14 is a constant determined empirically
        r_G = merge all regions in G;
        call 2-vital solver for r_G
    else
        for each region r ∈ G
            for each shared boundary block b between r and another region r_2 ∈ G
                do a 1-vital search for b in r_2;
            reduce liberty target for all successfully tested boundary blocks to 1
            take unproven (1-vital search not successful) blocks as special blocks;
            generate external moves for special blocks (for both attacker/defender);
            call 2-vital solver for r.
```

Fig. 5. Search for weakly dependent groups

3.4 Other Improvements to the Solver

The following further enhancements were made to the solver since the version described in [3].

Improved static solver – the static solver contains a more complete implementation of the concepts in [3] than the preliminary version used in the 1997 experiments. Its performance is about 3 to 4 % better on the same test set. See Subsection 5.1 for detailed results.

External eyes of blocks – if a boundary block of a region r has one sure liberty elsewhere, this information is stored and used in the search for r by lowering the liberty target for that block.

Time limit instead of depth limit – the previous experiments used a fixed-depth 6-ply search. The current solver uses a time limit instead, which allows it to search much deeper.

4 Search Enhancements

Below we discuss move generation and move ordering (in 4.1) and evaluation functions (in 4.2).

4.1 Move Generation and Move Ordering

In this work, we focus on proving that a region and its boundary blocks are safe. Therefore we have concentrated our efforts on generating and ordering the defender's moves. For the attacker, all legal moves in the region plus a pass move will be generated. When processing weakly dependent regions as described in Subsection 3.3, extra moves outside of the region might be generated for either attacker or defender. For details of this procedure please see [5]. The attacker is allowed to recapture immediately a ko. Therefore, the attacker will always win a ko-fight inside a region.

The solver utilizes EXPLORER's generic $\alpha\beta$ search engine which implements standard move-ordering techniques such as (1) trying the best move from a previous iteration first and (2) killer moves. There is currently no game-specific move ordering or pruning for the attacker. For the defender, the following safe forward-pruning technique is used: when a boundary block of a region is in atari, only moves that can possibly avert the capturing threat, such as extending the block's liberties or capturing the attacker's adjacent stones are generated. If no forced moves are found, all legal moves for the defender are generated.

For ordering the defender's moves, both a high-priority move-motivation detector and a normal scoring system are used. The motivation detector analyzes the purpose of the attacker's previous move, and classifies the situation as one of three priorities.

1. The attacker's move is close to one of the empty cutting points.
2. The attacker's move extends one or more cutting blocks.
3. Other attacker moves.

For priority-1 and priority-2 positions, a set of high-priority moves according to the attacker's motivation is generated first. For priority 1, most likely the attacker is trying to cut, so the cutting points close to this move, as well as the cutting points' 8 neighbor points, have high priority. For priority 2, most likely the attacker is trying to expand its own cutting block. Capturing this block is an urgent goal for the defender. Therefore, all liberties of this block are given high priority. The number of adjacent empty points is used to order liberties.

All moves in priority-3 positions and all remaining moves in priority-1 and priority-2 positions are sorted according to a score that is computed as a weighted sum

$$Move\ score = f_1 * LIB + f_2 * NDB + f_3 * NAB + f_4 * CB + f_5 * AP.$$

The formula uses the following five features.
1. Liberties of this defender's block (LIB).
2. Number of neighboring attacker's blocks (NAB).
3. Number of neighboring defender's blocks (NDB).
4. Capture bonus (CB): 1 if an opponent block is captured, 0 otherwise.
5. Self-atari penalty (AP): -1 if move is self-atari, 0 otherwise.

The following set of weights worked well in our experiments: $f_1 = 10, f_2 = 30, f_3 = 20, f_4 = 50, f_5 = 100$.

4.2 Evaluation Functions

Heuristic Evaluation Function – The evaluation function in [3] used only three values: *proven-safe*, *proven-unsafe* and *unknown*. Since most of the nodes during the search evaluate to *unknown*, we can improve the search by using a heuristic evaluation to differentiate nodes in this category. The heuristics are based on two observations.

1. An area that is divided into more subregions is usually easier to evaluate as *proven-safe* for our static evaluation function.
2. If the attacker has *active blocks* with more than 1 liberty, it usually means that the attack still has more chances to succeed.

Let NSR be the number of subregions and NAB be the number of the attacker's active blocks. Then the heuristic evaluation of a position is calculated by the formula

$$eval = f_1 * NSR + f_2 * NAB, \quad f_1 = 100, f_2 = -50$$

Exact Evaluation Function – The exact evaluation function recognizes positions that are *proven-safe* or *proven-unsafe*. A powerful function is crucial to achieve good performance. However, there is a tradeoff between evaluation speed and power. In our evaluation function there are two types of exact static evaluations, *HasSureLiberties()* and *StaticSafe()*. *HasSureLiberties()* is a quick static test to check whether all boundary blocks of a region have two sure liberties and the opponent cannot live inside the region. *StaticSafe()*, is a simplified static safety solver which takes the subregions created by the search into account. Because it has to compute regions, *StaticSafe()* is much slower than *HasSureLiberties()*. The relative speed of the two methods varies widely, but 5 to 10 times slower is typical. We use the following compromise rule: If the previous move changes the size of a region by more than 2 points, then *StaticSafe()* is used. Otherwise, the quicker *HasSureLiberties()* is used. In contrast, [3] used only a weaker form of *HasSureLiberties()*.

5 Experimental Setup and Results

The safety solver described here has been developed as part of the Go program EXPLORER [2]. To compare the performance of our current solver with the previous solver [3], our test set 1 is the same, the problem set *IGS_31_counted* from the Computer Go Test Collection [2]. The set contains 31 problems. Each of them is the final position of a 19 × 19 game played by human amateur players.

We also created an independent test set 2. It contains 27 final positions of games by the Chinese professional 9 dan player ZuDe Chen. Both sets are available at http://www.cs.ualberta.ca/~mmueller/cgo/general.html.

All experiments were performed on a Pentium 4 with 1.6 GHz and a 64MB transposition table. The following abbreviations for the solvers and enhancements are used in the tables.

Benson – Benson's algorithm, as in [3].
Static-1997 – static solver from [3].
Search-1997 – search-based solver, 6-ply depth limit, from [3].
Static-2004 – current version of static solver.
M1 – a basic 2-liberties search, similar to the one in [3].
M2 – M1 + consider external eyes of blocks as in Subsection 3.4.
M3 – M2 + region merging method as in Subsection 3.2.
M4 – M3 + move ordering and pruning as in Subsection 4.1.
M5 – M4 + improved heuristic and exact evaluation functions as in Subsection 4.2.
M6 – full solver, M5 + weakly dependent regions as in Subsection 3.3.

5.1 Experiment 1: Overall Comparison of Solvers

Table 1 shows the results for all methods listed above for test set 1. The set contains 31 full-board positions with a total of 31 × (19 × 19) = 11,191 points, 1,123 blocks and 802 regions. For methods M1–M6, a long time limit of 200 seconds per region was used. For results with shorter time limits, see Experiment 2.

Table 2 shows the results for all methods listed above for test set 2. This test set contains a total of 27 × (19 × 19) = 9,747 points, 1,052 blocks and 742 regions.

Table 1. Search improvements in test set 1

Version	Safe points	Safe blocks	Safe regions
Benson	1,886 (16.9%)	103 (9.2%)	204 (25.4%)
Static-1997	2,481 (22.2%)	168 (15.0%)	N/A
Search-1997	2,954 (26.4%)	198 (17.6%)	N/A
Static-2004	2,898 (25.9%)	212 (18.9%)	321 (40.0%)
M1	4,017 (35.9%)	326 (29.0%)	404 (50.4%)
M2	4,073 (36.4%)	330 (29.4%)	406 (50.6%)
M3	5,029 (44.9%)	444 (39.5%)	495 (61.7%)
M4	5,070 (45.3%)	451 (40.2%)	498 (62.1%)
M5	5,396 (48.2%)	484 (43.1%)	519 (64.7%)
M6 (Full)	5,740 (51.3%)	523 (46.6%)	548 (68.3%)
Perfect	11,191 (100%)	1,123 (100%)	802 (100%)

Table 2. Search improvements in test set 2

Version	Safe points	Safe blocks	Safe regions
Benson	1,329 (13.6%)	106 (10.1%)	160 (21.6%)
Static-2004	2,287 (23.5%)	188 (17.9%)	251 (33.8%)
M1	3,244 (33.3%)	273 (25.9%)	320 (43.1%)
M2	3,305 (33.9%)	278 (26.0%)	325 (43.8%)
M3	4,079 (41.9%)	380 (36.1%)	409 (55.1%)
M4	4,220 (43.3%)	394 (37.5%)	420 (56.7%)
M5	4,594 (47.1%)	440 (42.0%)	455 (61.4%)
M6 (Full)	4,822 (49.5%)	483 (45.9%)	481 (64.9%)
Perfect	9,747 (100%)	1,052 (100%)	742 (100%)

In results of test set 1, the current static solver performs similarly to the best 1997 solver. Adding search and adding region merging yield the biggest single improvements in performance, about 10% each. The heuristic evaluation function and weakly dependent regions add about 3% each. Other methods provide smaller gains with these long time limits, but they are essential for more realistic shorter times, as in the next experiment.

Results for test set 2 are a little bit worse than for test set 1, but that is true even for the baseline Benson algorithm. Our conclusion is that test set 2 is just a little bit harder, and the performance of the solver is comparable to its performance on test set 1.

5.2 Experiment 2: Detailed Comparison of Solvers

This experiment compares the six search-based methods M1–M6 in more detail on test set 1. The static solver can prove 321 out of 802 regions safe. Our best solver M6 can prove 548 regions with a time limit of 200 seconds per region. The remaining 254 regions have not been solved by any method.

A total of (548–321) = 227 regions can be proven safe by search. To further analyze the search improvements, we divide these regions into four groups of increasing difficulty, as estimated by the CPU time used.

Group 1, *very easy* (regions 322–346) – this group contains 25 regions. Most regions in this group have small size, less than 10. All methods M1–M6 solve all 25 regions quickly within a time limit of 0.1s (0.2s for M1).

Group 2, *easy* (regions 347–408) – this group contains 62 regions. Figure 6 shows two examples. Table 3 shows the number of regions solved by each method with different time limits. The number in braces is the difference between two methods. The performance of M1 and M2 is not convincing. By using region merging, M3 solves all 62 regions within 0.5s. The more optimized methods M4–M6 solve all within 0.1s. Region merging drastically improves the performance of solving these easy regions.

Group 3, *moderate* (regions 409–495) – this group contains 87 regions. Figure 7 shows two examples. Table 4 contains the test results. In this group, the search enhancements drastically improve the solver. M1 and M2 solve few problems. M3

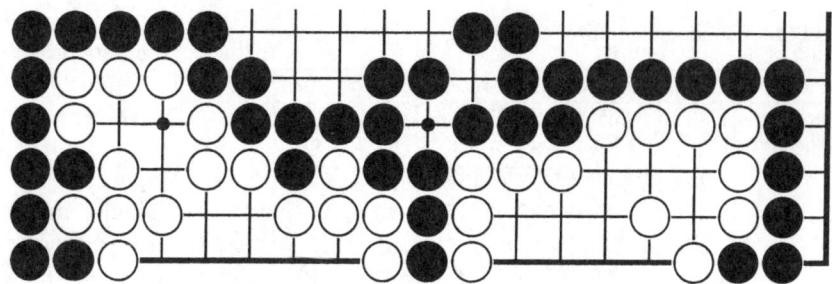

Fig. 6. Two examples of easy problems in group 2. Left: merged white region, size 10. Right: basic white region, size 11.

Table 3. Search results for Group 2, easy (62 regions)

Version	M1	M2	M3	M4	M5	M6
T=0.1s	0	23	38	62	62	62
T=0.5s	29 (+29)	31 (+8)	62 (+24)			
T=1.0s	39 (+10)	40 (+9)				
T=5.0s	43 (+4)	42 (+2)				
T=10s	43 (+0)	44 (+2)				
T=50s	43 (+0)	49 (+5)				
T=200 seconds	43 (+0)	49 (+0)				
Solved	43	49	62	62	62	62

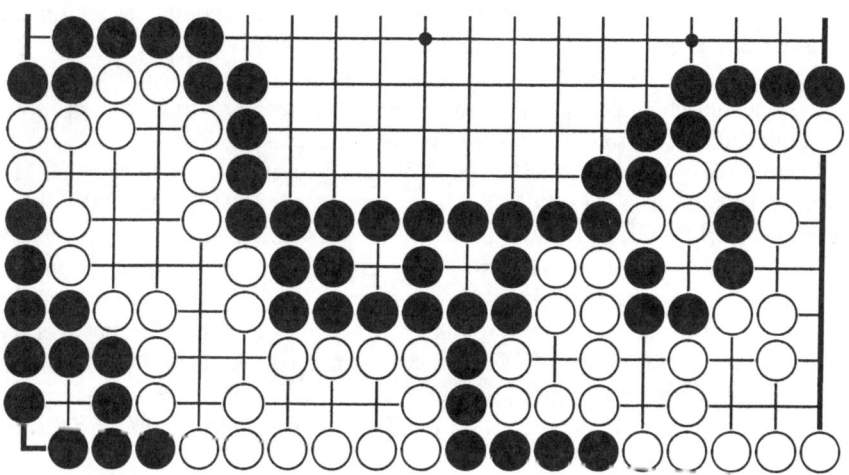

Fig. 7. Two examples of moderate problems in group 3. Left: merged white region, size 16. Right: basic white region, size 19. One white block has an external eye.

can solve 79 regions, but more than half of them need more than 10 seconds. The evaluation function drastically speeds up the solver. M5 solves all regions within 10 seconds. M6, using weakly dependent regions, solves 23 regions within 0.1s, as opposed to 0 for M5. All 87 regions are solved within 5s. In this category M6 outperforms all other methods.

Group 4, *hard* (regions 496–548) – this group contains the 53 regions that are solved in 5 to 200 seconds by M6. Figure 8 shows three examples. Table 5 contains the test results. This group includes 20 weakly dependent regions that cannot be solved by M1 to M5. Many of these problems take more than a minute even with M6. They represent the limits of our current solver.

Table 4. Search results for Group 3, moderate (87 regions)

Version	M1	M2	M3	M4	M5	M6
T=0.1s	0	0	0	0	0	23
T=0.5s	0	0	14 (+14)	14 (+14)	10 (+10)	37 (+14)
T=1.0s	0	6 (+6)	33 (+19)	33 (+19)	38 (+28)	59 (+22)
T=5.0s	0	6 (+0)	38 (+5)	38 (+5)	68 (+30)	87 (+28)
T=10s	0	8 (+2)	38 (+0)	40 (+2)	87 (+19)	
T=50s	0	10 (+2)	73 (+35)	79 (+39)		
T=200 seconds	13 (+13)	17 (+7)	79 (+6)	82 (+3)		
Solved	13	17	79	82	87	87

Table 5. Search results for Group 4, hard (53 regions)

Version	M1	M2	M3	M4	M5	M6
T=0.1s	0	0	0	0	0	0
T=0.5s	0	0	0	0	0	0
T=1.0s	0	0	0	0	0	0
T=5.0s	0	0	0	0	0	0
T=10s	0	0	0	0	11 (+11)	11 (+11)
T=100s	0	0	15 (+15)	17 (+17)	21 (+10)	28 (+17)
T=200 seconds	5 (+5)	5 (+5)	17 (+2)	20 (+3)	33 (+12)	53 (+25)
Solved	5	5	17	20	33	53

6 Conclusions and Future Work

The results of our work on proving territories safe are very encouraging. Using a combination of new region-processing methods and search enhancements, our current safety solver is significantly faster and more powerful than the solver in [3]. However, most large areas with more than 18 empty points remain unsolvable due to the size of the search space. Figure 9 shows an example. Although this region has only 18 empty points, our current solver cannot solve it within 200 seconds and a 14-ply search. In order to handle larger areas, it can be improved in the following areas.

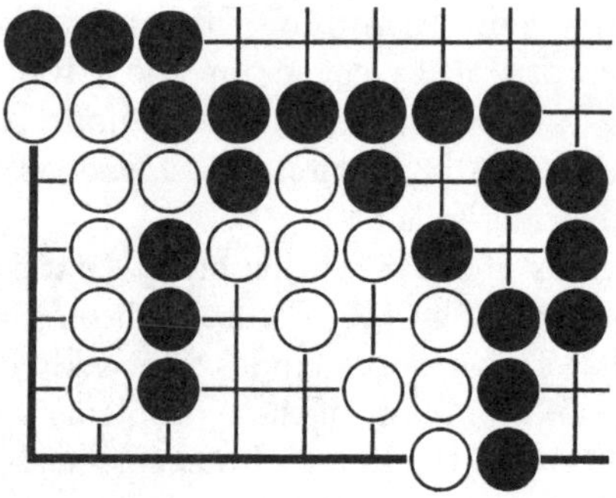

110   X. Niu and M. Müller

(a) Merged white region, size 17.

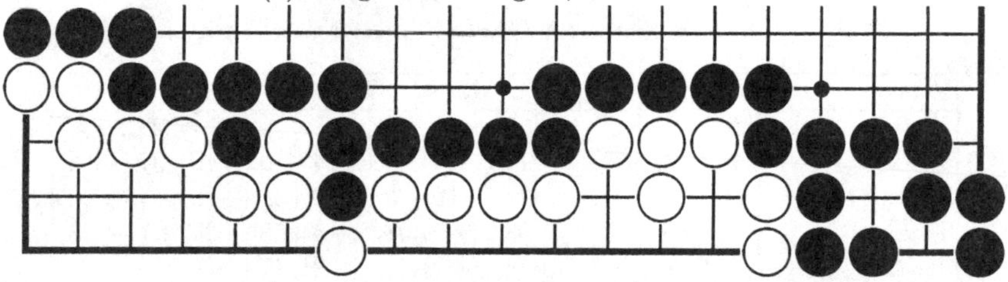

(b) Two weakly dependent white regions, size 11 and 9.

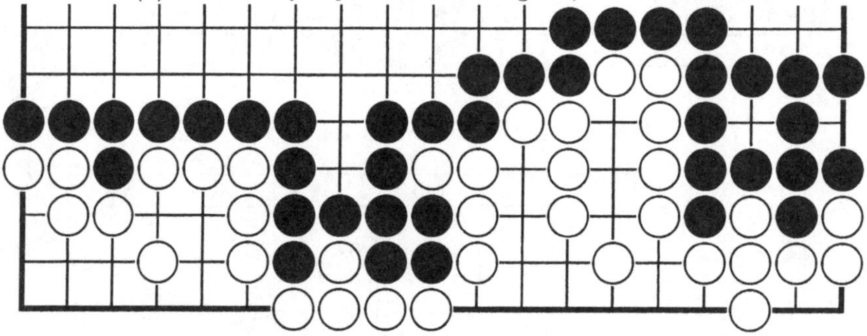

(c) Three weakly dependent white regions, size 13, 14 and 2.

Fig. 8. Three examples of hard problems in group 4

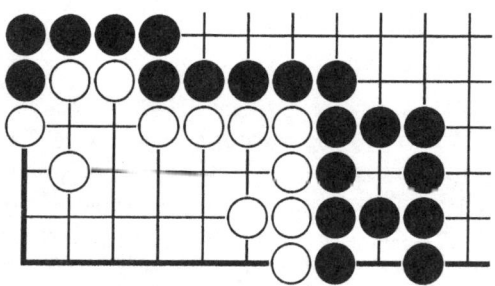

Fig. 9. Unsolved region, size 18

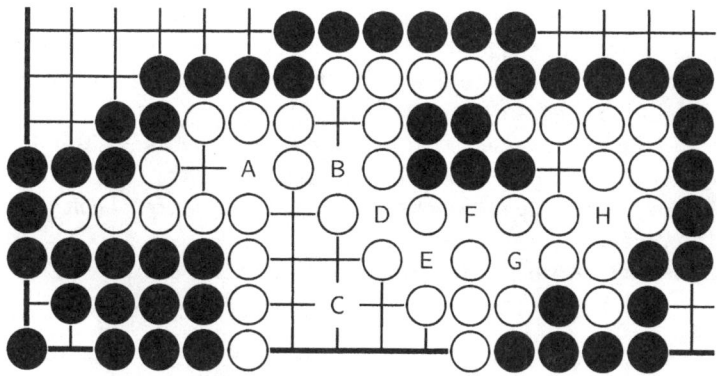

Fig. 10. An example of multiple related regions

Move generation – more Go knowledge could be used for safe forward pruning. Instead of generating all legal moves, in many cases the program could analyze the attacker's motivations and generate refutation moves. Move ordering and pruning for the attacker should also be investigated.

Evaluation function – our current exact evaluation function is all-or-nothing, and tries to decide the safety of the whole input area. If the area becomes partially safe during the search, this information is ignored. However, it would be very useful in order to simplify the further search. Also, more research on fine-tuning the evaluation function is needed.

Region processing – we can reduce the search space by treating a large region as several weakly dependent small sub-regions. Most sub-regions will be affected only by moves in the sub-region and possibly moves close to the boundary in other sub-regions. In addition, many strongly related regions could be treated as weakly related regions in practice. Figure 10, slightly simplified from position 16 of test set 1, shows an example. By our definition, regions $A \ldots H$ are strongly related, and are merged into a single region of size 25. However, if the partition were $A \cup C$ and $B \cup D \cup E \cup F \cup G \cup H$ then each merged region would be small and could be solved. In practice, this happens very often, for example in 7 out of the 31 test positions in test set 1. Better methods are needed to analyze the relationships between regions and to process regions more selectively.

Search method – in place of $\alpha\beta$, a modern search algorithm such as DF-PN [4, 8] would probably work well in this domain.

More future work ideas include the following six topics.

- Handle special cases such as seki, snapback, double ko.
- Use the solver in EXPLORER to prove regions unsafe and find successful invasions, or defend against them.

- Compare the performance against heuristic Go programs in borderline cases where it is hard to judge statically whether a defensive move is necessary. Such a test would indicate how much the method can improve the playing strength of Go programs.
- Develop a heuristic version that can find possible weaknesses in large areas.
- Compare the performance with Life and Death solvers such as GOTOOLS [9] in positions where the safety of territory problem is equivalent to a life-and-death problem.
- Build a solver for small board Go that utilizes this engine.

References

1. D.B. Benson. Life in the game of Go. *Information Sciences*, 10:17–29, 1976. Reprinted in Computer Games, Levy, D.N.L. (Editor), Vol. II, pp. 203-213, Springer Verlag, 1988.
2. M. Müller. *Computer Go as a Sum of Local Games: An Application of Combinatorial Game Theory*. PhD thesis, ETH Zürich, 1995. Diss. ETH Nr. 11.006.
3. M. Müller. Playing it safe: Recognizing secure territories in computer Go by using static rules and search. In H. Matsubara, editor, *Game Programming Workshop in Japan '97*, pages 80–86, Computer Shogi Association, Tokyo, Japan, 1997.
4. A. Nagai. *Df-pn Algorithm for Searching AND/OR Trees and Its Applications*. PhD thesis, University of Tokyo, 2002.
5. X. Niu. Recognizing safe territories and stones in computer Go. Master's thesis, University of Alberta, 2004. in preparation.
6. E. van der Werf, H.J. van den Herik, and J.W.H.M. Uiterwijk. Learning to score final positions in the game of Go. In H.J. van den Herik, H. Iida, and E.A. Heinz, editors, *Advances in Computer Games 10*, pages 143 – 158. Kluwer, 2004.
7. R. Vilà and T. Cazenave. When one eye is sufficient: a static classification. In H.J. van den Herik, H. Iida, and E.A. Heinz, editors, *Advances in Computer Games 10*, pages 109 – 124. Kluwer, 2004.
8. M.H.M. Winands, J.W.H.M. Uiterwijk, and H.J. van den Herik. An effective two-level proof-number search algorithm. *Theoretical Computer Science*, 313(3):511–525, 2004.
9. T. Wolf. The program GoTools and its computer-generated tsume go database. In H. Matsubara, editor, *Game Programming Workshop in Japan '94*, pages 84–96, Computer Shogi Association, Tokyo, Japan, 1994.

Searching for Compound Goals Using Relevancy Zones in the Game of Go

Jan Ramon and Tom Croonenborghs

Department of Computer Science,
Katholieke Universiteit Leuven, Leuven, Belgium
{Jan.Ramon, Tom.Croonenborghs}@cs.kuleuven.ac.be

Abstract. In complex games with a high branching factor, global alpha-beta search is computationally infeasible. One way to overcome this problem is by using selective goal-directed search algorithms. These goal-directed searches can use relevancy zones to determine which part of the board influences the goal. In this paper, we propose a general method that uses these relevancy zones for searching for compound goals. A compound goal is constructed from less complex atomic goals, using the standard connectives. In contrast to other approaches that treat goals separately in the search phase, compound goal search obtains exact results.

1 Introduction

In complex games with a high branching factor, such as the Asian game of Go, global alpha-beta search is computationally infeasible. Therefore, there has recently been some research towards local searches.

Local searches can be obtained by splitting the global search into several independent local searches. This approach is successfully adopted by Müller using decomposition search in the endgame [5]. Unfortunately, in the middle game it is rare to find isolated regions which are not influenced by other regions. Therefore, other notions of locality have been proposed which are more suitable for the middle game.

One way to preserve locality is by using selective goal-directed search algorithms [1,3,6]. This approach has been successfully adopted to several sub-games of Go, such as capturing a string of stones, connecting two strings of stones, and making life for a group. These goal-directed searches can use relevancy zones to determine which part of the board influences the goal. However, searches for such elementary goals cannot be used for solving more complex problems or for evaluating game positions. For instance, in Figure 1, there are two white stones with only one liberty. If Black is to move, he[1] can capture one of them, but then the other stone can escape. The question then arises how he should play in order to capture both stones. As we will see, Black can capture both stones by playing a ladder break on the crossing of both ladders. Cazenave and Helmstetter

---

[1] In this article, when 'he' and 'she' are both possible, 'he' is used.

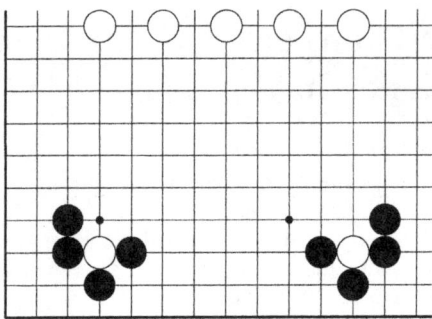

Fig. 1. A double ladder

[4] made a start on combining several goals into one search by considering the proof of the connection of two goals A and C using the proofs of the connections between A and B and between B and C.

In this paper, we propose a general method that searches with compound goals. A compound goal is a more complex goal, that is built from simpler atomic goals. Searching using compound goals is especially interesting when the atomic goals used are more or less independent, but not completely. In contrast to other approaches that treat local goals separately in the search phase (such as planning approaches [7]), compound goal search obtains exact results. The technique presented here is a generalization of [4] on transitive connections, as we treat compound goals which are arbitrary combinations (using conjunction, disjunction, and negation) of arbitrary atomic goals (not only connections but also, e.g., capturing and living).

The remainder of this paper is organized as follows. Section 2 reviews selective goal-directed search algorithms and introduces the terminology used in this paper. Next, in Section 3 we explain how we construct compound goals, using atomic subgoals. In Section 4 our experiments and analyses can be found. Finally, in Section 5 we present our conclusions and ideas for further work.

2 Searching for Atomic Goals

In this section the basics of selective goal-directed search algorithms are reviewed and the used terminology is introduced.

In games such as Go, locality is important, as the board is large and at the same time intermediate goals can be formulated using information of only a small local area. Special algorithms have been developed recently. The important idea in these algorithms is that one can be very selective on the candidate moves by only considering those that could potentially prevent a negative result that follows when no move (a nullmove) is played. To human Go players too, it is common knowledge that "my opponent's point is my point".

A goal is formulated for a particular player. He is called the attacker or MAX. The opponent is called the defender or MIN (he should try to prevent MAX from reaching the goal). For a particular goal G we will denote MAX by MAX(G) and MIN by MIN(G). Some common examples of goals in increasing order of complexity include the capture goal (MAX should capture a particular block of stones), the connection goal (MAX should connect two of his blocks), the life goal (MAX should make life for a given set of blocks) and the invade goal (MAX should make a living group somewhere in a given area).

The order of a move is an important concept. Generally speaking it indicates how directly the move can realize the goal. A MAX move of order n indicates that the goal can be reached, if it is followed by n successive moves of MAX[2]. By using this notion of the order of a move, the search can be kept local by limiting the order of the moves to be searched. A MAX move of order 0 reaches the goal directly. A move of a higher order n is a move such that if MIN passes, MAX can play a move of order at most $n-1$. For the capture goal, ladders are moves of order 1 and a geta (net) has order at least 2.

Generally speaking, a relevancy zone is a set of intersections that supports a proof. When a search engine needs to prove a tactical goal, it has to remember all the reasons that are responsible for the result of that search. This collection of intersections is called the relevancy zone or trace of that search.

To illustrate the use of relevancy zones, the double ladder example from the previous section is repeated. Figure 2 shows the same position as Figure 1, but this time the relevancy zones are included. A triangle indicates an intersection that belongs to the relevancy zone, obtained by the local search which goal is to capture the left white stone. The squares indicate the intersections from the relevancy zone of the local search for the right white stone. The intersection of both relevancy zones are the points marked by a circle.

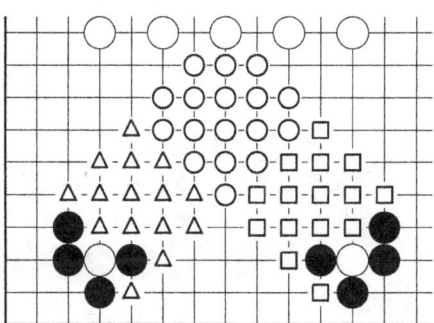

Fig. 2. A double ladder, including the relevancy zones

Relevancy zones have several uses. We mention two of them. First, as a proof remains valid as long as the intersections of its relevancy zone remain unchanged, it is natural for a playing program to store the relevancy zone of a proof together with the result, so it knows when it has to recompute the status of the goal. Second, if it is proved that for a certain order no move exists that reached the

---

[2] Definitions differ somewhat among the different articles in literature.

goal, one can use the relevancy zone of that proof as the set of candidate moves in a new search for moves of a higher order.

Different relevancy zones can be obtained by proving the same tactical goal, e.g., when proving that a block cannot be captured in one move, one only needs to add two liberties of that block (arbitrarily) to the relevancy zone. Ideally and not considering memory constraints, one could use a tree-like representation for a relevancy zone that entails the minimal relevancy zone, obtained by browsing through the tree. Currently, only one of the proofs is used as the relevancy zone for that search. This still obtains correct results, but it can be less efficient. (It can be necessary to recalculate the search, just because the "wrong" liberty is added.)

We distinguish between two subsets of a relevancy zone: the positive relevancy zone (denoted Rzone$^+$), containing all points such that when they remain unchanged, the result for the associated search cannot become better than the previously computed result for MAX and the negative relevancy zone (denoted Rzone$^-$) of all points such that when they remain unchanged the obtained result can not get worse for MAX. To illustrate this distinction, consider the position in Figure 3, where the goal is to capture the white group, so the black player is MAX and White is MIN. The outcome for the search is WON for the player to move. The markings indicate the relevancy zone, the circles indicate the Rzone$^+$-relevancy zone, which contain the intersections that can make the result better for Black (MAX) and the triangles indicate the Rzone$^-$-relevancy zone, intersections that, when changed, can make the result worse for Black (better for White). The two rectangles indicate the only intersections that belong to both Rzone$^+$ and Rzone$^-$.

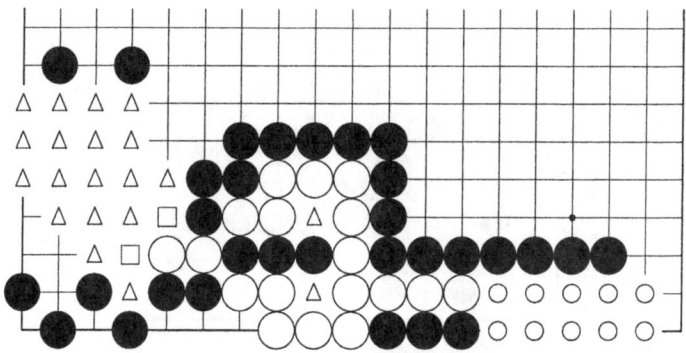

Fig. 3. Illustration of Rzone$^+$ and Rzone$^-$

One of the selective local search algorithms described in literature is lambda search [6]. Lambda trees are limited by the order of the moves (in lambda search, a λ^i attacking move threatens to do a λ^{i-1} move if the defender passes).

Abstract Proof Search [1] is based on similar ideas. However, the search can be both bounded by the depth and the order of the moves. Also, candidate moves cannot only be generated from the relevancy zones of lower order or lower depth searches, but also from goal and game specific hard-coded knowledge. The advantage of the latter is that often smaller sets of candidate moves can be used. Unfortunately, coding the knowledge turns out to be a quite cumbersome task which can only be done with some confidence for the lower order moves. For higher order moves, later versions of abstract proof search fall back on relevancy zones.

Several generalizations of abstract proof search have been proposed. Among these there is Iterative Widening and General Threats Search. Iterative widening [2] works in an analogous way as iterative deepening in that it performs an iteration of searches of increasing complexity by trying the most plausible candidate moves in the earlier searches and gradually considering more candidates resulting in 'wider' trees. Generalized Threats Search [3] is a generalization of the above algorithms.

In this paper we will use an implementation of abstract proof search (with the different generalizations) called TAIL, which was originally created by T. Cazenave and further extended in collaboration. The search algorithm is implemented in a generic goal-independent way. For each goal, a number of methods have to be specified; the most important ones are those listed in Table 1.

Table 1. Methods to specify for each goal

Method	Description
do_move($move, color$)	perform a move and necessary updates to the goal status
undo_move()	retract a move
eval($colortomove$,Rzone$^+$,Rzone$^-$)	evaluates the current position w.r.t. the goal and the player who moves first
max_moves($order$,Rzone$^+$,Rzone$^-$)	returns the candidate moves for MAX up to order $order$
min_moves($order$,Rzone$^+$,Rzone$^-$)	returns the candidate moves for MIN up to order $order$

We denote the application of a method M of some goal G in some position P with $G.M[P](args)$, where $args$ are the arguments. We will denote the relevancy zones of that call, i.e., the set of all intersections which may not change for the result of the method call to remain valid, with Rzone$^+\bigl(G.M[P](args)\bigr)$ and Rzone$^-\bigl(G.M[P](args)\bigr)$. One important method is the evaluation function. It either returns WON, KO, LOST, or UNKNOWN. The first three values are terminal, which means that the search can be stopped in nodes where these values are obtained. In this paper we do not make a distinction between the different kinds of ko. There can be several heuristic levels in UNKNOWN, according to the belief of the evaluation function in the possibility to reach the goal. We will denote these with UNKNOWN$_i$ ($i \in \mathbb{R}$) where higher values of i are more desirable for MAX.

3 Compound Goals

In the previous section we have discussed existing work on searching atomic goals. In this section we explain how one can search for compound goals, mainly using the same search algorithms.

We start with defining a language that allows one to build more complex goals from atomic ones. Therefore one can use the three standard connectives: conjunction, disjunction, and negation. These complex goals are constructed as new goals, by defining the functions listed in Table 1, so that any search algorithm (for instance those described in the previous section) can be used to compute the outcome. We start with the simplest operator: the negation.

3.1 Negation

The negation of a goal is the inverse goal, where a player tries to prevent his opponent from achieving the basic goal. We define it more accurately below.

Definition 1. *Let G be a goal. Then $not(G)$ is a goal, called the negation of G, such that* MAX$(not(G))$ = MIN(G) *and* MIN$(not(G))$ = MAX(G). *Moreover, for every position P, G.eval$[P](c)$ =* WON $\Leftrightarrow not(G)$.eval$[P](c)$ = LOST, G.eval$[P](c)$ = LOST $\Leftrightarrow not(G)$.eval$[P](c)$ = WON *and* G.eval$[P](c)$ = KO $\Leftrightarrow$ $not(G)$.eval$[P](c)$ = KO

The implementation of the necessary methods to be able to search for $not(G)$ given the methods for the basic goal G is in theory straightforward. Table 2 gives a summary. Note that because the roles of MAX and MIN are switched in the inverse goal, the roles of Rzone$^+$ and Rzone$^-$ also need to be reversed.

Table 2. Methods for the negation of a goal

Method	Implementation
$not(G)$.max_moves(*order*, Rzone$^+$, Rzone$^-$)	G.min_moves(*order*, Rzone$^-$, Rzone$^+$)
$not(G)$.min_moves(*order*, Rzone$^+$, Rzone$^-$)	G.max_moves(*order*, Rzone$^-$, Rzone$^+$)

However, there are a number of additional requirements on the basic goal in order for this schema to work. In particular, the evaluation function should be sufficiently accurate in returning the value LOST. Indeed, for the existing approaches, evaluation functions are often implemented asymmetric. It is important to return WON in positions where the goal is reached, but if UNKNOWN is returned instead of LOST in positions where the basic goal cannot be reached anymore, the search will take longer. However, it will still succeed in the same iterative deepening iteration. If the evaluation function is able to detect the LOST status early enough and the basic goal can be disproved by the original search algorithm then the negation of the goal can be proved efficiently. However, for a number of goals detecting the LOST status is more difficult than detecting the WON status. For instance, for the `capture` goal, the status is WON if the block to capture is removed from the board. Many playing programs assume that a block cannot be (easily) captured when its number of liberties is above some threshold, but even a block with tens of liberties can be dead. Hence, detecting that a block cannot be captured may require to recognize life.

3.2 Conjunction

A conjunction of two goals is the goal in which MAX is trying to achieve both goals. Below it is defined more accurately.

Definition 2. *Let A and B be goals such that* MAX(A) = MAX(B). *Then $A \wedge B$ is a goal, called the conjunction of A and B, such that* MAX$(A \wedge B)$ = MAX(A) *and* MIN$(A \wedge B)$ = MIN(A) = MIN(B). *Moreover, for every position P,*

- *if $(A \wedge B)$.eval$[P](c)$ = WON, then* MAX *can win both A and B if c moves first;*
- *if $(A \wedge B)$.eval$[P](c)$ = KO, then* MAX *can win both A and B by winning a ko, if c moves first;*
- *if $(A \wedge B)$.eval$[P](c)$ = LOST, then if c moves first, it is not possible for* MAX *to win both A and B, even not by winning a ko.*

We first propose a simple, static implementation of a compound goal construction for the conjunction goal that does not need search for the atomic goals independently (in 3.2.1). In a second step we will then discuss a dynamic implementation for $A \wedge B$ that performs local search for the goals A and B individually (in 3.2.2). Finally, we discuss the correctness of the conjunction goal search (in 3.2.3).

3.2.1 Static Functions

Below we discuss two types of static functions, viz. a static evaluation function and static candidate move generation.

A static evaluation function

We will denote the static evaluation function of the conjunction goal with $(A \wedge B)$.eval$_S$. For the case MIN moves first, we define it by

$$(A \wedge B).\text{eval}_S[P](\text{MIN}) = A.\text{eval}[P](\text{MIN}) \wedge B.\text{eval}[P](\text{MIN})$$

where Table 3 gives the conjunction of two position evaluations.

The values in this table are only valid in the case that the two evaluations are independent. With MIN to move, this means that the actual static evaluation differs from the values of the "raw" evaluation function when the evaluation functions for A and B detect WON or KO before this status has actually been realized on the board. The remaining "aji" (latent possibilities of

MIN in the remaining proof of the realizability of the evaluation) of one goal could prevent the winning by MAX of the other goal in the case the relevancy zones (Rzone$^-$)of A.eval$[P]$(MIN) and B.eval$[P]$(MIN) overlap. In that case $(A \wedge B)$.eval$_\mathrm{S}[P]$(MIN) should return a (high) UNKNOWN value to urge deeper search.

Furthermore, it is noteworthy that the mathematical intuition of conjunctions (taking the minimum of two values) is not applicable to the case of ko. Indeed, if both A and B evaluate to KO, then usually $A \wedge B$ evaluates to LOST as MAX can not win two ko fights at the same time, and $A \wedge B$ evaluates only to KO if both ko fights are in fact the same ko fight and MAX can win (lose) both A and B by winning (losing) this one ko. Even the conjunction of A.eval$[P]$(MIN) = KO and B.eval$[P]$(MIN) = WON can turn into $(A \wedge B)$.eval$_\mathrm{S}[P]$(MIN) = LOST instead of the expected KO in the (rare) case that B could only be won in double ko, providing an infinite source of ko threats for MIN to win the goal A.

We define the static evaluation function for the case where MAX moves first, by

$$(A \wedge B).\mathtt{eval_S}[P](\mathrm{MAX}) = \tag{1}$$
$$\bigl(A.\mathtt{eval}[P](\mathrm{MIN}) \wedge B.\mathtt{eval}[P](\mathrm{MAX})\bigr) \vee \bigl(A.\mathtt{eval}[P](\mathrm{MAX}) \wedge B.\mathtt{eval}[P](\mathrm{MIN})\bigr)$$

where the disjunction $E_1 \vee E_2$ of two evaluations E_1 and E_2 can be found in Table 3.

Table 3. Conjunction and disjunction of two independent evaluations

E_1	E_2	$E_1 \wedge E_2$	$E_1 \vee E_2$
WON	WON	WON	WON
WON	KO	KO (or LOST)	WON
WON	LOST	LOST	WON
WON	UNKNOWN$_j$	UNKNOWN$_j$	WON
KO	WON	KO (or LOST)	WON
KO	KO	LOST (or KO)	WON (or KO)
KO	LOST	LOST	KO (or WON)
KO	UNKNOWN$_j$	UNKNOWN$_{j-c}$	UNKNOWN$_{j+c}$
LOST	WON	LOST	WON
LOST	KO	LOST	KO (or WON)
LOST	LOST	LOST	LOST
LOST	UNKNOWN$_j$	LOST	UNKNOWN$_j$
UNKNOWN$_i$	WON	UNKNOWN$_i$	WON
UNKNOWN$_i$	KO	UNKNOWN$_{i-c}$	UNKNOWN$_{i+c}$
UNKNOWN$_i$	LOST	LOST$_i$	LOST
UNKNOWN$_i$	UNKNOWN$_j$	UNKNOWN$_{\min(i,j)}$	UNKNOWN$_{\max(i,j)}$

The actual static evaluation function also differs here from the "raw" case when one of the conjunctions evaluates to LOST and the relevancy zones of the local searches overlap. In this case UNKNOWN is returned, so that the "global" search for the compound goal continues.

Static candidate move generation

When the atomic goals are not searched individually, the union of the candidate moves generated by the atomic goals have to be taken as the candidate moves for the compound goal. So,

$$(A \wedge B).\mathtt{max\_moves}[P] = A.\mathtt{max\_moves}[P] \cup B.\mathtt{max\_moves}[P]$$
$$(A \wedge B).\mathtt{min\_moves}[P] = A.\mathtt{min\_moves}[P] \cup B.\mathtt{min\_moves}[P]$$

3.2.2 Dynamic Functions

Below we discuss tow types of dynamic functions, viz. a dynamic evaluation function and dynamic candidate move generation.

A dynamic evaluation function

The evaluation function for the conjunction goal can be much more accurate by using local search to the components of this compound goal. This allows to prune much faster.

For every node P in the search for $A \wedge B$, the first call to a method of the goal $A \wedge B$ performs internally a search for the goals A and B with c moving first. If $A.\mathtt{eval}[P](c)$ and $B.\mathtt{eval}[P](c)$ respectively return UNKNOWN, for both $c = \mathrm{MAX}$ and $c = \mathrm{MIN}$, this does not mean that potentially four independent searches have to be performed since to determine the set of candidate moves for MIN, the standard algorithm already does a search with MAX moving first (in order to obtain the relevancy zone). We will denote the results of these searches with $A.\mathtt{search}[P](colortomove)$ and $B.\mathtt{search}[P](colortomove)$ and the corresponding relevancy zones with $\mathtt{Rzone}^+(A.\mathtt{search}[P](colortomove))$, $\mathtt{Rzone}^-(A.\mathtt{search}[P](colortomove))$, $\mathtt{Rzone}^+(B.\mathtt{search}[P](colortomove))$, and $\mathtt{Rzone}^-(B.\mathtt{search}[P](colortomove))$. These local searches will either return a more informative value (WON, KO, or LOST) or will still return UNKNOWN. The latter means that a deeper local search is needed. If searching more deeply for the local goal is too expensive, so will the search for $A \wedge B$ be. This would mean that it is better to return UNKNOWN from the node P in the search for $A \wedge B$ rather than expanding it further. Of course, if $A \wedge B$ is searched using iterative deepening, in deeper iterations, more time can be allocated to the local searches.

Once both local searches are finished the function discussed in the previous paragraph is used to combine both results. As pointed out, if the relevancy zones of the two local searches overlap, the compound goal evaluates to UNKNOWN and deeper search is needed, by continuing the global search algorithm with the compound (conjunction) goal.

Using the information of the local searches, the appropriate intersections can be added to $\mathtt{Rzone}^+(A \wedge B.\mathtt{search}[P](colortomove))$ and $\mathtt{Rzone}^-(A \wedge B.\mathtt{search}[P](colortomove))$, so that one is able to construct hierarchies of goals.

Dynamic candidate move generation

If MAX moves need to be generated, this means that the evaluation function for MAX moving first (Equation 1) leads to UNKNOWN. In the game of Go, passing is allowed and hence eval[P](MAX) will always be higher or equal to eval[P](MIN), so one can concentrate on the eval[P](MIN) evaluation. Further, because the individual goals are searched, one can find the proofs for the outcomes of these individual goals in the different relevancy zones.

Based on the results of A.eval[P](MIN) and B.eval[P](MIN), the set of candidate MAX moves can be obtained. If both these atomic goals evaluate to LOST, this means that MAX has to play a move that changes the result of both sub searches and therefore has to play a move from the intersection of the Rzone$^+$ of each atomic goal. If one of the above searches evaluates to WON, the MAX moves are those played on the intersections from the Rzone$^+$ of the other goal.

Moreover, candidate moves for MAX that do not allow MAX to win either goal A or B individually will be pruned away immediately by the local searches for goals A and B. In this way, one can hope that only a very small set of candidate moves for MAX will remain.

In a MIN node, whatever MIN does, MAX can win both goals independently. This is a consequence of the fact that after the local search for the goals independently, the node has not been pruned. So, MIN has to try a move in

$$\text{Rzone}^-\big(A.\text{search}[P](\text{MIN})\big) \cup \text{Rzone}^-\big(B.\text{search}[P](\text{MIN})\big)$$

that makes miai of preventing MAX's win in A and B. In many cases, a move in Rzone$^-\big(A.$search$[P](\text{MIN})\big) \cap$ Rzone$^-\big(B.$search$[P](\text{MIN})\big)$ will work for MIN, so these moves can be tried first.

While in [4] for the case of transitive connections, it is argued that some candidates for MIN are not in the union of the relevancy zones as described above. But as will be shown in the next paragraph, the fact that this union of relevancy zones is sufficient follows from the definitions of relevancy zones and conjunction goal.

We illustrate this with the same example as in [4]. In the position in Figure 4a, we consider the connection of A and B and the connection of B and C. Figure 4b contains the Rzone$^-$-zones, the circles denote Rzone$^-\big(A - B.$search$[P](\text{MIN})\big)$, the triangles Rzone$^-\big(B - C.$search$[P](\text{MIN})\big)$, and the rectangles the intersection of the two. White 1 is a working move for MIN, which does not belong to the relevancy zone of the proof of either connection and subsequently is not considered as a candidate move. If White plays 1, Black cannot protect against 2 and 3 simultaneously. But this does not necessarily mean that white 1 should indeed be a candidate move. Point 1 will appear in the Rzone$^-(A \wedge B)$ though, when the search for the compound goal (transitive connection) is continued, 2 is tried and Black defends at 1 against this by capturing the stone. As [4] already points out, 3 is also a working move on itself belonging to the intersection of the relevancy zones of both connections. Moreover, 3 is the only (and real) move that prohibits the transitive connection. So locally, 1 is a sente move against the connection, but not a threat that should be included in the set of candidate moves for MIN. Figure 4c shows a slightly modified example, where the move at 3 in the original example is solved for Black and consequently white 1 is no longer a working move.

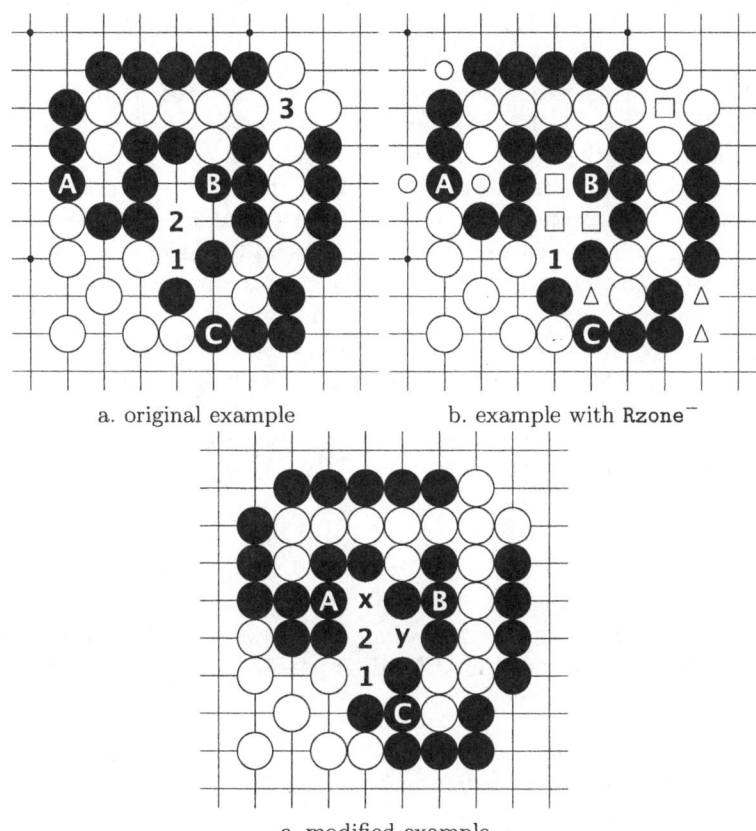

a. original example   b. example with Rzone⁻

c. modified example

Fig. 4. A counterexample for generating candidate moves for MIN

While it seems that a great deal of search should be performed to determine the status of a conjunction of goals, several optimizations can be applied. First, if a player moves outside the relevancy zone of one of both goals, the status of that goal does not change, and local search for that goal is not necessary in the next search node. Second, one can try to simplify the search by first trying those moves that minimize the intersection of the relevancy zones of the subgoals.

3.2.3 Correctness of the Conjunction Goal Search

Local searches are abstract proof searches. It only returns a terminal result (LOST,WON) if this is the correct result. When a result involving ko is returned, it is possible that a deeper search still obtains a 'better' result (WON instead of KOWON, or LOST instead of KOLOST).

To prove the correctness of the proposed compound search algorithm, we must show that all candidate moves are correctly generated and the evaluation function is correct. When the evaluation function returns a value which is not in the UNKNOWN range, it is correct that this result can be obtained (this follows from the definition and the properties of the local abstract proof searches). Still, it could be that a 'better' (non-ko instead of ko) result can be obtained if local searches are searched deeper and change from non-ko to ko.

We show that all necessary candidate moves are generated. With this we mean that there is no move which is not considered as candidate move and could yield a better result for either player. In the above subsection on dynamic candidate move generation using the relevancy zones of the local searches, we already showed that in the case none of the local searches evaluates to UNKNOWN, the set of candidate moves is correct. In the case one of the local searches evaluates to UNKNOWN, a correct and complete set of candidate moves cannot be derived. Still, if the global search gives a solution with a principal path of nodes where no subgoals evaluated to UNKNOWN, the result is correct. In the other case, one can get a more certain answer by deepening the local searches in the principal path that returned UNKNOWN, or by iterating the global search and adding in nodes with subgoals returning UNKNOWN, the appropriate relevancy zones of the search starting in this node in the previous global iteration, to the set of candidate moves.

3.3 Disjunction

A disjunction of two goals is a goal where MAX tries to realize at least one of both goals. Defining a disjunction of goals is possible using the classical laws of logic.

Definition 3. *Let A and B be goals such that* $\text{MAX}(A) = \text{MAX}(B)$. *Then $A \vee B$ is a goal, called the disjunction of A and B, such that*

$$A \vee B = not\bigl(not(A) \wedge not(B)\bigr)$$

This shows in fact that the properties of a disjunction of goals will be dual to the properties of a conjunction of goals. Of course, in practice a direct implementation without the negations turns out to be faster.

4 Experiments

In this section we will present experiments in order to evaluate our compound goal search and compare it with existing methods.

In order to make our implementation to solve compound goals we had first to construct basic goals producing good relevancy zones, also for goals that cannot be solved in a few moves. This provided an interesting opportunity to evaluate the use of relevancy zones as compared to hard coded knowledge. The use of relevancy zones does not mean that no hard coded knowledge is used. Some minimal goal-dependent knowledge is used. In the `capture` goal, for instance, the fact is used that one needs to play on a liberty to capture a block in 1 move. However, we use no move selection based on knowledge of the games or heuristics for move ordering.

4.1 Atomic, Relevancy-Based Games

In several experiments, it seems to turn out that search using hard coded knowledge is often faster, while search using relevancy zones solves more problems. Table 4 gives an impression of the results for different thresholds for the maximum nodes allowed for three datasets. These datasets were provided by T. Cazenave and contain problems from games amongst computer programs together with some artificial problems. For two of them the goal is to capture an opponent string or defend a friendly string from capturing. These datasets contain 144 and 75 problems. In the third dataset of 90 problems, the goal is to connect two friendly strings or disconnect two opponent strings.

Table 4. Comparing search using (mainly) relevancy zones to search using hard-coded knowledge. For different thresholds and for 'maximum nodes to traverse', the timings of both methods and the number of correctly solved problems is given.

goal	test set	nodes/problem	hard-coded		relevancy zones	
			time(s)	solved	time(s)	solved
capture	1	1000	0.57	57%	0.64	51%
		20000	1.18	57%	2.96	57%
		200000	1.85	58%	11.41	82%
	2	1000	0.27	33%	0.25	32%
		20000	0.50	35%	1.47	44%
		200000	0.77	35%	5.72	50%
connect	1	1000	0.56	58%	0.66	68%
		20000	1.24	60%	0.77	68%
		200000	3.98	68%	0.78	68%

4.2 Compound Goals

Now we move to compound goals. However, there seems to be no large dataset on compound goals available. We will therefore limit our discussion to a smaller set of hand-constructed and real-game problems. Most problems from games are transitive connection problems from [4]. In future work, we will collect a larger dataset of problems from games requiring compound goal techniques.

Table 5 lists a number of compound goals problems. We consider basically two methods to solve compound goals. The first one is the method described up to now. The second (naive) one is a method without search local to one of the subgoals. Candidate moves are just the union of the candidate moves proposed by the separate goals. This is about the best guess one can do without using relevancy zones and using no local search. As can be seen, for some problems

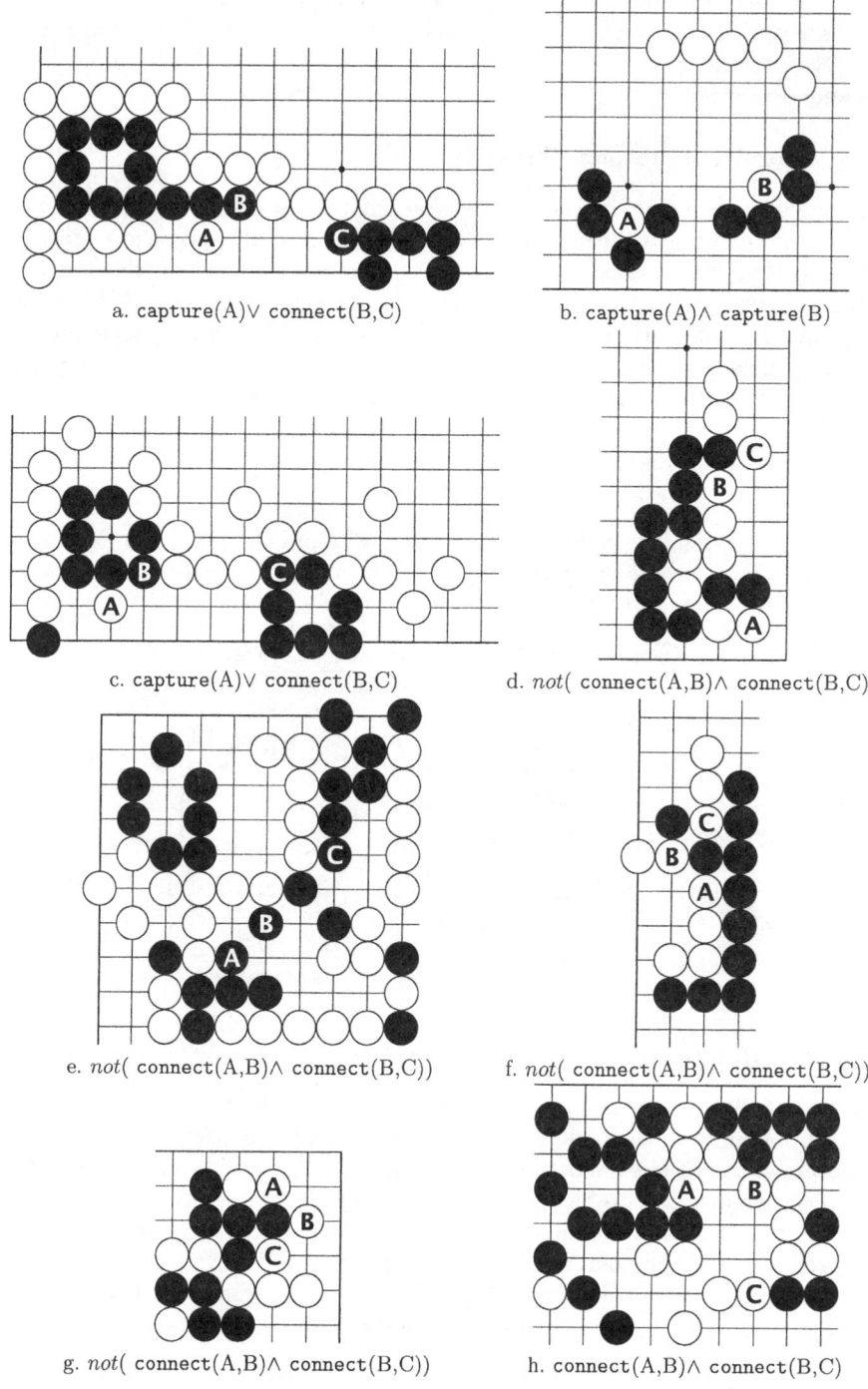

Fig. 5. Compound goal problems

the naive method works and is much faster. However, on several problems the naive method fails, i.e., it does not find a correct solution (a.o. due to missing candidate moves). Also, the naive method is not guaranteed to produce correct results. Of course, instead of this naive method one could try to take a safe approximation of the correct set of candidate moves, but this is very difficult because the complexity rapidly blows up as unnecessary candidates are added.

Table 5. Compound goals problems. Timings are given in seconds.

	goal	dynamic	static
1 (Figure 1)	capture(D3)∧capture(M3) (WON)	0.14	timeout
2 (Figure 5a)	capture(A)∨ connect(B,C) (WON)	0.39	timeout
3 (Figure 5b)	capture(A)∧ capture(B) (WON)	0.10	timeout
4 (Figure 5c)	capture(A)∨ connect(B,C) (WON)	0.39	timeout
5 (Figure 5d)	not(connect(A,B)∧ connect(B,C)) (WON)	1.12	0.01
6 (Figure 5e)	not(connect(A,B)∧ connect(B,C)) (WON)	1.12	0.01
7 (Figure 5f)	not(connect(A,B)∧ connect(B,C)) (WON)	1.62	0.01
8 (Figure 5g)	not(connect(A,B)∧ connect(B,C)) (LOST)	0.15	wrong
9 (Figure 5h)	connect(A,B)∧ connect(B,C) (WON)	0.55	0.03

5 Conclusions and Further Work

We proposed a general method for searching for compound goals. While existing techniques can guess good moves quite fast for such goals, our compound goal search technique is more accurate. Also, it is more general than existing techniques. Hence it allows one to determine accurately the status of a larger area on the board than existing methods. We considered atomic goals and three connectives to combine them: negations, disjunction, and conjunction. The binary connectives depend on correct computation of relevancy zones of their subgoals.

There are several directions of further work. First, a more accurate treatment of ko could make the search algorithm much more valuable in real games. When combining goals, it can be important to know what kind of ko is the local result of each of the subgoals.

Second, we have limited our discussion to the main logical operators (negation, conjunction, and disjunction). However, other operators may be useful. For instance, consider the problem of making a maximal number of points in some area (e.g., during an invasion). Then, the subgoal would return a real value instead of the logical values LOST, WON, and KO considered here. Combining such goals would be much more similar to decomposition search.

Finally, a more memory-intensive approach could avoid much recomputation. In our current approach, a global transposition table is used. However, it would be interesting to consider local transposition tables for every subgoal, which store also relevancy zones. This would probably save much computation, while if care is taken would still fit easily into current memory capacity.

Acknowledgements

Jan Ramon is a post-doctoral fellow of the Fund for Scientific Research (FWO) of Flanders. Tom Croonenborghs is supported by the Flemish Institute for the Promotion of Science and Technological Research in Industry (IWT). We thank Tristan Cazenave and his colleagues for the collaboration on the code used for this paper ("tail").

References

1. Tristan Cazenave. Abstract proof search. In T. Marsland and I. Frank, editors, *Proceedings of the Second International Conference on Computers and Games*, volume 2063 of *Lecture Notes in Computer Science*, pages 39–54, Hamamatsu, Japan, 2000. Springer-Verlag.
2. Tristan Cazenave. Iterative widening. In *Proceedings of the 17th International Joint Conference on Artificial Intelligence*, volume 1, pages 523–528, 2001.
3. Tristan Cazenave. A generalized threats search algorithm. In Schaeffer J, M. Müller, and Y. Björnsson, editors, *Proceedings of the Third International Conference on Computers and Games*, volume 2883 of *Lecture Notes in Computer Science*, pages 75–87, Edmonton, Alberta, Canada, 2002. Springer-Verlag.
4. Tristan Cazenave and Bernard Helmstetter. Search for transitive connections. *Information Sciences*, 175:284–295, 2005.
5. Martin Müller. Decomposition search: A combinatorial games approach to game tree search, with applications to solving go endgames. In *Proceedings of the International Joined Conference on Artificial Intelligence*, volume 1, pages 578–583, 1999.
6. Thomas Thomsen. Lambda search in game trees with applications to Go. In T. Marsland and I. Frank, editors, *Proceedings of the Second International Conference on Computers and Games*, volume 2063 of *Lecture Notes in Computer Science*, pages 57–80, Hamamatsu, Japan, 2000. Springer-Verlag.
7. Steven Willmott, Alan Bundy, John Levine, and Julian Richardson. An adversarial planning approach to Go. In H.J. van den Herik and H. Iida, editors, *Proceedings of the First International Conference on Computers and Games*, volume 1558 of *Lecture Notes in Computer Science*, pages 93–112, Tsukuba, Japan, 1998. Springer-Verlag.

Rule-Tolerant Verification Algorithms for Completeness of Chinese-Chess Endgame Databases

Haw-ren Fang

Department of Computer Science,
University of Maryland, Maryland, USA
hrfang@cs.umd.edu

Abstract. Retrograde analysis has been successfully applied to solve Awari [6], and construct 6-piece Western chess endgame databases [7]. However, its application to Chinese chess is limited because of the *special rules* about indefinite move sequences. In [4], problems caused by the most influential rule, *checking indefinitely*[1], have been successfully tackled by Fang, with the 50 selected endgame databases were constructed in concord with this rule, where the 60-move rule was ignored. A conjecture is that other special rules have much less effect on staining the endgame databases, so that the corresponding stain rates are zero or small. However, the conjecture has never been verified before. In this paper, a rule-tolerant approach is proposed to verify this conjecture. There are two rule sets of Chinese chess: an Asian rule set and a Chinese rule set. Out of these 50 databases, 24 are verified complete with Asian rule set, whereas 21 are verified complete with Chinese rule set (i.e., not stained by the special rules). The 3 databases, KRKCC, KRKPPP and KRKCGG, are complete with Asian rule set, but stained by Chinese rules.

1 Introduction

Retrograde analysis is widely applied to construct databases of finite, two-player, zero-sum and perfect information games [8]. The classical algorithm first determines all terminal positions, e.g., checkmate or stalemate in both Western chess and Chinese chess, and then iteratively propagates the values back to their predecessors until no propagation is possible. The remaining undetermined positions are then declared as draws in the final phase.

In Western chess, as well as many other games, if a game continues endlessly without reaching a terminal position, the game ends in a draw. However, in Chinese chess, there are special rules other than checkmate and stalemate to end a game. Some positions are to be treated as wins or losses because of these rules, but they are mistakenly marked as draws in the final phase of a typical retrograde algorithm. The most influential special rule is checking indefinitely.

---

[1] Another name of the concept of checking indefinitely is *perpetual checking*.

In [4], 50 selected endgame databases in concord with this rule were successfully constructed. In this paper, a rule-tolerant verification algorithm is introduced to find out which of these databases are stained by the other special rules.

The organization of this paper is as follows. Section 2 gives the background as the previous works. Section 3 describes the special rules in Chinese chess. Section 4 abstracts these special rules and formulates the problems. Section 5 presents the rule-tolerant algorithms for verifying the completeness of Chinese-chess endgame databases. Section 6 gives the conclusion and suggests two future lines of work. Experimental results are given in the Appendix.

2 Background

Retrograde analysis is applied to the two-player, finite and zero-sum games with perfect information. Such a game can be represented as a *game graph* $G = (V, E)$ which is directed, bipartite, and possibly cyclic, where V is the set of vertices and E is the set of edges. Each *vertex* indicates a position. Each *directed edge* corresponds to a move from one position to another, with the relationship of *parent* and *child* respectively. In Chinese chess, a *position* is an assignment of a subset of pieces to distinct addresses on the board with a certain player-to-move. Positions with out-degree 0 are called *terminal* positions.

2.1 A Typical Retrograde Algorithm

Definition 1. *A win-draw-loss database of a game graph* $G = (V, E)$ *is a function,* $DB : V \to \{\text{win}, \text{draw}, \text{loss}\}$. *Each non-terminal position* $u \in V$ *satisfies the following constraints.*

1. *If* $DB(u) = \text{win}$, *then* $\exists (u,v) \in E$ *such that* $DB(v) = \text{loss}$.
2. *If* $DB(u) = \text{loss}$, *then* $\forall (u,v) \in E$, $DB(v) = \text{win}$.
3. *If* $DB(u) = \text{draw}$, *then* $\exists (u,v) \in E$ *such that* $DB(v) = \text{draw}$, *and* $\forall (u,v) \in E$, $(DB(v) = \text{draw}) \vee (DB(v) = \text{win})$.

Definition 1 draws the most fundamental game-theoretical constraints that a win-draw-loss database must satisfy. A classical retrograde algorithm for constructing a win-draw-loss database consists of three phases: initialization, propagation, and the final phase.

1. In the initialization phase, the win and loss terminal positions are assigned to be **wins** and **losses**, respectively. They are checkmate or stalemate positions in Chinese chess.
2. In the propagation phase, these values are propagated to the their parents, until no propagation is possible.
3. The final phase is to mark undetermined positions as **draws**.

In the propagation phase, if an undetermined position has a child being a **loss**, it is assigned as a **win** to satisfy constraint (1) in Definition 1. If an undetermined position have all children as **wins**, it is assigned as a **loss** to satisfy constraint (2). The process continues until no update is possible. The remaining undetermined positions are marked as **draws** in the final phase. These marked draws satisfy constraint (3). This is a very high level description of retrograde algorithms. In practice, it is usually implemented as: whenever a position has its status determined, it propagates the result to its parents.

Retrograde analysis cannot apply to the whole game graph of Chinese chess, $G = (V, E)$, on a physical computer, because the graph is too big. Therefore, the algorithm is applied to a subgraph $G' = (V', E')$, which satisfies $\forall u \in V'$, $(u, v) \in E \Rightarrow ((v \in V') \wedge ((u, v) \in E'))$. The subgraph is typically partitioned into multiple endgame databases according to the numbers of different pieces remaining on the board. To simplify the notation and without losing generality, this subgraph is also called the Chinese-chess game graph throughout this paper.

2.2 Retrograde Analysis for Chinese Chess

The classical retrograde analysis requires that a game which does not end in a terminal vertex must end in a draw. Otherwise, problems may occur in the final phase. In Western chess, a game which continues endlessly without reaching a terminal vertex is judged to be a draw. In Chinese chess, however, such a game may end in a win or loss. These positions are sometimes mistakenly marked as draws in the final phase. As a result, the draw declaration can only safely be applied to the Chinese-chess endgame databases with only one player having attacking pieces [2,9].

Assuming both players play flawlessly, a position is called *stained* by some special Chinese-chess rule, if the game ends differently (i.e., win-draw-loss status changes) when this rule is ignored. A database is stained if it has one or more stained positions. In contrast, a database is *complete* if all the positions in it have the correct win-draw-loss information. The most influential special rule to stain the endgame databases is checking indefinitely. In [4], 50 endgame databases were successfully constructed in concord with this rule. All the recorded win and loss positions in the databases are verified as correct. However, the marked draws in the databases are possibly stained by other special rules. Therefore, a database is complete, if all the marked draws are not stained. The main theme of this paper is to investigate whether there are positions marked as draws in the databases and whether they are stained by the special rules.

3 Special Rules in Chinese Chess

In Chinese chess, the two sides are called Red and Black. Each side has one King, two Guards, two Ministers, two Rooks, two Knights, two Cannons, and five Pawns, which are abbreviated as K, G, M, R, N, C and P, respectively[2]. The pieces Rooks, Knights, Cannons and Pawns are called *attacking pieces* since they

---

[2] The English translation of the Chinese names differs by author.

can move across the *river*, the imaginary stream between the two central horizontal lines of the board. In contrast, Guards and Ministers are called *defending* pieces because they are confined to the domestic region[3].

In addition to checkmate and stalemate, there are various rules of indefinite move sequences to end a game. They are called *special rules* in this paper. An indefinite move sequence is conceptually an infinite move sequence. In real games, it is determined by the threefold repetition of positions in a finite move sequence [1, page 20, rule 23, page 65, rule 3].

3.1 Chinese and Asian Rule Sets

There are two rule sets of Chinese chess: an Asian rule set and a Chinese rule set. The differences are generally about the special rules other than checking indefinitely. In both Asian and Chinese rule sets, there are dozens of detailed special rules and sub-rules. Some rules are exceptions of some others. Because they are very complicated, we attempt to verify the completeness of a given endgame database via a *rule-tolerant* approach, instead of formulating all these rules.

All the special rules discussed in this paper refer to the rule book [1], in which pages 1–46 describe the Chinese rule set, and pages 47–119 describe the Asian rule set. Readers do not require this book to follow this paper. However, this rule book keeps to be cited to confirm that the theory is correct. The rules of checking indefinitely and mutual checking indefinitely are well studied in [3,4]. We focus on the other special rules.

3.2 The Rules of Chasing Indefinitely

In Chinese chess, *chasing indefinitely* is forbidden. The general concept is that a player cannot chase some opponent's piece continuously without ending [1, page 21, page 64]. The term *chase* is defined similarly to the term check, but the prospective piece to be captured is not the King but some other piece. For example in Figure 1(a) with Red to move, the game continues cyclically with moves Re0-e2 Ng2-f0 Re2-e0 Nf0-g2, etc. Red loses the game since he[4] is *forced* to chase the Black Knight endlessly. Here and throughout this paper, a player is said to be forced to play the indicated moves, if he will lose the game by making any other moves because of the rules of checkmate, stalemate or checking indefinitely.

In some cases, chasing is allowed. For example, the Kings and the Pawns are allowed to chase other pieces [1, page 22, rule 27, page 65, rule 9]. In Figure 1(b) with Red to move, the game continues cyclically with moves Rb0-c0 Pb1-c1 Rc0-b0 Pc1-b1, etc. Although Black chases the Red Rook endlessly, the game ends in a draw because the chaser is a Pawn. Besides, some types of chasing

---

[3] The notation and basic rules of Chinese chess in English can be found in the ICGA web page http://www.cs.unimaas.nl/icga/games/chinesechess/, and in FAQ of the Internet news group rec.games.chinese-chess, which is available at http://www.chessvariants.com/chinfaq.html.

[4] In this paper we use 'he' when 'he' and 'she' are both possible.

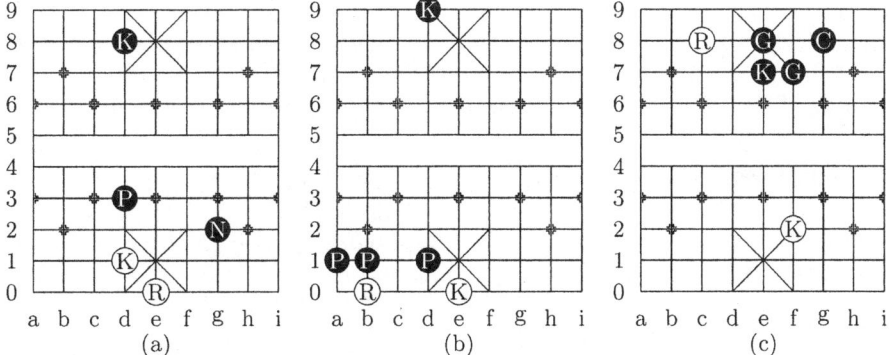

Fig. 1. Examples to illustrate the special rules

indefinitely is allowed under the Asian rule set, but forbidden by the Chinese rules. For example, it is allowed in the Asian rule set to endlessly chase one piece every other move and chase another piece at the moves in between [1, page 103, rule 32], whereas it is forbidden by the Chinese rules [1, page 24, rule 28.12]. An example is given in Figure 2(c). The rules of chasing indefinitely may also stain the endgame databases.

3.3 Summary of Special Rules

An indefinite move sequence is composed of two *semi-sequences*: one consists of the moves by Red and the other has the moves by Black. A semi-sequence is classified as being *allowed* or *forbidden* by the special rules. In the 50 endgame databases in concord with checking indefinitely, only marked draws are possibly stained. Therefore, special rules about allowed semi-sequences of moves can be ignored. In the Asian rule set, a forbidden semi-sequence of moves is either checking indefinitely or chasing indefinitely [1, page 64–65]. In the Chinese rule set, a forbidden semi-sequence of moves consists of three types of moves: checking, chasing, and *threatening to checkmate* [1, page 21]. A threatening-to-checkmate move means that a player *attempts* to checkmate his opponent by a semi-sequence of checking moves. A semi-sequence of moves is *allowed* if it is not forbidden. According to [1, page 20, rule 24, page 64, section 2], the special rules applied to an indefinite move sequence are summarized as follows.

1. If only one player checks the other indefinitely, the player who checks loses the game.
2. Otherwise, if only one semi-sequence of moves is forbidden and the other is allowed, the player who played the forbidden semi-sequence of moves loses.
3. Otherwise, the game ends in a draw.

With the above summary, both mutual checking indefinitely and mutual chasing indefinitely result in a draw. The example in Figure 1(c) with Red to move illustrates the difference between the Asian and the Chinese rule sets. The game continues cyclically with moves Rc8-c7 Ge8-d7 Rc7-c8 Gd7-d8, etc. Red is forced to check every other move and chases at the moves in between. It is allowed by the Asian rules [1, page 64, rule 2] but forbidden by the Chinese rules [1, page 21, rule 25.2].

4 Problem Formulation

Below we formulate the problem to be solved more precisely. In 4.1 we deal with abstracting the special rules and in 4.2 we describe a rule-tolerant approach.

4.1 Abstracting Special Rules

Denote the Chinese-chess game graph by $G = (V, E)$. For ease of discussion, we assume V contains the illegal positions, in which the own King is left in check, and E includes the moves to these illegal positions. Given a position $v \in V$, we define $\bar{v} \in V$ to have the piece assignment on the board the same as that of v but with a different player-to-move. The boolean function $check : E \rightarrow \{\text{true}, \text{false}\}$ is to indicate whether the edge (u, v) is a checking move. In other words, $check((u, v)) = \text{true}$ if and only if the next mover in $\bar{v}$ can capture the opponent's King immediately.

Definition 2. *In an infinite sequence of moves (v_0, v_1), (v_1, v_2), etc., the first mover loses the game because of the rule of checking indefinitely, if*

1. $\forall$ even i, $check((v_i, v_{i+1})) = \text{true}$.
2. $\forall n \geq 0$, $\exists$ odd $j > n$, such that $check((v_j, v_{j+1})) = \text{false}$.

In addition, the game results in a draw because of mutual checking indefinitely if, $\forall i \in N \cup \{0\}$, $check((v_i, v_{i+1})) = \text{true}$.

Note that this definition ignores the 60-move rule. Two semi-sequences of indefinite moves are (v_0, v_1), (v_2, v_3), etc. and (v_1, v_2), (v_3, v_4), etc.

We use the boolean function $threaten : E \rightarrow \{\text{true}, \text{false}\}$ to indicate whether a given move is threatening to checkmate. It is defined by $threaten((u, v)) = \text{true}$ if the own King in u is not in check and the next mover in $\bar{v}$ can checkmate his opponent with a semi-sequence of checking moves, assuming both players play flawlessly. If the next mover in u is in check, the move (u, v) is generally to get out of check and not counted as an attempt to checkmate the opponent.

Before defining a chasing move, we need to define the capturing move. A *direct capturing* move has a capturing piece and a captured piece other than the King[5]. Two direct capturing moves are treated the same if they have the same capturing piece and captured piece. A move (u, v) is called a chasing move, if $\bar{v}$ has a capturing move which u does not have and after the capturing move from $\bar{v}$, the own King of v is not in check (i.e., not an illegal position) [1, page 24, rule 29]. The boolean function $chase^* : E \rightarrow \{\text{true}, \text{false}\}$ is defined to indicate

---

[5] The cases of capturing a protected piece and indirect capturing are omitted here but will be discussed in Subsection 4.2.

whether a given edge (u,v) is a chasing move. The boolean function $chase: E \rightarrow$ {true, false} is to indicate whether a given edge is a forbidden chasing move. Note that $(chase((u,v)) = \text{true}) \Rightarrow (chase^*((u,v)) = \text{true})$. The definition of mutual and non-mutual chasing indefinitely is similar to that of mutual and non-mutual checking indefinitely in Definition 2, but replacing $check((u,v))$ by $chase((u,v))$. For the Asian rule set, we assume that the given database to be verified is in concord with checking indefinitely and mutual checking indefinitely, such as the 50 endgame databases in [4]. Therefore, we may focus only on chasing indefinitely in the verification algorithms for the Asian rules.

Definition 3. *Let the boolean function $forbid : E \rightarrow$ {true, false} indicate whether a given move is forbidden. Then,*

1. *Asian rules: $forbid((u,v)) := chase((u,v))$.*
2. *Chinese rules: $forbid((u,v)) := check((u,v)) \vee threaten((u,v)) \vee chase((u,v))$.*

If a semi-sequence of indefinite moves (v_0, v_1), (v_2, v_3), etc. is forbidden, then $\forall$ even i, $forbid((v_i, v_{i+1})) = \text{true}$.

4.2 A Rule-Tolerant Approach

There is a problem we need to face. The value of $chase((u,v))$ may depend on the other moves in a move sequence. Sometimes we cannot determine $chase((u,v))$ without inspecting the other moves in the move sequence. We call such a move *path-dependent*. For example, usually Pawns are allowed to chase; therefore, $chase((u,v)) = \text{false}$ if the chasing piece is a Pawn. However, with the Chinese rule set, if the opponent plays a forbidden semi-sequence of moves at the same time, then $chase((u,v)) = \text{true}$ with the piece to chase being a Pawn [1, page 32, rule 11]. Another example in the Asian rule set is: indefinitely chasing one piece every other move and chasing another piece at the moves in between is allowed [1, page 103, rule 32]. As a result, $forbid((u,v))$ is path-dependent. Nevertheless, $check((u,v))$, $threaten((u,v))$, $chase^*((u,v))$ and $forbid^*((u,v))$ are path-independent.

A rule-tolerant approach is proposed as follows. Let the boolean function $f(s)$ indicate whether a given semi-sequence of moves s is forbidden. Instead of programming the function $f(s)$, we look for another boolean function $f^*(s)$ satisfying $(f(s) = \text{true}) \Rightarrow (f^*(s) = \text{true})$. In other words, if we know $f^*(s) = \text{false}$, we can safely declare s is not a forbidden semi-sequence of moves. If a move sequence is composed of two semi-sequence of moves satisfying $f^*(s) = \text{false}$, the game is verified as a draw. If all positions marked as draws in a given database DB are verified as draws, then the database can be declared complete, assuming only marked draws are possibly stained.

Lemma 1. *Define the boolean function $forbid^* : E \rightarrow$ {true, false} as follows.*

1. *Asian rules: $forbid^*((u,v)) := chase^*((u,v))$.*
2. *Chinese rules: $forbid^*((u,v)) := check((u,v)) \vee threaten((u,v)) \vee chase^*((u,v))$.*

A semi-sequence of moves (v_0, v_1), (v_2, v_3), etc. is called suspiciously forbidden if, $forbid^*((v_i, v_{i+1})) = $ **true** for all even i. If a semi-sequence of moves is not suspiciously forbidden, it is an allowed semi-sequence of indefinite moves.

Proof. By Definition 3 and $(chase((u,v)) = $ **true**$) \Rightarrow (chase^*((u,v)) = $ **true**$)$, $(forbid((u,v)) = $ **true**$) \Rightarrow (forbid^*((u,v)) = $ **true**$)$. A forbidden semi-sequence of moves (v_0, v_1), (v_2, v_3), etc. satisfies that $\forall$ even j, $forbid((v_j, v_{j+1})) = $ **true**. If it is not suspiciously forbidden, then $\exists$ even i such that $forbid^*((v_i, v_{i+1})) = $ **false**, which implies $forbid((v_i, v_{i+1})) = $ **false**. A contradiction. □

In both Chinese and Asian rule sets, capturing a *protected* piece is usually considered as not a capturing move [1, page 24, rule 29.3, page 108, rule 35]. We call them *nullified* capturing moves in this paper. The general concept of a protected piece p is that, if p is captured by some opponent's piece q, the player of p can capture q back immediately *without exposing* the own King in check [1, page 24, rule 29.4, page 107, rule 34]. It has an exception in the Asian rule set: chasing a protected Rook by Knights and Cannons is forbidden [1, page 86, rule 20]. It is also true in the Chinese rule set because the chaser attempts to gain from capturing [1, page 21, rule 25.3, page 24, rule 29.1]. The key for being rule-tolerant is $(chase((u,v)) = $ **true**$) \Rightarrow (chase^*((u,v)) = $ **true**$)$. The function $chase^*((u,v))$ is defined by the chasing moves of u and $\bar{v}$. Ignoring nullified capturing moves of $\bar{v}$ remains rule-tolerant, because $chase^*((u,v))$ has a greater chance to be **true**. The problem occurs only when u has a nullified capturing move, which remains a capturing move but not nullified in $\bar{v}$. Figure 2(c) gives such an example. In the Appendix, nullified capturing moves are taken into account in the experiments. It is also shown that ignoring this rule in the verification algorithms does not change the conclusion from the experiments on the 50 endgame databases.

All the Asian rules are taken into account in the rule-tolerant approach, but we do ignore two Chinese rules about forbidden moves in the current experiments. (a) One is *indirect capturing*. It means that a move does not capture a piece immediately, but will capture some opponent's piece after a semi-sequence of checking moves made by the capturing player [1, page 24, rule 29.1]. (b) The other is capturing an *insufficiently* protected piece. It means that the player can *gain* after a sequence of capturing moves by both players [1, page 24, rule 29.3]. It is an exception of chasing a protected piece, i.e., capturing a protected piece is usually not counted as a capturing move, unless the capturing player can gain some piece after a sequence of capturing moves. The function $chase^*((u,v))$ is defined by the chasing moves of u and $\bar{v}$. With the Chinese rule set, problems may occur when $\bar{v}$ has a capturing move of case (a) or (b) which u does not have. These two rules require substantial programming. For case (a), all the finite semi-sequences of checking moves ending with capturing moves need to be considered. For case (b), the order of the capturing needs to be taken into account. Cases mixed with (a) and (b) are even more complicated.

If a player can win the game by either checkmate or stalemate, the position is already correctly marked by a typical retrograde algorithm. If a player can win the game by checkmate, stalemate, or the rule of checking indefinitely, then the position is already correctly marked by a checking-indefinitely-concordant retrograde algorithm by Fang [4]. Given a database from [4], our consideration deals with the positions in which some player can win the game only by special rules other than checking indefinitely, assuming both players play flawlessly. The experiments are on the 50 endgame databases in [4]; each has at most three attacking pieces (except KRKPPP which has four) remaining on the board and a database size less than 1GB. Usually, it requires several attacking pieces to form an indefinite move sequence of cases (a) or (b). Therefore, the effect because of ignoring the two Chinese rules in the experiments tends to be very minor.

5 Rule-Tolerant Verification Algorithms

Given a checking indefinitely concordant database of Chinese chess, in which the marked win and loss positions have correct information, we need to foresee if any of the marked draws may be mistaken because of the special rules. To achieve this goal, *suspicious move patterns* of special rules are defined (in 5.1), computed (in 5.2), and verified (in 5.3).

5.1 Suspicious Move Patterns of Special Rules

For the ease of discussion, we assume there is an *attacking* side and the other side is *defending*. The attacking side tries to win the game by forcing the defending side play the forbidden semi-sequence of moves, whereas the defending side tries to avoid such a semi-sequence.

Definition 4. *Given a win-draw-loss database DB of the Chinese-chess game graph $G = (V, E)$, a suspicious move pattern of special rules is a subgraph of G, denoted by $G^* = (V^*, E^*)$, with V^* being partitioned into V_A^* and V_D^* according to whether the next mover is attacking or defending. G^* satisfies the following constraints.*

1. $\forall u \in V^*$, $DB(u) = $ **draw**.
2. $\forall (u, v) \in E^*$ with $u \in V_D^*$, $forbid^*((u, v)) = $ **true**.
3. $\forall u \in V_D^*$, $((u, v) \in E) \wedge (DB(v) = $ **draw**$)) \Rightarrow ((u, v) \in E^*)$.
4. $\forall u \in V^*$, $\exists (u, v) \in E^*$, i.e., out-degree is at least 1.

Constraint (1) is because only positions marked as draws are possibly stained. Constraint (2) ensures the defending side play suspicious forbidden moves all the time inside the pattern. Constraint (3) makes the defending side unable to quit the pattern without losing the game. Constraint (4) keeps the pattern indefinite.

Lemma 2. *Given a win-draw-loss database for the Chinese-chess game graph G, and any two suspicious move patterns of special rules $G_1^* = (V_1^*, E_1^*)$ and $G_2^* = (V_2^*, E_2^*)$ having the same attacking and defending sides, $G^* = (V_1^* \bigcup V_2^*, E_1^* \bigcup E_2^*)$ is also a suspicious move pattern of special rules.*

Proof. The graph G^* clearly satisfies constraints (1) and (4). $\forall u \in V_1^*$ (or V_2^*), if the next mover in u is defending, no more edges going out of u are added in the union. Therefore, constraints (2) and (3) are satisfied. □

Theorem 1. *Given a win-draw-loss database of Chinese chess, there exist two unique maximum suspicious move patterns of special rules. In one of them, Red is the attacking side, whereas in the other, Black is the attacking side.*

Proof. Since the game graph $G = (V, E)$ of Chinese chess is finite, the number of suspicious move patterns of special rules is finite, with a given win-draw-loss database of Chinese chess. Denote these move patterns with Red as attacking side by $G_i^* = (V_i^*, E_i^*)$ for $i = 1, 2, \ldots, n$. By Lemma 2, $\overline{G^*} = (\bigcup_{i=1}^n V_i^*, \bigcup_{i=1}^n E_i^*)$ is a suspicious move pattern of special rules. It is maximum since $V_j^* \subseteq \bigcup_{i=1}^n V_i^*$ and $E_j^* \subseteq \bigcup_{i=1}^n E_i^*$ for $j = 1, 2, \ldots, n$. The proof is completed by swapping the attacking and defending sides. □

Given a win-draw-loss database of Chinese chess in concord with checking indefinitely from [4], each forbidden semi-sequence of indefinite moves is inside one of the two maximum patterns, assuming that both players, based on the given win-draw-loss database, play flawlessly to avoid playing and force each other to play a suspiciously forbidden semi-sequence of indefinite moves. By Lemma 1, a forbidden semi-sequence of moves must be suspiciously forbidden. Therefore, if these two maximum suspicious move patterns of special rules are empty, then the given database is complete. The discussion proves the following theorem.

Theorem 2. *Given a win-draw-loss database DB of Chinese chess in concord with checking indefinitely, if both maximum suspicious move patterns of special rules are empty, then database DB is complete.*

5.2 Computing the Maximum Suspicious Move Patterns

Lemma 3. *Given a win-draw-loss database for the Chinese-chess game graph $G = (V, E)$, the two maximum suspicious move patterns of special rules $\overline{G^*} = (\overline{V^*}, \overline{E^*})$ are induced subgraphs of G, i.e., $\forall (u, v) \in E$, $(u, v \in \overline{V^*}) \Rightarrow ((u, v) \in \overline{E^*})$.*

Proof. Given $(u, v) \in E$, if the next mover of u is the defending side, the statement is true by constraints (1) and (3) in Definition 4. If the next mover of u is the attacking side, the statement is true because the graph $\overline{G^*}$ is maximum. □

By Lemma 3, if we know $\overline{V^*}$ with a given win-draw-loss database DB, then $\overline{G^*} = (\overline{V^*}, \overline{E^*})$ can be determined as an induced subgraph of $G = (V, E)$. Define $\overline{V} = \{u : (u \in V) \land (DB(u) = \textbf{draw})\}$ and $\overline{G} = (\overline{V}, \overline{E})$ as an induced subgraph of G, i.e., $\overline{E} := \{(u, v) : ((u, v) \in E) \land (u, v \in \overline{V})\}$. Then $\overline{G^*}$ is a subgraph of $\overline{G}$ because of constraint (1) in Definition 4. The algorithm to compute $\overline{V^*}$ consists of two phases: initialization and pruning. The first phase is to compute suspicious win and loss candidate sets W and L respectively, such that $\overline{V^*} \subset W \bigcup L$. The pseudo-code of initialization phase is as follows.

$W \leftarrow \emptyset, L \leftarrow \emptyset$
for all $u \in \overline{V}$ with the next mover of u as the defending side **do**
    if $\forall (u,v) \in \overline{E}, forbid^*((u,v)) = \textbf{true}$ **then**
        $L \leftarrow L \bigcup \{u\}$
        $\forall (w,u) \in \overline{E}, W \leftarrow W \bigcup \{w\}$
    end if
end for

The second phase is to prune unqualified candidates in W and L by an iterative process, until no pruning is possible. If a suspicious loss candidate $u \in L$ has a child $v \in \overline{V}$ but $v \notin W$, then u does not satisfy constraint (3) in Definition 4 and therefore is pruned. If a suspicious win candidate in W does not have a child in L, then it does not satisfy constraint (4) and therefore is pruned. This observation suggests the following algorithm.

repeat
    {Prune unqualified suspicious loss positions.}
    for all $u \in L$ **do**
        if $\exists (u,v) \in \overline{E}$ such that $v \notin W$ **then**
            $L \leftarrow L - \{u\}$
        end if
    end for
    {Prune unqualified suspicious win positions.}
    for all $v \in W$ **do**
        if $\forall (v,u) \in \overline{E}, u \notin L$ **then**
            $W \leftarrow W - \{v\}$
        end if
    end for
until No more pruning is possible.

When no more pruning is possible, the subgraph of G induced by $W \bigcup L$ satisfies all constraints in Definition 4. Therefore, it is the maximum suspicious move pattern of special rules $\overline{G^*} = (\overline{V^*}, \overline{E^*})$, i.e., $\overline{V^*} = W \bigcup L$. Swapping the attacking and defending sides, we obtain the other maximum suspicious move pattern. Some strategies, such as children counting, can be applied to improve efficiency.

5.3 Verification Algorithms

By Theorem 2, if both the maximum suspicious move patterns of special rules are empty, then the given database DB is complete. This verification algorithm is called rule-tolerant because of the tolerance between boolean functions $chase^*((u,v))$ and $chase((u,v))$. If these two maximum suspicious move patterns are small enough, we may inspect the positions inside the patterns with a rule book to see if they are truly stained by the special rules. Besides, we may try trimming the maximum move patterns by reducing the tolerance.

Lemma 4. *Assume we are given a path-independent boolean function* $\overline{chase^*}$: $E \rightarrow \{\textbf{true}, \textbf{false}\}$, *which satisfies* $(chase((u,v)) = \textbf{true}) \Rightarrow (\overline{chase^*}((u,v)) = \textbf{true}) \Rightarrow (chase^*((u,v)) = \textbf{true})$ *for all* $(u,v) \in E$, *where* $G = (V,E)$ *is* the Chinese-chess game graph. Define the boolean function $\overline{forbid}^* : E \to$ {true, false} based on $\overline{chase}^*$ in a similar way to $forbid^*(u,v)$ in Lemma 1. The resulting maximum suspicious move pattern $\overline{G}^*$ is a subgraph of G^* in Theorem 1, with given the same win-draw-loss database. All forbidden indefinite semi-sequences of moves are inside $\overline{G}^*$, with the assumption that both players play flawlessly.

Proof. For all $(u,v) \in E$, $(\overline{forbid}^*((u,v)) = $ true$) \Rightarrow (forbid^*((u,v)) = $ true$)$, because $(\overline{chase}^*((u,v)) = $ true$) \Rightarrow (chase^*((u,v)) = $ true$)$. Therefore, a suspicious move pattern based on $\overline{chase}^*$ is also suspicious based on $chase^*$. The resulting maximum suspicious move pattern $\overline{G}^*$ based on $\overline{chase}^*$ is a subgraph of G^* based on $chase^*$. With the fact $(chase((u,v)) = $ true$) \Rightarrow (\overline{chase}^*((u,v)) = $ true$)$ for all $(u,v) \in E$, the last statement is true by the discussion similar to the proof of Lemma 1 and Theorem 2. □

The function $\overline{chase}^*((u,v))$ is obtained from $chase^*((u,v))$ by excluding some path-independent allowed moves. The algorithm in Subsection 5.2 requires the boolean function $forbid^*((u,v))$ being path-independent. Therefore, we need to keep function $\overline{chase}^*((u,v))$ being path-independent when reducing the tolerance. The closer $\overline{chase}^*((u,v))$ to $chase((u,v))$, the smaller the tolerance, and the smaller the maximum suspicious move patterns we may obtain.

Theorem 3. *Given a win-draw-loss database with marked draws possibly stained and the attacking side specified, the maximum suspicious move pattern of special rules with the Asian rule set is a subgraph of that with the Chinese rule set, assuming they use the same function $chase^*((u,v))$.*

Proof. By the definition of $forbid^*((u,v))$ in Lemma 1, a suspiciously forbidden semi-sequence of moves with the Asian rule set is also suspiciously forbidden with the Chinese rule set. Therefore, a suspicious move pattern of special rules with the Asian rule set is also suspicious with the Chinese rule set, which implies this theorem. □

By Theorem 3, if both maximum suspicious move patterns are empty with the Chinese rule set, then they are also empty with the Asian rule set, and therefore the given win-draw-loss database is complete with both the Asian and Chinese rule sets.

6 Conclusion and Future Work

Retrograde analysis has been successfully applied to many games. In Chinese chess, its application is confined to the endgames with only one player having attacking pieces. Other endgame databases were not perfectly reliable because of the existence of special rules. In the experiments with the 50 endgame databases in concord with checking indefinitely in [4], 24 and 21 endgame databases with both players having attacking pieces are verified complete with the Asian and Chinese rule sets, respectively. The 3 endgame databases, KRKCC, KRKPPP and KRKCGG, are complete with the Asian rule set but stained by the Chinese rules, as shown in the Appendix. Two suggested future directions of work are listed below.

1. **Knowledgeable encoding and querying of endgame databases.** For the endgame databases verified to be complete, we may extract and condense the win-draw-loss information into physical memory via the approach by Heinz [5]. The result can improve the present Chinese-chess programs.
2. **Construction of complete endgame databases.** For those stained by the special rules other than checking indefinitely, further work is required for the Chinese-chess endgame databases with complete information.

Acknowledgement

The author would like to thank Tsan-sheng Hsu for his suggestion to tackle this problem, and Shun-Chin Hsu for sharing his experience of programming threatening to checkmate.

References

1. China Xiangqi Association. *The Playing Rules of Chinese Chess.* Shanghai Lexicon Publishing Company, 1999. In Chinese.
2. H.-r. Fang, T.-s. Hsu, and S.-c. Hsu. Construction of Chinese Chess endgame databases by retrograde analysis. In T. Marsland and I. Frank, editors, *Lecture Notes in Computer Science 2063: Proceedings of the 2nd International Conference on Computers and Games,* pages 96–114. Springer-Verlag, New York, NY, 2000.
3. H.-r. Fang, T.-s. Hsu, and S.-c. Hsu. Indefinite sequence of moves in Chinese Chess endgames. In J. Schaeffer, M. Müller, and Y. Björnsson, editors, *Lecture Notes in Computer Science 2063: Proceedings of the 3rd International Conference on Computers and Games,* pages 264–279. Springer-Verlag, New York, NY, 2002.
4. H.-r. Fang, T.-s. Hsu, and S.-c. Hsu. Checking indefinitely in Chinese-chess endgames. *ICCA Journal,* 27(1):19–37, 2004.
5. E. A. Heinz. Knowledgeable encoding and querying of endgame databases. *ICCA Journal,* 22(2):81–97, 1999.
6. J.W. Romein and H.E. Bal. Awari is solved. *ICCA Journal,* 25(3):162–165, 2002.
7. K. Thompson. 6-piece endgames. *ICCA Journal,* 19(4):215–226, 1996.
8. H. J. van den Herik, J. W. H. M. Uiterwijk, and J. van Rijswijck. Games solved: Now and in the future. *Artificial Intelligence,* 134:277–311, 2002.
9. R. Wu and D.F. Beal. Fast, memory-efficient retrograde algorithms. *ICCA Journal,* 24(3):147–159, 2001.

Appendix: Experimental Results on the 50 Endgame Databases

Before verifying the completeness of a given endgame database as the Chinese-chess game graph is split, we need to verify all its supporting databases. If any of the two maximum suspicious move patterns of the special rules are non-empty, we may inspect whether these suspicious positions are stained by the special rules. An endgame database is not complete, if it has positions truly stained. If both maximum suspicious move patterns of the special rules of a given endgame database are empty or all the suspicious positions are inspected to be not stained, then the endgame database is verified as complete, assuming all its supporting databases are complete. The overall procedure to verify a given endgame database includes verifying all its supporting databases in bottom-up order.

The experiments are performed on the 50 endgame databases in concord with checking indefinitely in [4]. Table 1 lists the statistics for the Asian rule set. It provides the number of positions for each maximum suspicious move patterns. There

Table 1. Statistics of max suspicious patterns of 50 endgame databases, Asian rules

Database Name	Complete	Number of Positions		Ignore Protecting		Consider Protecting	
		legal	draw	Red attack	Black attack	Red attack	Black attack
KRCKRGG	N	252077421	147261943	6152	517	5573	170
KRCKRG	N	141001563	66005451	90775	54	88632	156
KRCKRM	N	209431807	103902489	22361	974	21948	1214
KRCGGKR	N	256862617	34156099	271110	23970	269839	23970
KRCGKR	N	142232812	14112951	54040	354	54324	354
KRCMKR	N	210030190	23470723	58314	8865	29958	8865
KRCKR	N	31012335	9108171	4132	0	3917	0
KRNKRGG	N	251181481	103598842	20739	2428	16096	2196
KRNKRG	N	138660209	15453492	7877	977	7356	1142
KRNKRM	N	204263530	29396541	11220	2254	9342	2915
KRNKR	N	30118362	2936309	453	192	449	192
KRPKRG	N	84330363	35495583	31558	58	25360	187
KRPKRM	N	124050578	44636393	23584	1218	18256	1334
KRPKR	N	18443469	2635006	975	0	930	0
KRKNN	N	16300026	4621246	357	0	363	344
KRKNC	N	33568194	596093	296	162	296	162
KRKCC	Y	17300976	1409308	0	0	0	0
KRKNPGG	N	168307887	2123952	427	382	427	382
KRKNPG	N	92456806	112434	0	237	0	237
KRKNPM	N	136200539	466627	0	492	0	492
KRKNP	N	20011890	20026	0	79	0	79
KRKPPP	Y*	103676439	1179271	111	72	111	24
KRKPPGG	Y*	52598998	947571	0	61	0	61
KRKPPMM	Y	122221940	2211873	0	0	0	0
KRKPPG	Y*	28498574	53263	0	26	0	26
KRKPPM	Y	41658907	107613	0	0	0	0
KRKPP	Y	6084903	8187	0	0	0	0
KNPKN	N	21682338	4889447	17760	0	17746	0
KNPKCG	N	100076040	40930099	1917	0	1899	0
KNPKCM	N	149719630	67281695	7970	0	7142	0
KNPKC	N	22364304	6387677	268	0	301	0
KCPKC	Y	22956705	20094132	0	0	0	0
KRGGKR	Y	2997932	1840528	0	0	0	0
KRGKR	Y	1628603	966336	0	0	0	0
KRMKR	Y*	2389472	1370793	29	144	29	144
KRKR	Y	348210	193950	0	0	0	0
KRKNGGMM	N	63684381	33604025	83	0	83	0
KRKNGGM	Y	21879507	719210	0	0	0	0
KRKNGMM	N	35195142	4703606	1605	0	1622	0
KRKNGG	Y	3221138	6498	0	0	0	0
KRKNGM	Y	12032732	27288	0	0	0	0
KRKNMM	Y	7654095	121216	0	0	0	0
KRKNG	Y	1762807	3275	0	0	0	0
KRKNM	Y	2605497	5200	0	0	0	0
KRKN	Y	380325	609	0	0	0	0
KRKCGG	Y	3322727	1379102	0	0	0	0
KRKCMM	Y	7913097	2969808	0	0	0	0
KRKCG	Y	1820350	2462	0	0	0	0
KRKCM	Y	2694400	91748	0	0	0	0
KRKC	Y	393327	7479	0	0	0	0

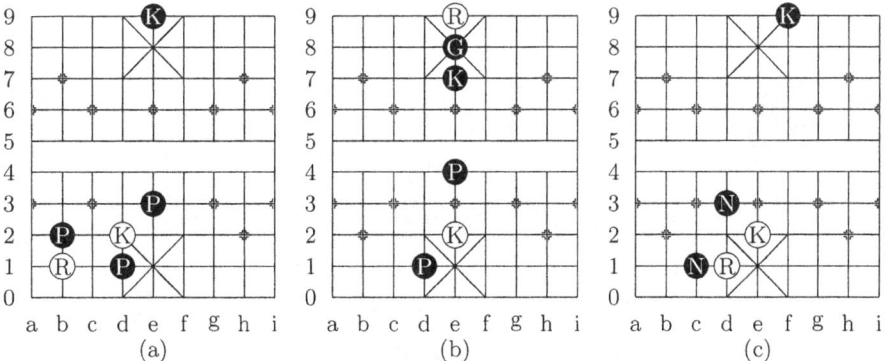

Fig. 2. (a)(b) Suspicious positions not stained by the special rules. (c) A game ends differently under the Asian and Chinese rule sets.

are two experiments on each database. One ignores the rule of capturing a protected piece discussed in Subsection 4.2, and the other takes it into account. Although the statistics differ, the conclusion does not change with this rule ignored. An endgame database with a non-empty maximum suspicious move pattern but verified as complete is denoted by Y*. The statistics exclude illegal positions. Two conjugate positions (i.e., piece assignment on the board is the same as each other in the mirror) are treated as the same one and counted only once.

In Figure 2(a), the game continues cyclically with moves Rb1-a1 Pb2-a2 Ra1-b1 Pa2-b2, etc. In Figure 2(b), the game continues indefinitely with moves Ke2-d2 Pd1-e1 Kd2-e2 Pe1-f1 Ke2-f2 Pf1-e1, etc. In both games, Red is forced to chase indefinitely. However, both are allowed because the chaser is a King or a Pawn [1, page 22, rule 27, page 65, rule 9]. In Figure 2(c), the game continues indefinitely with moves Ke2-d2 Nc1-b3 Kd2-e2 Nb3-c1, etc. Red is forced to chase the two Black Knights iteratively. It is allowed under the Asian rule set but forbidden by the Chinese rules [1, page 24, rule 28.12, page 103, rule 32].

A forbidden semi-sequence of moves under the Asian rule set is generally also forbidden under the Chinese rule set. To verify the completeness in Chinese rules, we may focus on the endgame databases complete with the Asian rule set. Table 2 lists the statistics of the experimental results for the Chinese rule set. It includes all the endgame databases complete with the Asian rule set. An endgame database complete with the Asian rule set but stained by the Chinese rules is denoted by N*. Examples are KRKCC, KRKPPP, and KRKCGG. Figure 3(a) illustrates a KRKCC endgame stained only by the Chinese rules. The game continues cyclically with the moves Ke2-f2 Ce6-f6 Kf2-e2 Cf6-e6, etc. Black is forced to check every other move, and threatens to checkmate at all moves in between. In Figure 3(b), the game continues indefinitely with the moves Rb8-c8 Pc1-b1 Rc8-c9 Kf9-f8 Rc9-b9 Pb1-c1, etc. Red is forced to play a semi-sequence of indefinite moves which consists of checking and chasing moves. In Subsection 3.3, Figure 1(c) illustrates another example of a KRKCGG endgame stained only by the Chinese rules.

In Chinese chess, Cannon requires an additional piece to jump over to capture. In KNPKC endgames, Red seems unlikely to force Black to play forbidden chas-

Table 2. Statistics of max suspicious patterns of 35 endgame databases, Chinese rules

Database Name	Complete	Number of Positions		Ignore Protecting		Consider Protecting	
		legal	draw	Red attack	Black attack	Red attack	Black attack
KRKNN	N	16300026	4621246	372	76	378	344
KRKCC	N*	17300976	1409308	602	0	602	0
KRKNPGG	N	168307887	2123952	1187	382	1188	382
KRKNPG	N	92456806	112434	348	237	348	237
KRKNPM	N	136200539	466627	725	519	725	519
KRKNP	N	20011890	20026	116	79	116	79
KRKPPP	N*	103676439	1179271	129	781	129	268
KRKPPGG	Y*	52598998	947571	0	68	0	68
KRKPPMM	Y	122221940	2211873	0	0	0	0
KRKPPG	Y*	28498574	53263	0	35	0	35
KRKPPM	Y	41658907	107613	0	0	0	0
KRKPP	Y	6084903	8187	0	0	0	0
KNPKN	N	21682338	4889447	18116	0	18092	0
KNPKCG	N	100076040	40930099	12979	0	11227	0
KNPKCM	N	149719630	67281695	10407	0	9410	0
KNPKC	N	22364304	6387677	2838	0	1060	0
KCPKC	Y	22956705	20094132	0	0	0	0
KRGGKR	Y	2997932	1840528	0	0	0	0
KRGKR	Y	1628603	966336	0	0	0	0
KRMKR	Y*	2389472	1370793	29	144	29	144
KRKR	Y	348210	193950	0	0	0	0
KRKNGGMM	N	63684381	33604025	83	0	83	0
KRKNGGM	Y	21879507	719210	0	0	0	0
KRKNGMM	N	35195142	4703606	1687	0	1708	0
KRKNGG	Y	3221138	6498	0	0	0	0
KRKNGM	Y	12032732	27288	0	0	0	0
KRKNMM	Y	7654095	121216	0	0	0	0
KRKNG	Y	1762807	3275	0	0	0	0
KRKNM	Y	2605497	5200	0	0	0	0
KRKN	Y	380325	609	0	0	0	0
KRKCGG	N*	3322727	1379102	0	891	0	891
KRKCMM	Y	7913097	2969808	0	0	0	0
KRKCG	Y	1820350	2462	0	0	0	0
KRKCM	Y	2694400	91748	0	0	0	0
KRKC	Y	393327	7479	0	0	0	0

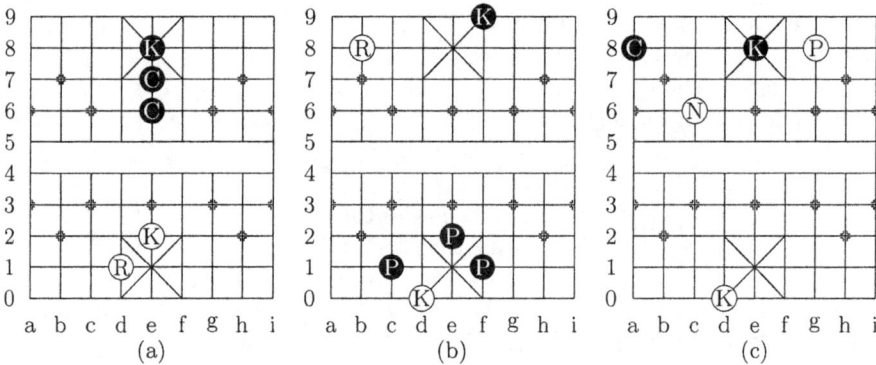

Fig. 3. Positions (a)(b) stained only by the Chinese rules and (c) stained by both rule sets

ing moves indefinitely, because Black has only one Cannon and the King. Note that Kings are allowed to chase. However, Figure 3(c) illustrates a surprising example. The game continues cyclically with the moves Nc6-b8 Ke8-e7 Nb8-c6 Ke7-e8, etc. Black loses the game because he chases a red Pawn indefinitely.

An External-Memory Retrograde Analysis Algorithm*

Ping-hsun Wu[1], Ping-Yi Liu[2,**], and Tsan-sheng Hsu[1,***]

[1] Institute of Information Science,
Academia Sinica, Taipei, Taiwan
{davidw, tshsu}@iis.sinica.edu.tw
[2] Department of Computer Sciences,
University of Texas, Austin, USA
pyliu@cs.utexas.edu

Abstract. This paper gives a new sequential retrograde analysis algorithm for the construction of large endgame databases that are too large to be loaded entirely into the physical memory. The algorithm makes use of disk I/O patterns and saves disk I/O time. Using our algorithm we construct a set of Chinese-chess endgame databases with one side having attacking pieces. The performance result shows that our algorithm works well even when the number of positions in the constructed endgame is larger than the number of bits in the main memory of our computer. We built the 12-men database KCPGGMMKGGMM, the largest database reported in Chinese chess, which has 8,785,969,200 positions after removing symmetrical positions on a 2.2GHz P4 machine with 1 GB main memory. This process took 79 hours. We have also found positions with the largest DTM and DTC values in Chinese chess so far. They are in the 11-men database KCPGGMKGGMM; the values are 116 and 96, respectively.

1 Introduction

There have been a great deal of studies into the construction of endgame databases for a variety of games [17]. The algorithms for constructing these databases are done by retrograde analysis. Several previous algorithms required storing the entire database in the physical memory in order to access randomly the database during the entire computation [15].

When the target endgame gets too large to fit into the main memory, fast memory-efficient algorithms are devised. There are several possible variations that could be used. We list some of them below. The first one is to find parallel machines with adequate amounts of total physical memory [2,14]. However, it is

---

* Research supported in part by NSC grants 91-2213-E-001-027 and 92-2213-E-001-005.
** The work was done while this author was with the Institute of Information Science, Academia Sinica, Taiwan.
*** Corresponding author.

difficult to find large machines to cope with the increasingly larger endgames. The next approach is to compress the database and use fewer bits to represent the partial information for positions during construction. Thompson first uses two bits per position for the construction of 6-piece databases for Western chess [16]. In a recent work, one bit was used to represent a position in Chinese chess [19]. Since at least one bit is needed to represent a position, if complicated and time-consuming encoded schemes are to be avoided, it is difficult to further compress the databases using this scheme. It is also not time-economical to build endgames with more positions than the number of bits in the main memory. The third approach is to devise elegant engineering solutions to reduce the amount of disk I/O usage. One such example is in the game of awari [10]. By using good index functions, in which only positions whose indexes are in a close interval are randomly accessed during construction, the page fault rate is significantly reduced. Another successful example is in the game of checkers [13]. By using an advanced compression scheme that supports real-time decompression and a specialized disk swapping algorithm, endgames each containing up to 170 billion positions, the current largest endgame databases reported for any game, are built. However, this approach is mostly game dependent.

The last approach uses the idea of partitioning a large database into several sub-databases, with this method, a sub-database built later only uses information contained in a few, but not all of the previous built sub-databases. One of the earliest works using this procedure was the partitioning of positions according to the current player in Thompson's 1986 work [15] for Western chess. Later, Thompson partitioned databases according to the number of pieces in the endgames to build 6-piece Western chess endgame databases [16]. In [7], the endgame KQQKQP was built by partitioning the databases according to the positions of the pawn. Other games also employed similar approaches, such as checkers [9]. By applying this approach, Hsu and Liu [6] devised a general graph-theoretical framework for the verification of Chinese-chess endgame databases, with possible usage in other similar games.

In this paper, we introduce a new approach, namely *external memory algorithms*, to construct endgame databases. Recently, this approach has been used, independently, to remove duplicated nodes in breath-first search [8]. In such an algorithm, final results are stored entirely on this disk at all times. During computation, intermediate results are also stored on the disk. Intermediate results are *merged* into the final result each time they are available. To merge intermediate and final results, which are both stored on the disk, we need to perform disk I/O operations. The time complexity of an external memory algorithm depends on the number and pattern of I/O accesses, in addition to the number of instructions executed. In this paper, all timing data are measured on a 2.2GHz P4 machine with 1 GB memory under FreeBSD 4.5-RELEASE. The largest database that we have built is KCPGGMMKGGMM, which contains 8,785,969,200 positions after removing symmetrical ones. The number of positions is larger than the number of bits in the physical memory of our computer. We also found positions with the currently reported largest DTM and DTC values in the 11-men database KCPGGMKGGMM whose values are 116 and 96, respectively.

The structure of this paper is as follows. Section 2 gives four major categories of retrograde analysis algorithms. Section 3 describes a new approach of retrograde analysis, using external memory algorithms. Section 4 shows our experimental results. Finally, Section 5 gives concluding remarks.

2 Previous Algorithms

Retrograde analysis is a well-known technique for constructing endgame databases for a variety of games. We list below four major categories of sequential algorithms[1] for its implementation, depending on the selection of candidates in each propagation iteration, and how to tell whether a position value can be changed for the rest of the propagations.

We call a position *stable* if its value cannot be changed for the rest of the iterations. A position is *terminal* if its position value can be directly decided, e.g., it is checkmate or stalemate. A position is *unknown* (respectively, *win* in i plies and *lose* in j plies for some i and j) if its current value is unknown (respectively, win-in-i and lose-in-j). Given a position p, let $current(p)$ be its current best value during the construction, and $nchild(p)$ be the number of its children whose values are not stable. Let W_i be the set of positions whose current values are win-in-i-moves. Let L_i be the set of positions whose current values are lose-in-i-moves. Let D be a set of positions whose current values are draw. For convenience, a position in W_i (respectively, L_i or D) is called a W_i- (respectively, L_i- or D-) position. A W_i-position, for some positive integer i, is also called a *win-position*. We define similarly *loss-positions* and *unknown-positions*. By definition, a stalemate position is an L_0-position and a checkmate position is a W_1-position respectively. A position p is a *child* of another position p' if p' can reach p in one ply. The position p' is also called a *parent* of p. Given a position p, the function UPDATE$_b(p)$ updates the current best value of p's parents using the current best value of p as follows:

1. if $current(p)$ is lose-in-i, $current(p')$ is lose, unknown, or win-in-j and $j > i + 1$, then $current(p')$ is win-in-$(i + 1)$;
2. if $current(p)$ is win-in-i, $current(p')$ is unknown or lose-in-j and $j < i$, then $current(p')$ is lose-in-i.

The function UPDATE$_f(p)$ changes $current(p)$ to the currently best possible value by checking the current best values of its children. The function VERIFY$_{loss}(p)$ verifies whether the values of all of p's children are win-in-some-i-moves. Given an endgame A, let α_A be the maximum number of distinct position values, including win-in-i, lose-in-j, draw, and unknown, in A. Let β_A be the maximum

---

[1] We note that there are several parallel/distributed implementations of retrograde analysis algorithms [2,14]. Most recently, a parallel/distributed solution was used to solve awari [12]. This paper addresses only sequential algorithms.

number of children that a position may have. Without considering any complex encoding scheme for position values, we need at least $\lceil \log_2(\alpha_A) \rceil$ bits for storing the value of a position.

We list the algorithms next. Note that each algorithm may need a different initialization, as well as a final cleaning phase. Here we only focus on the main part of the algorithms, namely the propagation phase. For each algorithm, we state the cost of coding, space needed to store partial information of a position during computation, and time complexity.

2.1 Repeatedly Forward Checking

The first category, which we call *repeatedly forward checking*, scans the database again and again until none of the database values is changed during an entire scan. During each scanning, a position p updates its current best value by probing the current best values of its children. That is, during the scan we apply UPDATE$_f(p)$ to each position p.

{∗ I Repeatedly Forward Checking ∗}
I.1 **repeat**
    I.1.1 **for** each position p in the endgame **do**
        I.1.1.1 UPDATE$_f(p)$;
    until no position values is changed for the entire **for** loop;

This algorithm is simple and easy to be implemented. Nalimov [11] reported an implementation of such an algorithm with some fine-tuning for efficiency. The amount of space required for each position is $\lceil \log_2(\alpha_A) \rceil$. However, depending on the orders used to scan the database, a position may need to apply several times before it becomes stable. Each call to UPDATE$_f$ needs to invoke a time-consuming *move-generator*. Since each call to UPDATE$_f$ requires accessing positions spread out in the database, good memory management and compression schemes need to be used to reduce the amount of disk I/O operations.

2.2 Layered Backward Propagation with Forward Checking

The second category, which we call *layered backward propagation with forward checking*, backward propagates the value of a position to its parents using UPDATE$_b$. The propagation is done in *layers*, namely L_i-positions propagate to W_{i+1}-positions in one batch and W_{i+1}-positions to L_{i+1}-positions in another batch. The algorithm first finds the set of terminal positions, i.e.,W_1 and L_0, would be the initial set of candidates. We initially set $i = 0$. For each L_i-position p, we apply UPDATE$_b(p)$. The parents whose values are changed by values propagated from p are put to W_{i+1}. For each parent p' of a W_{i+1}-position in the current candidate set, we apply VERIFY$_{loss}(p')$ and put it to L_{i+1} if VERIFY$_{loss}(p')$ returns true. Thus, we have a new set of L_{i+1}-positions. We then increase the value of i. This iterative process continues until the candidate set is empty.

{∗ II Layered Backward Propagation with Forward Checking ∗}
II.1 $i \leftarrow 0$;
II.2 **while** $L_i \neq \emptyset$ or $W_{i+1} \neq \emptyset$ **do**
   II.2.1 **for** each L_i-position p **do**
      II.2.1.1 UPDATE$_b(p)$;
      II.2.1.2 put the parents whose values are changed by p to W_{i+1};
   II.2.2 **for** each W_{i+1}-position p **do**
      II.2.2.1 UPDATE$_b(p)$;
   II.2.3 **for** each parent p' of a W_{i+1}-position **do**
      II.2.3.1 **if** VERIFY$_{loss}(p')$ is true **then** put p' to L_{i+1};
   II.2.4 $i \leftarrow i + 1$;

The correctness of this algorithm is guaranteed by Lemma 1 and Lemma 2.

Lemma 1. *For a loss-position p, the database entry is stable if and only if $nchild(p) = 0$. For a win-position p, the database entry is stable if and only if (1) $nchild(p) = 0$, or (2) $current(p)$ is win-in-some-moves and has the best possible value among all unpropagated entries in the database.*

Proof. We prove the "if" part. The "only if" part follows trivially. This lemma is trivially true if $nchild(p) = 0$ since its value cannot be further updated. Note that during propagation, if $current(p)$ is win and it is the best value among all entries in the database, it cannot be updated by any propagated value. Hence, this lemma is trivially true. □

Note that it is trivial to know if $nchild(p) = 0$, p is stable. Without the second condition in the lemma, it is possible to have a position whose current best value is win in some moves, but cannot propagate because some of its children are not stable. These children may be in a cycle in the state graph and none of them can become stable because of the cycle.

Lemma 2. *A position is stable the first time it becomes a win-position during the iterative propagation if all terminal positions are L_0- and W_1-positions.*

Proof. The proof is by induction on the propagation with i. When $i = 0$, L_0-positions are terminal positions and stable. W_1-positions are terminal positions or propagated from L_0-positions. Since L_0-positions are stable, and win-in-1-move is the best value among all entries in the database, W_1-positions are stable. For the induction hypothesis, we assume that W_j-positions are all stable for $j = 1$ to k. When $i = k$, an L_k-position p is a parent of W_k-positions, and all of the children of p are win-positions. Since all known win-positions are stable, p is stable. For a parent p' of p, if p' was a win-position before being propagated from p, $current(p')$ was win-in-j-moves where $j = 1$ to k, and $current(p')$ is not changed by p. If $current(p')$ is changed by p, and p' is put to W_{k+1}, p' was not a win-position but an unknown-position or a loss-position. Hence, it is the first time p' becomes a win-position during the propagation. Since win-in-$(k+1)$ is the best value among unpropagated entries, by the second condition in Lemma 1, p' is stable. Thus the lemma is true. □

This algorithm uses the fact that a position p can propagate its value to its parents to avoid the need to call UPDATE$_f$ for all of the parents of p. Lemma 1 indicates that a loss-position needs to propagate its value to its parents only when all of its children are win-positions. Lemma 2 indicates that a win-position propagates its value to its parents exactly once, that is, the first time it becomes "win" during the iterative propagation. Each position put to L_i or W_{i+1} is stable. Each position propagates its value only once. However, the algorithm needs to call an *unmove-generator* to find the parents of a position. The algorithm also needs to call a *move-generator* for each parent of a win-position.

The amount of space required for each position is $\lceil \log_2(\alpha_A) \rceil$. The algorithm is reasonably easy to implement. However, it accesses random positions in the endgame when calling the *move-* and *unmove-generators*. To solve the problem of requiring the endgame to be fully resided in the physical memory, several researchers [16,19] discovered memory-saving techniques, based on the observations that during the i-th iteration only the L_i- and W_{i+1}-positions are accessed. Hence, a bit-map of the endgame for the positions in L_i or W_{i+1} is used. As a result, Thompson [16] used two bits per position in the physical memory during the propagation phase. Wu and Beal [19] reduced it to one bit per position.

2.3 Backward Propagation with Unknown-Children-Counting

The third category, which we call *backward propagation with unknown-children-counting*, uses a counter to keep the number of unknown child positions if all of its current known children are win-positions. This algorithm also backward propagates the value of a position. However, the propagation is not done in layers. It keeps in a queue all positions whose values are changed and need to be propagated. The positions in the queue are propagated in an FIFO order. If a position value is updated due to propagated values from its children, then this position is put into the queue. The process stops when the queue is empty. Initially, the set of terminal positions is in the candidate queue. In each iteration, we apply UPDATE$_b(p)$ for a candidate p. If p is a loss-position, the parents whose values are changed by p become new candidates. If p is a win-position, we decrease $nchild(p')$ by 1 for each parent p' of p. Therefore, p' becomes a new candidate when $nchild(p') = 0$. This iterative process continues until the candidate queue is empty.

Lemma 3. *A position is stable the first time it becomes a win-position during the propagation if the queue initially contains only terminal positions, i.e., L_0- and W_1-positions, and L_0-positions are put into the queue before W_1-positions.*

Proof. It is trivially true that a W_1-position is stable. Consider the candidate queue only contains L_i- and W_{i+1}-positions. L_i-positions propagate to W_{i+1}-positions, and W_{i+1}-positions propagate to L_{i+1}-positions. An L_{i+1}-position will not be propagated before an L_i-position by the queue structure. After all of L_i-positions are propagated, there will be no more L_j-positions, $j \leq i$, to be placed in the queue in the future. Similarly, the queue contains only L_{i+1}-positions and W_{i+2}-positions after all of the W_{i+1}-positions are propagated. There will be no more W_j-positions, $j \leq i + 1$, to be placed in the queue in the future.

Let the value of a position p be set to be win-in-i, $i > 1$, because of propagation from an L_{i-1}-position. The only chance for the value of p to be changed is to be set win-in-k, where $k < i$. However, there will be no more L_{k-1}-positions to be placed in the queue after an L_{i-1}-position is propagated. Hence, this lemma is true. □

{∗ III Backward Propagation with Unknown-children-counting ∗}
III.1 **while** the candidate queue is not empty **do**
    III.1.1 Remove position p from the candidate queue;
    III.1.2 UPDATE$_b(p)$;
    III.1.3 **if** p is a loss-position **then**
        III.1.3.1 put the parents whose values are changed by p into the candidate queue;
    III.1.4 **if** p is a win-position **then**
        III.1.4.1 **for** each parent p' of p **do**
            III.1.4.1.1 $nchild(p') \leftarrow nchild(p') - 1$;
            III.1.4.1.2 **if** $nchild(p') = 0$ **then** put p' into the candidate queue;

The algorithm needs additional codes for maintaining the unknown-children-counts and thus requires $\lceil \log_2(\alpha_A) \rceil + \lceil \log_2(\beta_A) \rceil$ bits per position. However, this algorithm does not need to call *move-generators* during the propagation phase. A loss-position propagates its value to its parents only once using its parents' unknown-children-count. Gasser [5] and Wirth [11] both implemented variations of this algorithm. Fang et al. [3,4] implemented a variation of this algorithm by replacing the candidate queue with repeatedly scanning the database for unstable and needed to be propagated positions.

2.4 Layered Backward Propagation with Unknown-Children-Counting

The fourth category, which we call *layered backward propagation with unknown-children-counting*, is a combination of the layered approach used in the second category and the children-counting approach used in the third category. Each position keeps its remaining number of unknown child positions. Initially, the set of terminal positions, i.e., W_1 and L_0, are the initial set of candidates. We initially set $i = 0$. For each L_i-position p, we apply UPDATE$_b(p)$. The parents whose values are changed by p (called *propagating packages*) are put to W_{i+1}. For each parent p' of a W_{i+1}-position p (also called a *propagating package*) in the current candidate set, we apply UPDATE$_b(p)$ and decrease $nchild(p')$ by 1. If $nchild(p') = 0$, put p' to L_{i+1}. We then increase the value of i. This iterative process continues until the candidate set is empty.

{∗ IV Layered Backward Propagation with Unknown-children-counting ∗}
IV.1 $i \leftarrow 0$;
IV.2 **while** $L_i \neq \emptyset$ or $W_{i+1} \neq \emptyset$ **do**
    IV.2.1 **for** each L_i-position p **do**
        IV.2.1.1 UPDATE$_b(p)$;
        IV.2.1.2 put the parents whose values are changed by p to W_{i+1};

IV.2.2 **for** each W_{i+1}-position p **do**
    IV.2.2.1 UPDATE$_b(p)$;
    IV.2.2.2 **for** each parent p' of p **do**
        IV.2.2.2.1 $nchild(p') \leftarrow nchild(p') - 1$;
        IV.2.2.2.1 **if** $nchild(p') = 0$ **then** put p' to L_{i+1};
IV.2.3 $i \leftarrow i + 1$;

The algorithm needs additional code for maintaining the unknown-children-counts and requires $\lceil \log_2(\alpha_A) \rceil + \lceil \log_2(\beta_A) \rceil$ bits per position. The algorithm also does not require the calls to *move-generators*. Further, each position propagates its value to its parents exactly once. The algorithm also needs additional codes for partitioning the endgame into layers according to their win or lose depths. Allis et al. [1], and Bal and Allis [2] implemented variations of this algorithm.

3 External Memory Implementation

We implement a variation of the algorithm in the fourth category. Our implementation handles the case when the size of the endgame is much larger than the physical memory size. Note Steps IV.2.1.1 and IV.2.2.2 need to access the parents of a position. The parents of a position are spread out in the endgame. Thus, if the endgame is not stored entirely in the physical memory, we need to perform a time-costly disk random access to obtain the values of the parents. We save time in disk I/O using the following key observation.

3.1 Observation

Assume we have a file with N fixed-sized records stored entirely on the disk. The records are numbered from 1 to N. We want to *update* a list of records in the file with the list of indexes $L = I_1, I_2, \ldots, I_n$. Note that it may be the case that $I_i = I_j$ for some $i \neq j$. We refer to this problem as the *disk updating* problem. To *update* a record, we read the content of this record, put it into the memory, perform some changes to it, and then write it back to the disk. We also assume if a record is updated twice, we can combine the two updates into one. It is well-known that each disk read or write operation is performed on a *block* basis with B records read and write at a time. We refer to B as the *blocking factor*. Here, we assume N can be evenly divided by B. We have N/B blocks in the file. Hence, if the two records with indexes I_i and I_{i+1} are stored in the same block, we can update the two records using one disk read and one disk write.

Lemma 4. *Assume each record is updated with equal probability. If we update the blocks with indexes in L in the order of $I_1, I_2, \ldots, I_n$, then on average we need to perform $n - (n-1) \cdot B/N$ disk reads and $n - (n-1) \cdot B/N$ disk writes.*

Proof. The chance of I_i and I_{i+1} belonging to the same block is B/N. Let $E(n)$ be the expected number of blocks updated for a list of n records. Then, $E(n) = E(n-1) + (N-B)/N = E(n-1) + 1 - B/N$. Note $E(1) = 1$. Thus, $E(n) = n - (n-1) \cdot B/N$. □

A disk accessing sequence described in Lemma 4 is called *random disk updating sequence*.

A reasonable alternative for visiting the records in L is to sort the indexes in L first. The records are visited in the sorted order. During sorting, we can also combine multiple updates of the same record into one update. Let n_d be the number of distinct indexes in L. Let the sequence $L' = I'_1, I'_2, \ldots, I'_{n_d}$ such that $I'_i < I'_{i+1}$ for all $1 \leq i < n_d$, and each index in L is in L'.

Lemma 5. *Assume I_i's are uniformly distributed, and I_i and I_j are independent.*

1. *The expected number of $n_d = N(1 - (1 - 1/N)^n)$, and*
2. *on average there are $\frac{N}{B}(1 - \frac{\binom{N-B}{n_d}}{\binom{N}{n_d}})$ distinct blocks in L'.*

Proof. Let $K(n)$ be the expected number of distinct elements if the n elements are chosen with equal probability. Thus the chance of picking a new element that is different from a set of already chosen $n - 1$ elements is $1 - K(n-1)/N$. Then $K(n) = K(n-1) + (1 - K(n-1)/N)$ and $K(1) = 1$. Thus $K(n) = 1 + (1 - 1/N) + (1 - 1/N)^2 + \cdots + (1 - 1/N)^{n-1}$. We have $n_d = K(n) = \frac{1-(1-1/N)^n}{1-(1-1/N)} = N(1 - (1 - 1/N)^n)$.

We now prove the rest of the lemma. Let the blocks in the file be $B_1, B_2, \ldots, B_{N/B}$. The chance of updating a record in B_i is independent of the chance of updating a record in B_j if $i \neq j$. The probability of having no record in L' coming from B_i is $P = \frac{\binom{N-B}{n_d}}{\binom{N}{n_d}}$. Note the numerator of the formula is the number of distinct combinations that the n_d elements come from blocks not in B_i. The denominator is the number of distinct combinations that the n_d elements come from all blocks. Hence, the expected number of distinct blocks in L' is $\frac{N}{B}(1 - P)$. □

Note that when $\lim_{N\to\infty}(1 - 1/N) = 1$. If $N \gg n$, then $n_d \simeq n$. The disk accessing sequence described above is called *sequential disk accessing sequence*. Moreover, note that a sequential updating of the records can combine multiple updates to the same record into just one update. The sequential updating process can also reduce the number of disk blocks being accessed. In the worst case, the entire file is read only once regardless of n and n_d. For the random updating process, the number of disk blocks being read is linear in n.

In addition to the above advantage, note that the disk operations favour access to records with the same *locality*. That is, we can reasonably assume adjacent blocks of a file are physically located in close proximity. Each disk maintains a current disk head location. A disk read or write to a location that is far away from its current head location suffers severe overload in moving the disk head. It is easy to see that a sequential updating process moves the disk head from the top of the file to the end of the file exactly once, while a random updating process spends excess extra time in moving the disk heads. Hence, the total time spent in a sequential disk accessing sequence is much less than the total time spent in a random disk accessing sequence even if $n = n_d$.

Table 1. T_{update} is the average total time to randomly update n records in a file of N records with $B = 10,240$ and each record has 1 bytes. B_{update} is the average number of blocks updated.

N	$25 \cdot 2^{20}$ (26214400)		$40 \cdot 2^{20}$ (41943040)		$100 \cdot 2^{20}$ (104857600)	
n	$T_{update}(sec)$	B_{update}	$T_{update}(sec)$	B_{update}	$T_{update}(sec)$	B_{update}
100	0.01	100	0.01	100	0.01	100
1,000	0.52	1,000	2.01	1,000	3.84	1,000
10,000	5.72	9,997	20.16	9,998	41.81	9,999
100,000	55.89	99,960	199.72	99,974	408.62	99,990
1,000,000	556.60	999,613	1974.91	999,759	3787.17	999,905

Table 2. T_{update} is the average total time to sequentially update n records in a file of N records with $B = 10,240$ and each record has 1 bytes. B_{update} is the average number of blocks updated.

N	$25 \cdot 2^{20}$ (26214400)		$40 \cdot 2^{20}$ (41943040)		$100 \cdot 2^{20}$ (104857600)	
n	$T_{update}(sec)$	B_{update}	$T_{update}(sec)$	B_{update}	$T_{update}(sec)$	B_{update}
1,000	0.44	828	1.41	884	2.63	952
10,000	1.72	2,512	3.13	3,747	7.35	6,395
100,000	1.66	2,560	2.62	4,096	6.69	10,239
1,000,000	1.22	2,560	2.52	4,096	6.46	10,240
10,000,000	0.72	2,560	1.88	4,096	5.88	10,240

See Table 1 and Table 2 for the timing data using random accessing sequences and its equivalent sequential accessing sequence. Note that in measuring the timing data for the sequential accessing sequences, we first generate a random sequence of indexes and then sort the sequence. Hence, if an index is in the sequence twice, it is only updated once. For a random accessing sequence, the record with two appearances of its index is updated twice. From the timing data, we can see that the process can be speeded up more than 500 times.

3.2 Refined Implementation

By using the above observations, we refine the implementation of Step IV.2.1 as follows.

{∗ Revised Step IV.2.1 ∗}
{∗ Expanding. ∗}
1. **for** each L_i-position p **do**
    1.1 find the parents of p and put them into an updating list U;
{∗ Sorting and Combining. ∗}
2. sort U according to the indexes of the positions in the endgame;
3. remove duplicated positions in the sorted U;
{∗ Updating. ∗}
4. **for** each position p' in U **do**
    4.1 **if** $current(p')$ changed by p **then** {put p' in W_{i+1}; mark p' stable;}

The revised implementation of Step IV.2.2 is as follows.

{∗ Revised Step IV.2.2 ∗}
{∗ Expanding. ∗}
1. **for** each W_{i+1}-position p **do**
    1.1 find the parents of p and put them into an updating list U;
{∗ Sorting and Combining. ∗}
2. sort U according to the indexes of the positions in the endgame;
3. **if** a position p' occurs p'_d times in U **then** remove duplications of p' and record p'_d in p';
{∗ Updating. ∗}
4. **for** each position p' in U with p'_d duplications **do**
    4.1 $nchild(p') \leftarrow nchild(p') - p'_d$;
    4.2 **if** $nchild(p') = 0$ **then** {put p' in L_{i+1}; mark p' stable;}

We first run the *expanding* process by finding the parents of propagatable positions. Using sorting, we convert a sequence of random disk accesses to a sequence of sequential disk accesses without accessing the same disk location twice. The process is called *combining*. The combining process not only reduces the number of disk I/O's and disk head moving distances, but also "combines" multiple updating operations of a record into one disk updating operation. In the revised steps of Step IV.2.1 if a position is a parent of more than one L_i-positions, then this position is updated only once instead of multiple times, as in the original implementation. Finally we update the disk sequentially according to the sorted and combined updating list. The process is called *updating*.

One problem with the above implementation is that it takes space and time to store and sort the parents. For the endgames we considered, the list L before sorting and compacting is usually larger than the endgame being constructed for the first few iterations. Hence, we cannot use internal-memory sort. In the literature, there are many so called *external-memory* sorting algorithms for sorting a list that cannot be entirely stored in the physical memory [18]. The list L is stored in the disk with n records. Each record represents an index. Let B be the disk blocking factor. That is, a disk block contains B records. Let M be the number of records that can be stored in the physical memory. Let $R = \lceil n/M \rceil$ i.e., the number of sublists that L is divided into in order for each sublist to be stored in the physical memory. Let $M' = \lceil n/R \rceil$, i.e., the size of a sublist if L is to be evenly divided into R sublists. Note that the size of each sublist is either M' or $M'-1$. Since M' is usually large, we assume the size of each sublist is M'. Let $W = \lfloor M/B \rfloor$, i.e., the number of blocks that can be stored in the physical memory at the same time.

Lemma 6 ([18]). *Assume $n > M$, $W \geq R$ and the list L is entirely stored in the disk. Sorting L can be done using $O(n \cdot \log_2 n)$ amount of computation and sequentially reading and writing the list L stored in the disk twice.*

Proof. We assume the list is to be sorted in non-decreasing order. The idea for the algorithm is to evenly divide the list L into R sublists, each with size M'. Each sublist is loaded into the physical memory and then uses an internal

Table 3. T_{srt}^4 is the average total time to sort n 4-bytes elements whose indexes are between 0 and $2^{31} - 1$. T_{mrg}^4 is the average total time spent in external merge while sorting. T_{srt}^4 includes T_{mrg}^4.

M	$50 \cdot 2^{20}$		$100 \cdot 2^{20}$		$150 \cdot 2^{20}$		$175 \cdot 2^{20}$	
n	$T_{srt}^4(sec)$	$T_{mrg}^4(sec)$	$T_{srt}^4(sec)$	$T_{mrg}^4(sec)$	$T_{srt}^4(sec)$	$T_{mrg}^4(sec)$	$T_{srt}^4(sec)$	$T_{mrg}^4(sec)$
2^{20}	0.61	0.00	0.61	0.00	0.62	0.00	0.62	0.00
$10 \cdot 2^{20}$	7.28	0.00	7.27	0.00	7.27	0.00	7.26	0.00
$100 \cdot 2^{20}$	108.68	31.54	89.99	0.00	89.88	0.00	90.01	0.00
$200 \cdot 2^{20}$	258.42	80.47	261.76	80.74	263.05	80.86	262.30	80.96
$300 \cdot 2^{20}$	396.05	128.15	396.31	124.09	397.24	122.27	397.87	121.85
$400 \cdot 2^{20}$	535.52	178.80	533.88	170.69	533.82	166.29	532.07	165.05
$500 \cdot 2^{20}$	674.31	227.21	675.10	220.18	671.39	214.39	668.15	206.91

Table 4. T_{srt}^5 is the average total time to sort n 5-bytes elements whose indexes are between 0 and $2^{31} - 1$. T_{mrg}^5 is the average total time spent in external merge while sorting. T_{srt}^5 includes T_{mrg}^5.

M	$40 \cdot 2^{20}$		$80 \cdot 2^{20}$		$120 \cdot 2^{20}$		$140 \cdot 2^{20}$	
n	$T_{srt}^5(sec)$	$T_{mrg}^5(sec)$	$T_{srt}^5(sec)$	$T_{mrg}^5(sec)$	$T_{srt}^5(sec)$	$T_{mrg}^5(sec)$	$T_{srt}^5(sec)$	$T_{mrg}^5(sec)$
2^{20}	0.83	0.00	0.83	0.00	0.82	0.00	0.83	0.00
$10 \cdot 2^{20}$	9.39	0.00	9.36	0.00	9.38	0.00	9.41	0.00
$100 \cdot 2^{20}$	140.27	37.39	151.63	37.91	116.31	0.00	116.58	0.00
$200 \cdot 2^{20}$	319.06	90.65	321.55	89.76	322.21	87.68	321.90	87.45
$300 \cdot 2^{20}$	485.60	143.55	487.06	138.79	487.50	135.32	488.38	134.47
$400 \cdot 2^{20}$	654.50	196.97	657.07	191.29	656.29	186.68	654.05	180.47
$500 \cdot 2^{20}$	823.34	252.27	826.05	244.04	825.22	236.55	825.80	232.65

memory routine to sort. It takes $O(M' \log_2 M')$ time to sort a sublist of size M'. After a sublist is sorted, it is stored in a file. The total time used in sorting sublists is thus $O(R \cdot M' \log_2 M') = O(n \cdot \log_2 M')$.

These R sublists are then merged. During merging, we only need to load one block of each sublist in order to perform the operation. Hence, we can only load W blocks. Since we assume $W \geq R$, we can merge the sublists in one phase. We use a standard heap data structure to store the current smallest elements of sublists that are not yet selected. The size of the heap is R. The top of the heap is the next smallest element to be selected. Once an element from a sublist is removed from the heap, the next element in this sublist is inserted. The heap is then re-organized. It takes $O(\log_2 R)$ time to re-organize the heap. Hence the amount of computation needed in merging is $O(n \log_2 R)$. The total amount of computation is thus $O(n \cdot \log_2(M' \cdot R)) = O(n \cdot \log_2 n)$.

The above computation obviously sequentially reads and writes L stored in the disk twice. □

For our implementation, $W \geq R$ for all of the cases is considered. We list in Tables 3 and 4, the time needed to sort items with 4 and 5 bytes per item,

Table 5. Timing data in seconds. A $*$ means it takes to long to measure. T_{srt}^4 is the average total time to sort n 4-byte indexes referencing between 0 and $N-1$. T_{upd_s} is the average total time to sequentially update n 1-byte records in a file of N records. T_{s+u} is the average total time to sort n indexes then to sequentially update n referenced records. T_{upd_r} is the average total time to randomly update n referenced records. $M = 50 \cdot 2^{20}$ indexes.

N	$25 \cdot 2^{20}$ (26214400)				$40 \cdot 2^{20}$ (41943040)				$100 \cdot 2^{20}$ (104857600)			
n	T_{srt}^4	T_{upd_s}	T_{s+u}	T_{upd_r}	T_{srt}^4	T_{upd_s}	T_{s+u}	T_{upd_r}	T_{srt}^4	T_{upd_s}	T_{s+u}	T_{upd_r}
10^5	0.06	1.66	1.72	55.89	0.06	2.62	2.68	199.72	0.06	6.69	6.75	408.62
10^6	0.58	1.22	1.80	556.60	0.58	2.52	3.10	1974.91	0.58	6.46	7.04	3787.17
10^7	6.94	0.72	7.66	$*$	6.94	1.88	8.82	$*$	6.94	5.88	12.82	$*$
10^8	103.65	2.71	106.36	$*$	103.65	3.82	107.47	$*$	103.65	7.58	111.23	$*$

respectively. For the largest sorted list containing 500 million items, it takes about $1.3 \cdot 10^{-6}$ seconds to sort a 4-byte item on average. It takes about $1.6 \cdot 10^{-6}$ seconds to sort a 5-byte item on average. If the list can be stored entirely in the physical memory, it takes about $7.6 \cdot 10^{-7}$ (respectively, $9.7 \cdot 10^{-7}$) seconds to sort a 4-byte (respectively, 5-byte) item on average. Hence, sorting a large list stored in the disk is roughly 2 to 3 times slower than sorting a list stored in the physical memory. The timing data is consistent with the analysis in Lemma 6 with negligible deviation. Our experiments show that this method is economical to sort lists that are too large to be stored in the physical memory.

In Table 5, we list the total cost of converting a disk random accessing sequence into a disk sequential accessing sequence. We also compare the cost with a disk random accessing sequence. From the data, we can expect the procedure to be 30 to 300 times faster than using the disk random accessing sequence when $N = 25 \cdot 2^{20}$. The speed is even more impressive for larger N — up to 600 times faster. Note that the sorting time increases twenty-fold when the data size is 10 times larger from $n = 10^7$ to 10^8 because of the extra overload needed in merging for the case $n = 10^8$. Therefore, we expect our implementation of the retrograde analysis algorithm to be efficient.

4 Experimental Results

We use the above refined algorithm to construct Chinese-endgame databases with only one side having attacking pieces. We have constructed a total of 396 endgame databases totalling 70 GB. The endgame databases constructed are listed in Table 6. Endgame databases that have had one, or more, non-king pieces removed are not listed. We use the letters K, G, M, R, N, C and P to represent pieces King, Guard, Minister, Rook, Knight, Cannon and Pawn, respectively.

On average, our program computes the value of a position in $2.6 \cdot 10^{-5}$ seconds. About a quarter of the time is spent on unmove generating. A tiny fraction of the time (about 4%) is spent on sorting.

Table 6. Statistics of sample endgame databases

| Database | | Number of | RTM | BTM | Max | Construction |
Name	Size (bytes)	Positions	Win %	Draw %	DTM	Time
KRRKGGMM	56,662,740	88,474,680	100.00	2.37	32	15:57
KRNKGGMM	114,598,800	176,949,360	99.97	7.59	32	26:37
KRCKGGMM	114,598,800	176,949,360	99.99	7.47	32	43:46
KRPKGGMM	70,032,600	105,327,000	99.99	8.28	32	14:23
KNNKGGMM	56,662,740	88,474,680	99.27	12.75	31	9:54
KNCKGGMM	114,598,800	176,949,360	99.34	12.81	36	25:17
KNPKGGMM	70,032,600	105,327,000	23.57	94.69	65	7:46
KPPPKGGMM	371,172,780	497,448,000	60.64	60.08	46	1:11:52
KCCPKGGMM	3,116,450,700	4,318,407,000	99.25	18.24	67	32:32:26
KCCGMMKGGMM	5,087,968,020	7,400,808,000	99.59	12.83	37	61:56:27
KCCGGMKGGMM	3,045,369,960	4,385,664,000	99.59	12.85	38	35:39:23
KCPGMKGGMM	2,143,092,600	3,122,501,400	67.29	50.78	114	11:59:08
KCPMMKGGMM	1,414,373,400	1,032,915,240	18.87	96.11	56	3:17:32
KCPGGKGGMM	551,113,200	799,470,000	39.56	86.47	58	1:35:13
KCPGMMKGGMM	6,288,499,800	9,030,768,000	68.26	50.15	73	42:31:30
KCPGGMKGGMM	3,763,940,400	5,382,816,000	66.92	51.17	116	24:39:19
KCPGGMMKGGMM	8,785,969,200	15,705,990,000	67.87	50.55	81	78:36:54

5 Concluding Remarks

In [16], Thompson declared that he had failed to implement an alternative algorithm that is similar to the second or fourth category. He noted the following three major problems in the implementation of this type of algorithm.

- The extra time needed to run the initialization phase.
- The drastical increase in the number of random disk accesses.
- The inability to resume the program.

Our current implementation solves all three problems. The initialization problem can be solved by using the partitioning scheme as described in [6]. The second problem appears while updating the database. Instead of probing a bitmap of the database being constructed by probing the disk randomly, Thompson's algorithm needs to compute the number of non-stable children. Our implementation solves this problem by batching disk accesses in a layer and then converting them to sequential disk accesses. The last problem is also solved by the layered approach. If the program is stopped during the i-th iteration, we can restart it from the beginning of the i-th iteration, once we record the status of the database after the $(i-1)$-th iteration is finished.

From the performance data, our implementation is reasonably fast and has built the currently known largest Chinese-chess endgame. We further note that if the endgame to be constructed can be placed entirely into the main memory, then our algorithm may be slower than previous algorithms done by others, due to the extra time spent in sorting. However, the performance of the code is still efficient.

Our algorithm can be used to construct databases either in DTM or DTC metrics. In comparison, a 10-men KCPGMKGGMM endgame database is constructed in about 13 hours on a 1GHz P3 machine with 256 MB memory [19]. Our largest constructed database contains more than 4 times the number of positions. The number of positions is more than the number of bits in the physical memory. Hence, the algorithm proposed in [19] cannot be used to construct this database efficiently. We have also discovered positions with larger DTM and DTC values. The previous largest values were 114 and 94, respectively. In KCPGGMKGGMM, we discovered positions with values 116 and 96, respectively. In the future, we will use our disk approach to build endgames with both sides having attacking pieces, which were addressed recently [3,4].

Acknowledgement

We thank Professor Shun-chin Hsu for his assistance, Haw-ren Fang for helpful discussions and making his retrograde analysis code in [3,4] available to us, members of IIS Computer Chinese-Chess Research Club for comments, and an anonymous referee for pointing out references [8] and [13].

References

1. L. V. Allis, M. van der Meulen, and H. J. van den Herik. Databases in awari. In D.N.L. Levy and D.F. Beal, editors, *Heuristic Programming in Artificial Intelligence 2: The Second Computer Olympiad*, volume 2, pages 73–86. Ellis Horwood, Chichester, England, 1991.
2. H. Bal and L. V. Allis. Parallel retrograde analysis on a distributed system. In *Proceedings of the 1995 ACM/IEEE Supercomputing Conference*, 1995. http://www.supercomp.org/sc95/proceedings/463_HBAL/SC95.HTM.
3. H.-r. Fang, T.-s. Hsu, and S.-c. Hsu. Construction of Chinese chess endgame databases by retrograde analysis. In T. Marsland and I. Frank, editors, *Lecture Notes in Computer Science 2063: Proceedings of the 2nd International Conference on Computers and Games*, pages 96–114. Springer-Verlag, New York, NY, 2000.
4. H.-r. Fang, T.-s. Hsu, and S.-c. Hsu. Indefinite sequence of moves in chinese chess endgames. In J. Schaeffer, M. Müller, and Y. Björnsson, editors, *Lecture Notes in Computer Science 2883: Proceedings of the 3rd International Conference on Computers and Games*, pages 264–279. Springer-Verlag, New York, NY, 2002.
5. R. Gasser. Solving nine men's morris. In R. Nowakowski, editor, *Games of No Chance*, volume 29 of *MSRI*, pages 101–113. Cambridge University Press, Cambridge, 1996. ISBN: 0-521-57411-0.
6. T.-s. Hsu and P.-Y. Liu. Verification of endgame databases. *ICGA Journal*, 25(3):132–144, 2002.
7. P. Karrer. KQQKQP and KQPKQP≈. *ICCA Journal*, 23(2):75–84, 2000.
8. R. E. Korf. Delayed duplicate detection: Extended abstract. In *Proceedings of the 18th International Joint Conference on Artificial Intelligence*, pages 1539–1541, 2003.

9. R. Lake, J. Schaeffer, and P. Lu. Solving large retrograde-analysis problems using a network of workstations. In H. J. van den Herik, I. S. Herschberg, and J. W. H. M. Uiterwijk, editors, *Advances in Computer Chess*, volume 7, pages 135–162. University of Limburg, Maastricht, The Netherlands, 1994. ISBN: 9-0621-6101-4.
10. T. R. Lincke and A. Marzetta. Large endgame databases with limited memory space. *ICCA Journal*, 22(4):131–138, 1999.
11. E. V. Nalimov, C. Wirth, and G. McC. Haworth. KQQKQQ and the Kasparov-world game. *ICCA Journal*, 22(4):195–212, 1999.
12. J. W. Romein and H. E. Bal. Awari is solved. *ICGA Journal*, 25(3):162–165, 2002.
13. J. Schaeffer, Y. Björnsson, N. Burch, R. Lake, P. Lu, and S. Sutphen. Building the checkers 10-piece endgame databases. In H. J. van den Herik, H. Iida, and E. A. Heinz, editors, *Advances in Computer Games: Many Games, Many Challenges*, volume 10, pages 193–210. Kluwer Academic Publishers, 2003. ISBN: 1-4020-7709-2.
14. L. Stiller. Parallel analysis of certain endgames. *ICCA Journal*, 12(2):55–64, 1989.
15. K. Thompson. Retrograde analysis of certain endgames. *ICCA Journal*, 9(3):131–139, 1986.
16. K. Thompson. 6-piece endgames. *ICCA Journal*, 19(4):215–226, 1996.
17. H. J. van den Herik, J. W. H. M. Uiterwijk, and J. van Rijswijck. Games solved: Now and in the future. *Artificial Intelligence*, 134:277–311, 2002.
18. J. S. Vitter. External memory algorithms and data structures: Dealing with MASSIVE DATA. *ACM Computing Surveys*, 33(2):209–271, 2001.
19. R. Wu and D. F. Beal. Fast, memory-efficient retrograde algorithms. *ICCA Journal*, 24(3):147–159, 2001.

Generating an Opening Book for Amazons

Akop Karapetyan and Richard J. Lorentz

Department of Computer Science,
California State University, Northridge, USA
{akop.karapetyan, lorentz}@csun.edu

Abstract. Creating an opening book for the game of Amazons is difficult for many reasons. It is a relatively new game so there is little accumulated knowledge and few expert games to draw on. Also, the huge branching factor, over 1000 available legal moves during most of the opening, makes selecting good moves difficult. Still, one or two bad moves in the opening can doom a player to passive play and an almost certain loss, so it is essential for a good program to play the opening well. We discuss a number of possible methods for creating opening books, focusing mainly on automatic construction, and explain which seem best suited for games with large branching factors such as Amazons.

1 Introduction

Amazons is a fairly new game, invented in 1988 by Walter Zamkauskas of Argentina and is a trademark of Ediciones de Mente. There is still much to learn about how to play this game well, especially in the opening. There are few expert players to learn from and few expert games from which to study openings. In fact, there are currently only very primitive strategic concepts known that Amazons players agree are important for playing the game well. One of the reasons progress is slow is that the branching factor in Amazons is huge, especially during the opening. There are 2,176 legal first moves and during the first 10 moves of the game the average number of legal moves is well over 1,000 [1]. It is still not well understood how to determine which of these many choices of moves is best. We discuss these issues and how they relate to creating an opening book in more detail in Section 2.

There have been a number of recent attempts to automatically create opening books in other games like Othello [2, 3] and Awari [3]. Though these ideas cannot be immediately applied to Amazons because of its extremely high branching factor, they have provided the framework and the inspiration for our work. In Section 3 we discuss various algorithmic ideas for automatically generating an opening book, see how they relate to the methods mentioned above, and discuss implementation issues and experimental results.

We conclude in Section 4 with some analysis of the various methods described and suggestions for further work.

2 The Game of Amazons and Opening Books

Amazons is usually played on a 10 by 10 board where each player is given 4 amazons initially placed as shown in Figure 1(a). Although the game may be played on

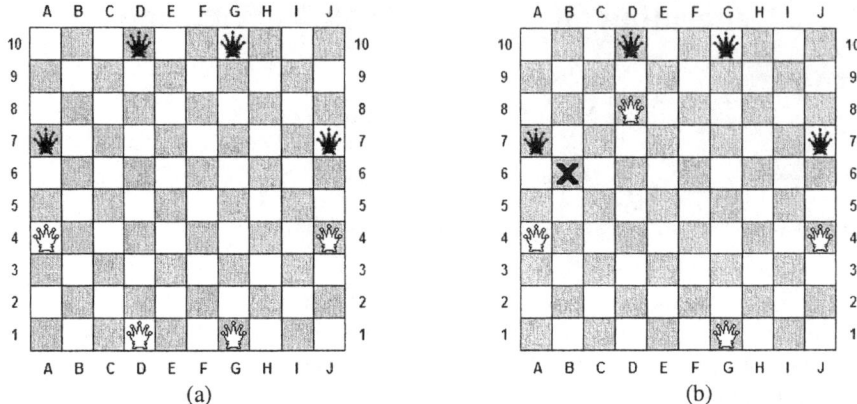

Fig. 1. The initial position and a typical first move

other sized boards and with different numbers and initial placements of amazons, it is this version of the game that is most often played and so will be the version that will be discussed here.

A move comprises two steps: first an amazon is moved like a chess queen through unoccupied squares to an unoccupied square. Then from this new location the amazon "throws an arrow" again like a chess queen moves across unoccupied squares to an unoccupied square where it stays for the remainder of the game. For example, from the initial position in Figure 1 (a), one of white's 2,176 legal moves is to move the amazon from square D1 to square D8 and then throw an arrow to square B6, where the arrow must stay for the remainder of the game. This move is denoted D1-D8(B6) and is shown completed in Figure 1 (b). In fact, this move has recently become a popular opening move for White because it helps distribute White's amazons evenly throughout the board, it gives the moved amazon increased mobility, and it restricts Black's amazons on A7 and on D10.

White moves first and passing is not allowed. The last player to make a legal move wins. Since the idea is to make more moves than your opponent, three strategic concepts emerge. (1) Keep your amazons mobile, that is, place them in positions so they are able to make as many moves as possible. We refer to this as *mobility*. (2) Position your amazons in such a way that squares can be reached quickly, e.g., a square that one of your amazons can reach in two moves but the opponent requires three moves to reach will more likely provide you rather than your opponent with a legal move later. We call this *accessibility*. (3) Using your amazons and your arrows, enclose large regions so that these regions will later provide you with ample moves. This is called *territory*.

Early Amazons programs tended to emphasize accessibility and so this was usually the major component of their evaluation functions. Hence, a very common opening move used to be D1-D7(G7). However, it was later discovered that mobility is at least as important in the opening because if a player manages to completely surround an opponent's amazon, the game essentially becomes a battle of three amazons against four, and the four amazons, if not too behind in the other areas usually have an easy win. Jens Lieberum's program, AMAZONG [4], was the first program to bring this to our attention by easily winning the Seventh Computer Olympiad Amazons tournament [5]. Details concerning AMAZONG's evaluation function can be found in [6].

A system for creating opening books for Amazons, or any other game for that matter, must possess at least the following three features. (1) *Automatic*. Since our understanding of the game is still quite rudimentary construction of an opening book must be flexible, allowing us to quickly and automatically rebuild the book or effortlessly modify portions of the book. (2) *Interactive*. Amazons programs can still be terribly wrong in their move making decisions, especially in the opening. Hence, the opening book should be easy for a human to examine, traverse, and modify, overriding the choices that were automatically formulated. (3) *Symbiotic*. Once an opening book is created it should be simple to incorporate it into an Amazons-playing program. Similarly, a strong Amazons-playing program is necessary to provide the engine that allows a book to be built and it should be easy to incorporate this engine into the opening-book creator. For the purposes of this research the opening book was designed to interact with the program INVADER, a strong Amazons-playing program developed at California State University, Northridge [7]. We will discuss point (1), the automatic construction of opening books, in detail in Section 3.

As for point (2), we have found a number of features to be necessary to allow us to properly hand tune an opening book. Creating an opening book is typically an iterative process. First we automatically create a book. Next we examine the book to find its strengths and weaknesses. A weakness might mean that the book tends to avoid or not even consider a move that we consider to be strong, or conversely tends to use a move that we consider to be weak. We adjust the book accordingly and then automatically expand the size of the book, especially around the changed nodes and continue iterating in a similar fashion. So, to be able to perform these steps easily the opening-book system should make it simple to examine the moves in the book. Figure 2 is a screen shot of our opening-book program called SCRIBE and shows how we deal with some of these issues.

The top-left pane of the output shows the opening book expressed in the usual "outline" format where all moves at a particular depth are indented the same amount and the moves are sorted with the better moves on top. Clicking on any of the moves displays the position on the board to the right and below the board data relevant to that node of the book is shown.

Various tools allow us to add and remove nodes, change values of nodes, and flag nodes as being the best move regardless of their value. Also, we can force the book to expand from a certain node. So, for example, if we notice the book does not give enough attention to a move it considers weak but we consider strong, we can both increase the value of that move so that the book will be forced to choose it from the corresponding position and we can then force the book to expand around this node.

Finally, the pane in the bottom is used to keep a record of the search information that was performed in the process of calculating the book moves. This allows us to examine the search data of moves that seem to have incorrect values in the book.

Regarding point (3), symbiosis, it is obvious that a book generator needs to be integrally connected with the game player. In fact, it is the search engine of the game player that is used to perform lookahead during opening-book construction. Also,

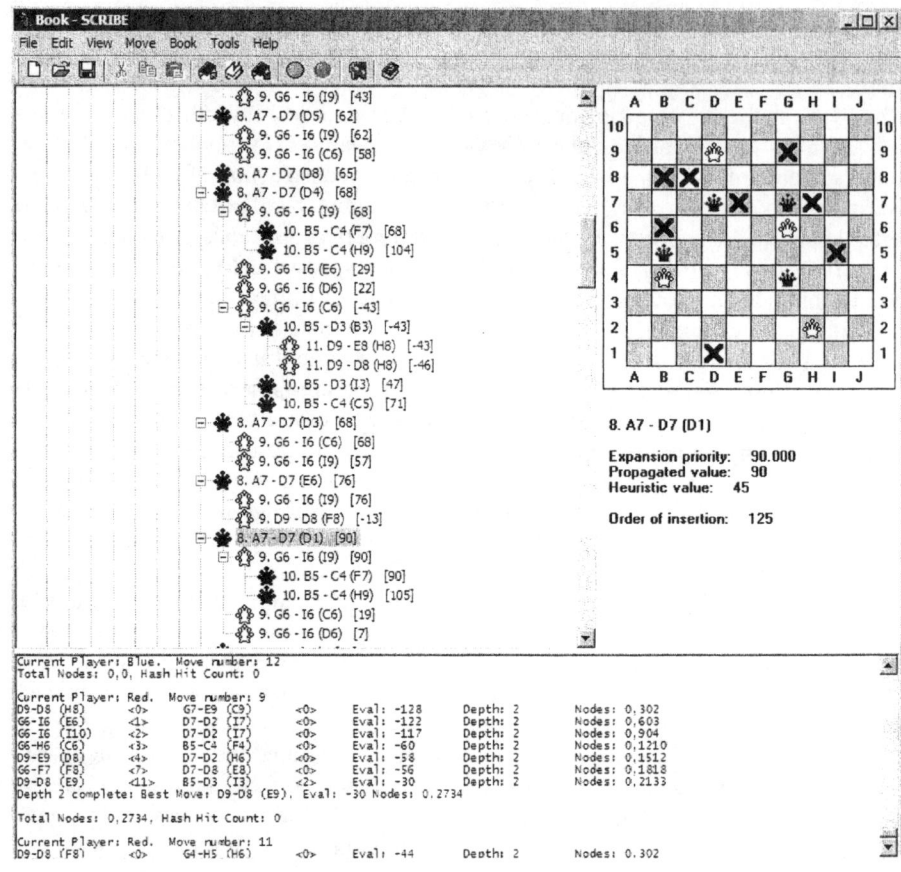

Fig. 2. Screen shot of SCRIBE

once a book has been constructed, it needs to be quickly and seamlessly integrated back into the game player. For Amazons this is especially important since the evaluation function, and even the search engine, are continuously changing and the opening book needs to be able to reflect the most current knowledge about the game. Hence, SCRIBE can easily rebuild the book as INVADER evolves and INVADER can easily accept new books created by SCRIBE.

Of course the most important aspect of creating an opening book is the quality of the moves themselves, but the importance of designing a system that is easy to use and modify should not be ignored.

3 Automatic Generation of an Opening Book

An opening book can conceptually be viewed as a tree, just like a game tree, where the root is the initial game position; nodes in general correspond to positions that occur later in the game and also record information such as the heuristic evaluation of that node and the negamax value of the position; edges represent legal moves from one position to the next; and leaves represent positions at the frontier of the book, that is, positions for which the book has no suggested moves. Since transpositions occur very rarely in Amazons games, we choose not to view the opening book as a more complicated graph structure, and SCRIBE actually represents the book internally in this way and stores it entirely in main memory.

Lincke distinguishes between passive and active book construction [3]. Passive book construction involves adding moves to the opening book based on information gleaned from experts, either from their games or from their knowledge of the game, while active construction means constructing the book automatically. In Section 2 we argued that a mixed approach is the best approach and briefly explained how we implemented the passive aspects of the book's construction. However, for Amazons the heart of a good opening book builder is the part that builds and expands the book automatically.

Automatically building an opening book is an iterative process [2]. First, choose a node to *expand*, that is, add a new child, a leaf, to this node. In other words, a new book move is added to the position corresponding to the node that is expanded. Next, calculate the heuristic evaluation value of this newly added position. Then do a negamax propagation of its value up the tree. Since this step can change the values of nodes in the tree, it can change the structure of the book resulting in a new book that may now choose different moves than it did before this move was added. We repeat this process to increase the size of the book. In practice, in the first step when adding a new child we may actually add more than one node. This will be discussed in more detail below.

Obviously, choosing which node to expand is the crucial step. We would like our opening book to be populated with moves that are actually likely to be made. A book full of moves that are never made is of no use to us. We refer to the process of choosing nodes to expand as the *expansion algorithm* and discuss four possibilities below.

We make a few assumptions about the implementations independent of the particular algorithm. First, viewing the book as a tree, we assume that every node in the tree contains a value associated with it called the *propagated value* [3]. Leaves have propagated values corresponding to an evaluation done of that position and all other nodes have the usual negamax value. Note that the value associated with leaves is not necessarily simply the result from some evaluation function. Of course it can be, but we usually calculate this value by doing a normal INVADER lookahead from that position down to some fixed depth. The actual depth we choose is user selectable in SCRIBE.

Also, INVADER's evaluation function oscillates in a way that, all else being equal, the last player to move will have a noticeable advantage in evaluation. This seems to be a common feature of Amazons programs. It is quite normal to see fairly large oscillations in evaluation values as lookahead progresses from one level to the next. This is especially true in the opening where it is usually obvious that by simply making a reasonable move you have improved your position. This can be a serious problem for an opening book because leaves are equally likely to appear at even and at odd levels and those evaluated at odd levels will have disproportionately large values, incorrectly skewing the book towards moves along such paths. To deal with this *odd-even effect* we attempt to normalize evaluations so that they are more consistent across the various depths. We have tried a number of different methods with varying degrees of success.

One idea is to attempt to return a value in between what is returned at the odd and even levels. An obvious way to do this would be to return the average of the leaf evaluation and the evaluation of its parent. Experiments with this approach proved disappointing. It seems that if the parent node had an evaluation that was extreme (i.e., larger or smaller than most of its siblings) it would dominate the result that was returned and so even if the next move, the leaf, tended to move the evaluation back away from the extreme, the parent's contribution would be too great and would incorrectly skew the result. Attempts to scale the average by doing things like weighting the leaf higher would mitigate this problem, but would then reduce the needed correction for the odd-even effect. We were unable to find a proper balance.

Another method to normalize the evaluation is to simply "correct" the evaluation function. In the case of INVADER's evaluation we must take into account the fact that we add various bonuses to the evaluation (for such things as amazon mobility, amazon separation, etc.) and the amount of the bonus depends on the move number. Also, by the nature of the game, the advantage of the extra move is greater earlier in the game. So, the only reasonable way we have to correct the evaluation is to subtract a fixed amount from the evaluation at every odd level, where this amount decreases as the level increases, and eventually vanishes. With our examples we found that after ply nine there was no longer a need to apply a correction. Choosing the exact values of these corrections is done by hand and then hand tuned until they are working the way we want. This is a decidedly inexact, but reasonably effective approach.

Finally, we tried another approach similar to the first one. As was mentioned earlier, evaluations for the book are obtained by doing a, say, two-ply INVADER search. If this search ends at an odd level (and, therefore, begins at an odd level) we must perform a correction. One choice would be to simply not allow this and require all searches beginning at odd levels (or beginning at even levels if INVADER is asked to do an odd length search) to go one level deeper. This, however, would require too much extra time and would produce its own skewing of results as all such searches would have more information than searches commencing at even levels. Instead, we keep track of the leaf of the principal variation of the search and rather than use that for our evaluation, we do a one-move lookahead from that node and use the value returned from that search. The drawback to this method is that any search that goes only one deep is going to be prone to error and so sometimes this new result overcorrects and provides a value that now favours the other player too much. Nevertheless, we have found it to be quite effective and certainly far superior to simply ignoring the odd-even effect. For all empirical data that follows this last approach was used, though using the second approach yields similar results.

3.1 Best-First

The best-first method was first proposed in [2] and was further analyzed in [3] where some weaknesses of the approach were pointed out. In the case of best-first expansion we always expand the leaf at the end of the book's principal variation. The idea is that assuming the book is accurate, both players are likely to make moves along the principal variation and so by expanding this line we increase the number of book moves we will likely be able to make during a single game. However, it may turn out that when a line is examined to a sufficient depth it is discovered to be weaker than originally thought, thus changing the PV. So we must then be prepared to continue the expansion down a different line. To allow for this (likely) contingency we need *deviations* [2] available at each node so that they may replace lines that have been found to be weak. The easiest way to do this is to be sure that every non-leaf node contains at least one child that is a leaf. This way we will always expand at a leaf, simplifying the expansion algorithm, and these leaves provide ready deviations for lines that prove to be weak. The process of expanding a node first creates a child, a leaf, for that node. But then the expanded node is no longer a leaf so, we also create a new leaf node as a sibling to the expanded node to replace this lost leaf. This new leaf can serve as a deviation should the principal variation lose value. Hence, at each expansion we always create two new nodes.

The graph below summarizes a 1000-move Amazons book created using best-first expansion.

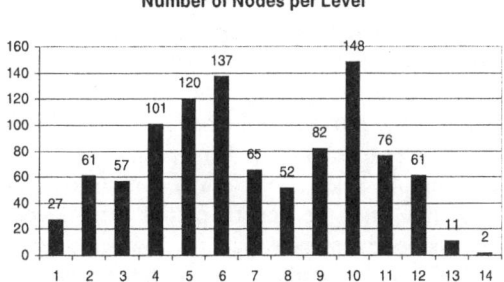

Fig. 3. 1000-Node Best-First Book

Figure 3 reveals that the distribution is far from ideal. In summary, the tree is too narrow and deep and should be wider and shallower. For example, well over one third of the moves in the book are at levels 10 and above, yet the likelihood of an actual game staying in the book so long is extremely low.

Long branches in an opening Amazons book are not very practical. In Amazons not only are there many different moves available at every turn but the number of plausible moves (both in terms of human understanding and in terms of INVADER evaluation) is also high, making it quite unlikely we can actually predict moves down to level 10. What is not clear from the figure, however, is the fact that the book is extremely unbalanced, making things even worse. Among the 27 possible first moves in the book, 25 of them are subtrees with 5 or fewer nodes. This means that the vast majority of the tree is concentrated under only two different first moves. Worse still, these do not appear to be the two best moves, though they are quite reasonable.

The above situation corresponds exactly to the situation described in [3]. Even if these two moves were in fact the best possible initial moves, an opponent could choose (by luck or because of knowledge of the book) the third best move, which is still a perfectly good move, and then very quickly get the program out of book, defeating the purpose of having a (large) opening book. But this is the nature of best-first expansion. As long as one move evaluates higher than any other, no matter by how small a margin, that node will continue to be expanded, at the expense of other reasonable looking moves. Compounding the problem is that eventually, when a large subtree reaches a certain size, the evaluation may drop to the point where another move with a very tiny tree becomes part of the PV. If book creation stops soon after this happens we have the annoying situation where the book's PV is actually the smallest subtree in the book. Continuing to build the book beyond this point would partially correct this problem, but in general, deciding when to stop building the book is a difficult question.

Even though the tree is too deep, surprisingly, there are times when the tree becomes too wide, also. This can happen very deep in the tree, where it is particularly disadvantageous. For example, in our book constructed above there is a move at level 8 that has 34 children. The reason this can happen has to do in part with the volatility of our evaluation and the fact there is still quite a bit of flux at this stage of the game. For example, it may appear that one of Black's amazons is nearly trapped and so the score evaluates high for White. But in just one or two moves it might be proven that that the amazon can free itself and so the evaluation can suddenly make a large switch. If such a discovery can be made with just one extra level of lookahead, then one node can generate many children, as described below.

Assume a white leaf with no siblings is expanded and this node evaluates high thinking it has Black in a bind. Further assume that when this node is expanded to create a child the extra level of lookahead now shows that White is not as well off as believed so the child evaluates much lower. But, as described above, when expanding this node we must also supply a new sibling, a deviation leaf, to the expanded node and this new leaf does not have the benefit of the extra level of lookahead. So, assuming it is a similar move to the one expanded, it will also evaluate high. Because of its high evaluation there is a good chance it will be chosen next for expansion and the process will continue until the deviation is so weak that it no longer becomes the PV. In the example mentioned above it must expand 34 nodes before this happens.

3.2 Drop-Out

Implicit in the discussion above is the fact that moves in the book that are nearly as good as the optimal moves, but have much smaller subtrees, are desirable moves for an opponent as he can then get out of book with minimal penalty. So, moves with small subtrees and with near optimal values should be given some expansion priority over the basic best-first expansion node, where, of course, "near optimal" and "small subtree" need to be defined more formally and then need to be tuned for the game in question. This is the main idea behind drop-out expansion [3]. Also, to help visualize the structure of a book, and to help explain later the drop-out expansion algorithm, the drop-out diagram was introduced. A drop-out diagram shows the depth and value of leaves assuming the opponent can make any move in the book and INVADER, say, makes optimal moves. In other words, it shows under what conditions an opponent might drop out of the book. The drop-out diagram for the 1000-node Amazons book described in the last section is shown in Figure 4. Evaluation values are scaled in such a way that 20 points is meant to represent one extra square of territory, that is, at the end of the game an evaluation of 60, say, would mean that the player has 3 extra

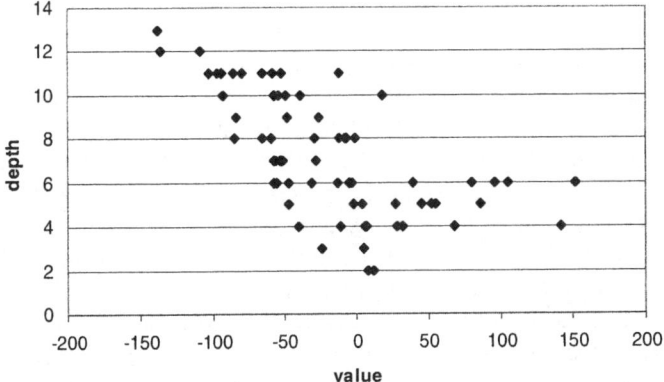

Fig. 4. Drop-out Diagram for the 1000-Move Best-First Expansion Book

squares and should win by 3 points. To get a feel for positional bonuses assigned by INVADER early in the game, it can be noted that completely surrounding an enemy amazon, for example, is awarded a 500 point bonus. That is, all else being equal (which, of course, it never is), completely enclosing an enemy amazon should lead to a 25 point win.

We see that if both players make optimal moves according to the book the opponent will drop out of the book at move 13 giving INVADER an evaluation of approximately -135. Looking further down the diagram, however, we see that the opponent can choose suboptimal moves, exit the book at move 3 and INVADER ends up with an evaluation of about -35. Given the fact that Amazons evaluations are quite imprecise at this early stage of the game, if the opponent were to discover or know about this line of play in the book, it would probably be a wise path to take. One, therefore, attempts to create a book where the cost of leaving the book early is sufficiently high as to discourage the opponent from doing so, since it is to the advantage of the program to stay in the book for as long as possible. In terms of the drop-out diagram we would like the slope of the left frontier of the dots to be sufficiently horizontal so that the opponent is reasonably unlikely to want to drop out of the book early. A more vertically sloped frontier implies the opponent can leave the book early and achieve nearly as good a score, thus defeating the purpose of having a deep opening book. In contrast, a slope that is too flat and spreads out too far to the right will have nodes that are unlikely to be chosen because of their low evaluation, but will still have large subtrees. This is a waste of both time, in creating these unneeded subtrees, and space, in saving moves in the book that are unlikely to be needed. The goal is to find the right balance.

Drop-out expansion is designed to try to produce books with an ideal slope in the drop-out diagram – one that properly discourages leaving the book, but does not waste too much time or space in the book on moves that are weak. Using this algorithm, all moves are considered for expansion where priority is given to moves with small subtrees and that have values that are close to the optimal value. The idea is to remove nodes that provide cheap and early exits from the book.

There are a couple of reasons why it is currently not feasible to use drop-out expansion for automatic Amazons opening-book construction. The main reason is the method was designed with small width games in mind, games that can construct deep books like Othello and Awari that have opening books to move 20 and beyond. The goal of drop-out expansion is to push out the drop-out diagram from bottom left to top right. Unfortunately, Amazons is not ready for this. Amazons needs a book that is compressed from the top. For practical purposes any moves in the book beyond depth 6 are so unlikely to be used that they are next to useless. Until we are either able to build opening books with billions of positions or we have evaluation functions that are much more accurate in the opening stages of the game we are not ready for books that can go much deeper than 6 or so moves. Currently, the ideal drop-out diagram for an Amazons book will have a frontier slope that is nearly horizontal at about level 4 and has virtually no entries above level 6. One could get this kind of behavior using drop-out expansion, but the factor dealing with evaluation values (the ω-factor [3]) is far less important than the depth of the tree, so a method that emphasizes depth is to be preferred.

As a side note, the basic drop-out algorithm asks that all siblings of every node be placed in the book during construction. Though this simplifies the expansion process, when dealing with a game like Amazons with over 1000 moves at each node it is extremely wasteful, so it would be best not to do this but instead add such nodes when needed as is done in the basic best-first algorithm. Dealing with these issues leads us to our next expansion algorithm.

3.3 Best-First with Depth Penalty

Since the main problem we have been facing in constructing an opening book for Amazons is the unnecessarily large depths achieved, we simplify the idea of drop-out expansion and simply penalize expansion nodes based on how deep the node is in the tree. In essence what we do is build the tree the same as we would using the basic best-first algorithm by propagating up heuristic values in a negamax fashion, but now node values are penalized only according to how far away the expanding leaf is from the node. We distinguish these values, which we call *expansion values*, from the normal propagated values. The deeper the leaf is that provides the expansion value for a node, the more that is subtracted from the normal propagated value and we then propagate up these expansion values to determine the *principal expansion variation* and then expand accordingly. In terms of implementation some care is necessary as the expansion value of a node is now both a function of the child with the maximum expansion value and the distance to the leaf of the principal expansion variation for that node. This means that the expansion value of a node will actually differ from the expansion value of the child propagating the value – by an amount equal to the depth penalty. As always, a certain amount of tuning is necessary, but without too much effort we are able to get pretty good results.

Figure 5 shows the results of two 1000-node books created using depth penalties of 50 and 100.

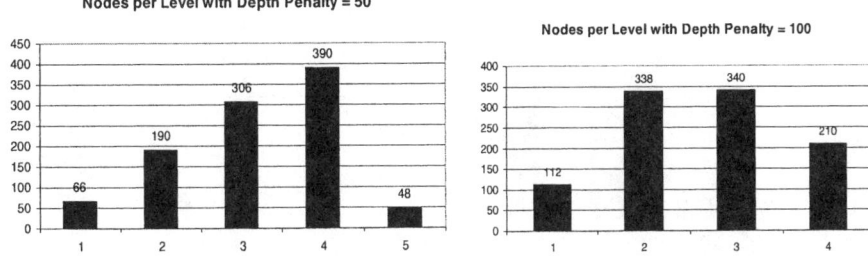

Fig. 5. 1000-Node Depth-First Books with Depth Penalty

Both obviously have much better move distributions than the book from Figure 3. But what is not obvious from the diagram is that a depth penalty of 50 still did not give the kind of distribution we were looking for. Among the 66 different positions at level 1, only 8 of them contained subtrees with 3 or more nodes in them. So even though we did not waste a lot of the book on very deep moves, most of the tree was still under only 10% of the first moves.

The book that was created with a depth penalty of 100 had a very nice distribution. A full one third of the level 1 positions had subtrees that went at least to depth 3. And one third of these positions, mostly those with the highest evaluations, had quite large subtrees, as one would hope for the moves that evaluate the highest. Also, the quality of the top moves in the tree agreed well with what would be expected from our current understanding of the game. Finally, most of the positions with subtrees consisting of two or fewer children came from moves that did not seem to have much promise and so could easily be pruned by hand, in preparation for further expansion of the tree as discussed in Section 2.

3.4 Breadth-First with Width Penalty

Since our main concern has been to limit the depth of the book another obvious solution would be to build the book in a breadth-first fashion. Of course given the huge move counts in Amazons we do not want to do a full-width construction, so for the purposes of these experiments we limited the width of the book at each level according to evaluation values. That is, we selected threshold difference values such that when generating children of a node, generation ceases when the difference between the heuristic value of a generated node and the heuristic value of its best sibling passes a certain threshold. This is what we refer to as the *width penalty*. The idea is that if a particular position has many moves that appear to be nearly equally good we do not want to exclude arbitrarily any of them but only exclude those moves that seem to be sufficiently weaker than the top moves. Since the volatility of heuristic evaluations seems to decrease as the move number increases, we decrease the threshold value as we descend the tree. Keeping the threshold the same produces a tree where nodes at lower levels have far too many children. As always, finding appropriate values for these thresholds proves to be an interesting exercise in tuning relative to the particular evaluation function being used.

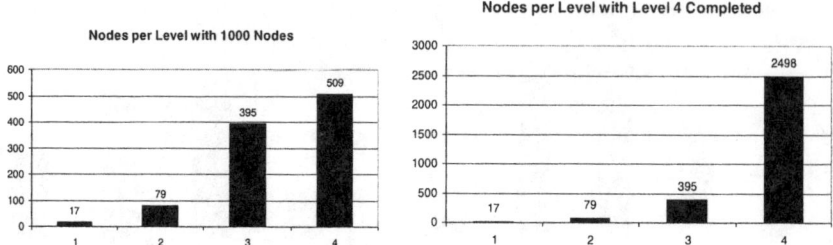

Fig. 6. Breadth-First Opening Books with Width Penalty

The first part of Figure 6 shows the node distribution after creating a 1000-node opening book using breadth-first expansion priority with width penalty, with threshold values that were tuned to provide a tree that was not too broad at the lower levels. Still, it is hard to avoid building a book that has many nodes at the lowest levels because of the exponential growth inherent to a full tree. Also, since level 4 was not completed the distribution is a bit misleading and the book is unsatisfactory since some level-3 moves have a full array of children and some have none. So, to help us better understand the nature of breadth-first expansion we allowed the book to continue building until level 4 was complete.

Looking at the second part of Figure 6 we see that the exponential nature of the tree hits us hard as now the vast majority of moves are at level 4. This is a bit worrisome as many of these moves have very low probability of ever being visited. The problem is that breadth-first expansion treats every node equally, so all nodes get expanded to the same depths and widths that every other node does, regardless of the "importance" of that node. The advantage is that the book looks more like a regular game-tree expansion, so we have more confidence that the moves at the top of the tree are the correct moves.

4 Conclusions and Future Work

We have studied four different expansion techniques for automatically building opening books for the game of Amazons. Best-first expansion and drop-out expansion seem most suited for games that do not have too many good choices of moves at each level and that are sufficiently well understood that deep evaluations actually have a reasonable probability of being played. This is not the case with Amazons, a game that has over 1000 possible moves at each level during the opening and literally hundreds of reasonable choices.

Since we have seen that deep opening books are not of much use to an Amazons player, a natural choice for opening-book construction would seem to be some modification of breadth-first (e.g., with width penalty) where deep lines cannot be created. We have found that this tends to create trees that deemphasize the first few moves, providing books that are very heavy near the leaves. Perhaps adding some other features to this algorithm can make it more viable. For example, modifying the width penalty at each level so that both the heuristic evaluation values and the number of siblings are taken into account might help. This way we can restrict the number of nodes at the lower levels where the book is less likely to need them.

It appears the best choice available to us is a modified version of best-first, such as the simple best-first with depth penalty. Properly tuned, this expansion algorithm produces trees that seem to have the right mix of moves, with a reasonable variety of choices at the top levels and sufficient, but not too much depth for the critical lines. Perhaps combining this algorithm with some of the breadth-first ideas will prove best. For example, a best-first algorithm with both a depth penalty and a breadth penalty that depends on evaluation values and node counts might produce an ideal book for Amazons.

A question not to be ignored is "how useful is a good opening book for an Amazons program?" Especially if we only stay in book for an average of fewer than four moves? This is a difficult question to answer. However, we have two pieces of evidence, though not completely convincing, suggest it is useful. First, by playing games where we generate variety by either slightly modifying the evaluation function, modifying the time available to each player, or by forcing a certain opening move we find that the version with the opening book wins about 60% of the games.

Secondly, we appeal to our human understanding of the game. We notice not too surprisingly that games played using the opening book start out better while still in the book. However, they seem to maintain some advantage for a number of moves after leaving the book. Unfortunately, either because of the nature of the program INVADER, or less likely, because of some not well understood property of the game of Amazons, small opening advantages seem to eventually wash out and land the game in positions that become very difficult to evaluate. Nevertheless, the opening book does prevent INVADER from making catastrophic moves in the opening and by all appearances does appear to give some added advantage. We suspect that INVADER is not yet ready for a good opening book, but when its play becomes steadier and stronger the opening book will make it even more so.

Finally, automatically constructed opening books, no matter how cleverly designed, have serious limitations. The most important lines might not get sufficient attention, that is, their subtrees may be too small or good moves might not appear in the book at all. Moves might end up with propagated values that a human can easily recognize are too large or too small. For these reasons it is important that a system that automatically generates opening books also allows for easy human intervention. At a minimum, the following three features are necessary. (1) A convenient method of traversing and visualizing the opening book. (2) The ability to override the automatically generated evaluation values, including forcing a move to the top or bottom of the candidate list. (3) To force the book to be expanded around certain nodes. With tools like these, the computer then becomes a powerful tool for aiding a human in creating a strong and useful opening book.

References

1. H. Avetisyan, and R. Lorentz. Selective Search in an Amazons Program. In Schaeffer, J., Müller, M, Björnsson, Y., editors, *Proceedings of the Third International Conference on Computers and Games (CG 2002)*, volume 2883 of *Lecture Notes in Computer Science*. Edmonton, Alberta, Canada. Springer-Verlag (2003) 123-141.

2. M. Buro. Toward Opening Book Learning. *ICGA Journal*, volume 22, no. 1 (1999) 98-102.
3. T. Lincke. Strategies for the Automatic Construction of Books. In Marsland, T, Frank, I., editors, *Proceedings of the Second International Conference on Computers and Games (CG 2000)*, volume 2063 of *Lecture Notes in Computer Science*. Hamamatsu Japan. Springer-Verlag (2001) 74-86.
4. http://swiss2.whosting.ch/jenslieb/amazong/amazong.html
5. R. Lorentz. First-time Entry Amazong wins Amazons Tournament. *ICGA Journal*, volume 25, no. 3 (2002) 182-184.
6. J. Lieberum. An Evaluation Function for the Game of Amazons. In van den Herik, H. J., Iida, H., Heinz, E.A., editors, *Advances in Computer Games: Many Games, Many Challenges, Proceedings of the 10th Advances in Computer Games Conference (ACG-10)*. Kluwer Academic Publishers, Boston (2004) 297-306.
7. http://www.csun.edu/~lorentz/amazon.htm

Building a World-Champion Arimaa Program

David Fotland

Smart Games,
San Jose, CA, USA
Fotland@smart-games.com

Abstract. Arimaa is a new two-player strategy game designed by Omar Syed to be difficult for computer players. Omar offers a $10,000 prize to the first program to beat a top human player. My program, BOT_BOMB, won the 2004 computer championship, but failed to beat Omar for the prize. This paper describes the problems with building a strong Arimaa program and details of the program's design.

1 Introduction

Arimaa is a new two-player perfect-information strategy game designed by Omar and Amir Syed. Their goal was to design a game that was fun to play, and very difficult for a computer to play well. The game has free placement of pieces in the initial position to foil opening books. It has a huge branching factor and long-term strategy, which should make full width search impractical. The game ends with most of the pieces still on the board, eliminating any benefit from endgame databases.

Omar offers a prize of $10,000 for the first program that can beat a strong player (selected by Omar), in a multi-game match with long time limits. The first computer championship was in January, 2004, and was won by my program BOT_BOMB. The computer vs. human championship was played against Omar. He won all eight games, although none was easy, and the average length of the games was 55 moves. Typical Arimaa games last 30 to 40 moves.

Omar contacted me in January, 2003, and suggested I might want to write a program. While I agreed that Arimaa is a more difficult game for computers than chess, I felt I could win the contest. My opinion is that Arimaa is more difficult than chess, but still much easier than Go. It is more like Shogi or Amazons. Even though Arimaa is difficult for computers, Arimaa is a new game, and people are not very good at it yet.

I started in February, and by May BOT_BOMB was the highest rated player at the web site. This May version of the program is still available at the web site under the name BOT_ARIMAANATOR, and is rated about 225 points below the current version. This gain was due to the addition of a goal evaluation and search extensions, and adding the pin evaluation. I stopped working on it over the summer, and in September discovered that the human players had become much stronger. Several new strategic concepts were discovered, and the human players were not blundering away pieces. Several players were rated higher than BOT_BOMB. I worked on the program until the end of December, but was not able to close the gap. As in computer chess, the program is relatively stronger than people at short time controls. At 30 seconds per move average, the program's rating is about 100 points higher than the tournament version, with 3 minutes per move.

Arimaa can be played on-line at http://arimaa.com or using software available from Smart Games at http://www.smart-games.com.

2 Rules of Arimaa

Arimaa is played on an 8x8 chess board, and can be played with a standard set of chess pieces, although Arimaa renames the pieces Elephant, Camel, Horses (2), Dogs (2), Cats (2), and Rabbits (8), in order from strongest to weakest. Gold (white) starts by placing the 16 pieces in the two rows closest to him[1], in any arrangement, as his first move, then Silver (Black) places pieces on his side of the board. Human play has shown that there are several popular initial arrangements, with different strategic consequences. Silver can place his pieces to counter Gold's arrangement, which compensates for Gold's first move advantage. Because all pieces are placed at once, the first move for each side has a branching factor of almost 65 million, so making the first move part of the search is infeasible. There are over 10^15 possible opening positions, making an opening book infeasible.

After placement, Gold moves first. For each player, a move consists of 4 steps. Each step moves a piece one square horizontally or vertically, except that Rabbits cannot move backwards. The four steps in a move can be used to move one piece or multiple pieces. Any step but the first in a move can be a pass, but the move must change the board position. The goal of the game is to get one of your Rabbits to the 8th rank.

The board has 4 traps, at the 3-3 squares. After any step, if a piece is on a trap, and there is no friendly piece in one of the four squares next to the trap, that piece is captured, and removed from the board.

Stronger pieces can pull, push, or freeze adjacent weaker enemy pieces. To pull a piece, the player moves a piece one step, then uses another step to move the adjacent enemy piece into the square he just vacated. To push a piece, the player moves an adjacent enemy one square in any direction, then moves his stronger piece into the open square. An adjacent weaker enemy piece is frozen unless it has an adjacent friendly piece. Frozen pieces cannot be moved, although they can still be pulled or pushed.

Repeating a position a third time loses the game. If there are no legal moves available, the player to move loses. The only way to draw is for both players to lose all eight Rabbits.

---

[1] In this paper we use 'he' when 'he' or 'she' are both possible.

3 Why is Arimaa Hard?

The key issue for Arimaa and computers is the huge branching factor. Some positions have only one legal step (after a push), but most have between 20 and 25 legal steps. The four steps in a move lead to about 300,000 4-step sequences. Not counting transpositions, there are typically between 20,000 and 30,000 distinct four step moves.

Because the pieces move slowly compared to chess, an attack can take several moves to set up, so the program must look ahead at least five moves (20 steps) to compete with strong players. This is too deep for a simple iterative deepening alpha-beta searcher. Forcing sequences are rare, so deep searching based on recognizing forced moves (such as PN search or shogi endgame search) are not effective.

Sacrificing material to create a threat to reach the goal, or to immobilize a strong piece, is much more common than sacrifices for attacks in chess, so programs that focus on material are at a disadvantage.

Evaluation of an Arimaa position is difficult, and very unlike a chess evaluation. Control of the traps is important, but control of the center is not, since it is easier to move a Rabbit to the goal near the edge of the board than the center. Pieces can be immobilized for many moves defending a trap, or preventing a Rabbit from reaching a goal, which affects the balance of strength on the rest of the board.

3.1 Two Example Positions

Below I provide two example positions in the Figures 1 and 2.

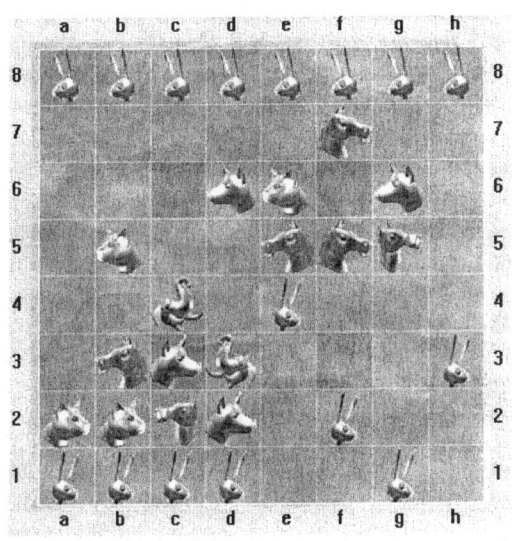

Fig. 1. Example position 1

In Figure 1, the Dog at c3 is on a trap, with the Elephant preventing it from being captured. The adjacent silver pieces are stronger than the Dog, so it cannot push them out of the way. The gold Elephant is pinned, defending the Dog. If it moves away, the Dog will be captured.

In Figure 2, the trap at f6 is dominated by strong nearby gold pieces. Once the gold Camel pushes the Cat at g6 out of the way, Gold will control 3 sides of the trap, and can start pulling pieces in. If the silver pieces move away to avoid being captured, Gold will be able to advance the Rabbit at h3 and reach the goal.

Silver has sacrificed two pieces to get an advanced Rabbit at g3. The gold Camel is frozen by the silver Elephant, so it cannot help defend the goal. If the gold Elephant moves away, Silver will capture the gold Camel in the trap at f6. Silver has a very strong position, and is threatening to reach the goal next turn.

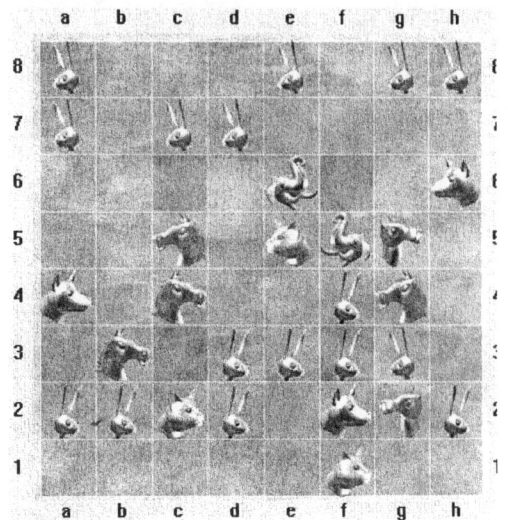

Fig. 2. Example position 2

4 The Program

My Arimaa program is called BOT_BOMB (which is the name used on the Arimaa web site, http://arimaa.com), but it also plays under the names BOT_SPEEDY, BOT_BOMBCC2004, and BOT_ARIMAANATOR. It is written in C++, derived from a chess program I wrote years ago. The rough specifications are:

4400 lines:	board representation and evaluation;
1800 lines:	search.

4.1 Board Representation

I use 64-bit bit-boards [5]. It was a good choice for a chess program, and even better for Arimaa, since the pieces move one space horizontally or vertically and there are many local evaluation terms. The bit-board class has members for logical operations, shifts, expand, count-bits, and iteration.

There is one bit board for each piece type, one for empty squares, and one for frozen pieces. There is an array which gives the piece at each square, and another which gives the strongest adjacent piece to each square, by each color. Some flags track if there is a pull or push in progress. Additional board class members track the side to move, the step number, hash values, and the material balance. All of this data is maintained incrementally when a move is made, and copied and restored to take back a move. The C++ operator "=" copies the core board data (373 bytes) using memcpy(), without copying any of the temporary data used during evaluation. This is much faster and simpler than writing an unmove function.

4.2 Evaluation

Below we discuss eight items of the evaluation function, viz., material, piece-square tables, Rabbit evaluation, mobility, trap evaluation, goal evaluation, pin evaluation, and center evaluation.

Material – In theory, material values in Arimaa should be completely relative, since if an Elephant is taken, the Camel becomes the new invulnerable piece. In practice, Elephants are never lost, so I use a fixed set of material values.

Rabbits	1.0, 1.5, 2.0, 2.5, 3.0, 4.0, 5.0, 12.0
Cat	2.5
Dog	3.0
Horse	6.0
Camel	11.0
Elephant	18.0

The rabbit value is adjusted depending on how many Rabbits are left. Once the last Rabbit is captured, the game cannot be won. You need two Rabbits to make goal threats on opposite sides of the board, and you need several Rabbits to defend your own goal. As the number of Rabbits goes down, the value of the remaining Rabbits goes up. The first captured Rabbit is worth 1.0, and the final one is worth 12.0.

Piece-square tables – This is a minor part of the evaluation, encouraging Rabbits to stay back to defend the goal, Dogs and Cats to stay near our traps to defend them, and the strong pieces to stay near the center and off of traps. Rabbits are encouraged to advance at the edges, and stay out of the center. For pieces, the values range from +0.1 to −0.5.

Rabbit evaluation – There is a bonus if a Rabbit has no enemy Rabbits on its file or adjacent files ahead of it (0.1 to 1.0 depending on how advanced it is). There is a bonus (0.1) to encourage Rabbits to have a piece to either side in the three points adjacent, behind or ahead. This tends to keep a solid wall of Rabbits and pieces across the board to make it harder to reach the goal. Advanced Rabbits are evaluated based on the relative strength of nearby pieces. This evaluation can be positive or negative, between about −0.3 and +5.0.

Mobility – Frozen pieces are penalized 0.02 for Rabbits, 0.1 for Cats and Dogs, 0.15 for Horses and 0.2 for the Camel. Mobility is very important for the stronger pieces. I found that if I make these penalties too large, the strong pieces get tied down freezing the weaker pieces, and become immobilized themselves.

The basic evaluation above is inadequate to prevent weak players from trapping the program's pieces. Weak players can also easily force a Rabbit to the goal. Searching at least 16 steps is required, but is not possible due to the high branching factor. To make the program competitive with strong players, the evaluation must do a static analysis that replaces several ply of look ahead.

Trap evaluation – Trap evaluation statically evaluates how many steps (1 to 6, or more) it will take to trap each piece on the board, assuming no enemy defensive moves. For any piece that could be trapped in six or fewer steps, it statically estimates the number of enemy steps it takes to defend that piece. There are about 50 cases, evaluated with a decision tree, about 900 lines of code. The evaluation function combines the individual values to estimate the future material balance, and identify threats. The algorithm is run once for each trap, and looks at the pattern of nearby pieces.

Goal evaluation – Goal evaluation statically evaluates how many steps (1 to 8, or more) it will take each Rabbit to reach the goal, assuming no intervening moves by the opponent. This is a tricky 700 lines of code, since there are many cases. It allows the search to find goals four steps earlier, and enables a highly pruned search to find defenses against goal threats. When this was implemented, weak players could no longer sacrifice pieces to force a goal, and strong players complained that the program defended tenaciously against goal threats. The strong players shifted to new strategies that immobilized pieces, won material, and did not try for the goal until there was a large material advantage.

I test the goal evaluation with 50 problems that have a forced goal in 10 to 12 steps, taken from actual games. With the goal evaluation enabled, it solves all 50 problems in an average of 1.2 seconds and 190 K search nodes. With the goal evaluation disabled, and a 10 minute time limit, it only solves 21 problems in an average of 436 seconds and 94 M nodes. The test positions are available at http://www.smart-games.com/mate12.ZIP

Pin evaluation – It is possible to use one piece to pin two enemy pieces near a trap, giving you a material advantage on the rest of the board. You can also use several weak pieces to pin an enemy piece on a trap. The pin evaluator handles these situations. This is the most difficult part of the evaluation, since it is hard to judge the relative values correctly.

Center evaluation – Strong pieces (Horse, Camel, and Elephant) are encouraged to have access to the center. The value depends on the number of steps it would take each piece to move onto one of the four center squares. For the Elephant, it ranges from 0.35 for being on the center to zero for being 4 or more steps away. The peak value for the Camel is 0.10.

The Camel is encouraged to stay away from the enemy Elephant, unless that Elephant is immobilized, or the Camel is near a trap with several friendly pieces nearby. If the enemy Elephant is not pinned, the Camel is encouraged to stay on its own side of the board. If the Camel is advanced, the enemy Elephant gets a bonus for being behind it. The bonuses are in the range of 0.1, and are sufficient to prevent strong players from pulling a camel to their trap in the opening, but are not a general solution.

4.3 Search

The search is fail-soft alpha-beta negamax principal variation search (PVS, also called NEGASCOUT) [6] with iterative deepening and transposition table. Each iteration and call to negamax extends the search by a single step, which makes move generation simple. Because the side to move only changes every four ply, alpha and beta are only exchanged every four ply, and the code for PVS and cutoffs is a little different. PVS typically has some code that looks like:

```
if (first move)
        score = -negamax(depth-1, -beta, -alpha);
else {
        score = -negamax(depth-1, -alpha-1, -alpha);
        if (score > alpha && score < beta)
                score = -negamax(depth-1, -beta, -alpha);
}
```

Arimaa code looks like:

```
if (step != 4)
        score = negamax(depth-1, alpha, beta);
else if (first move)
        score = -negamax(depth-1, -beta, -alpha);
else {
        score = -negamax(depth-1, -alpha-1, -alpha);
        if (score > alpha && score < beta)
                score = -negamax(depth-1, -beta, -alpha);
}
```

There can be no cutoffs in the first four ply so at the root, PVS is only used for iterations that search five steps or more. In negamax, PVS is only used at the 4th step of each move. Using it at the other three steps only slows down the search. The first iteration searches three steps, since with extensions, many interesting lines go four or more steps, complete a move, and help sort the next iteration.

Null move [7] is used to prune uninteresting branches. A null move can happen on any step except the first in a turn, and causes a pass for all the remaining steps in that turn. The search depth is reduced by four steps. A null move is only tried if beta is low enough that a cutoff is likely.

If all remaining steps are by the same color, the position is evaluated to see if it can get an early cutoff.

4.4 Search Extensions

If the goal evaluation shows three or fewer steps remaining to the goal, and the player to move has that many steps left in his turn, the position is scored as a mate without further search. By an experiment it is shown that the goal evaluation is accurate up to three steps. If there are four steps remaining, the search is extended one step to verify the goal.

If the opponent has a Rabbit four or fewer steps from the goal, the search is extended to find a defense. While searching for a defense, generated moves are pruned to only moves that are close enough to the Rabbit to affect the outcome. When there is a push, the next step is forced, so if that forced step is at depth zero, the search is extended one step.

4.5 Move Generation and Sorting

The move generator generates pulling and pushing moves first, then all others. Piece captures usually involve a push or pull, so these moves are more likely to cause a cutoff. The move generator never generates passes, although they are legal. Generating pass moves slows down the search, and does not seem to make the program stronger. The game ends with many pieces still on the board, so *zugzwang* is unlikely.

During a goal search, moves are sorted according to the distance from the piece moved to the Rabbit threatening to reach the goal. During the regular search, it first tries the move suggested by the transposition table, then two killer moves [1], then three moves from the history heuristic [2], then the rest of the moves. I tried using just the three history moves, and no killer moves, but the performance was worse, unlike the results in [3]. Perhaps this is because many Arimaa moves are quiet, so moves from one part of the tree are less likely to be effective elsewhere. Or perhaps it is because I only used the top 3 history moves, rather than doing a full sort.

4.6 Transposition Table

There is a 2 million entry transposition table [5] using Zobrist keys [4]. Half the entries are used for the primary table, and are replaced if the new value is closer to the root. Replaced entries are saved in the other half of the table.

5 Performance

On a 2.4 GHz Pentium, in the middle game, the program searches between 200K and 250 K nodes per second, where a node counts every time a move is made during the search. At 3 minute per move time control it completes 10 or 11 full steps of search, with extensions to a maximum of 14 to 18 steps.

The goal search can find 12 step forced goal sequences in under 1.5 seconds, and usually finds 20 step goal sequences within the 3-minute tournament time control. The program spends about 15 to 20% of the time generating moves, and 10 to 15% in search and move sorting. Most of the time is spent in the evaluation function. Early versions of the program with simple evaluations searched about 600 K nodes per second.

At the Arimaa web site, BOT_BOMB is currently rated 125 rating points below the top human. There are only 4 human players more highly rated. The next strongest computer opponent, BOT_OCCAM, is rated about 400 rating points below BOT_BOMB.

6 Future Work

The goal search, goal evaluation, and trap evaluation work very well, but there are still some bugs and missing cases to add. The evaluation of pieces stuck on traps or defending traps still has big problems, and leads to the loss of many games. Finally, the program has little sense of how to deploy its pieces, especially when one or more of the traps is tied up with strong pieces.

After the computer vs. human championship, the players discovered a way to sacrifice a piece to immobilize an Elephant forever, and easily win see Figure 3). Clearly mobility if the strong pieces needs more work.

The gold Elephant has just been immobilized, since it can not push anything or cross the trap. Silver can substitute weak pieces for the stronger ones involved, and get an advantage on the rest of the board. It is not easy to free the Elephant.

Fig. 3. An immobilized Elephant

7 Conclusion

Omar has succeeded in creating a game that is both fun for people to play and difficult for computers. In order to win you have to advance your Rabbits and pieces, but that puts them in danger of being pulled into traps in enemy territory. Pieces can be immobilized defending traps, giving one side a long-term strategic advantage.

A year ago it was not very hard to make a program to beat the strongest people, but the quality of top human play has improved faster than the computer programs. Any strong Arimaa program will have to evaluate accurately difficult strategic issues and have a deep selective search. Nevertheless, I have found that iterative deepening alpha-beta search is effective against most players. There does not seem to be a need for local search within the evaluation function, as in computer Go. But like Go, there are long-term positional effects that must be captured in a complex evaluation, that search cannot find.

References

1. S.G. Akl and M.M Newborn, The Principle Continuation and the Killer Heuristic, *ACM Annual Conference*, 466–473, 1977.
2. J. Schaeffer, The History Heuristic, *ICCA Journal*, Vol. 6, No. 3, pp. 16–19, 1983.
3. J. Schaeffer, The History Heuristic and Alpha-Beta Enhancements in Practice, *IEEE Transactions on Pattern Analysis and Machine Intelligence archive*, Vol. 11, No. 11, pp. 1203–1212, 1989.
4. Zobrist, A. L., A Hashing Method with Applications for Game Playing, *Technical Report 88*, Computer Science Department, University of Wisconsin Madison 1970, reprinted in *ICCA Journal*, Vol. 13, No. 2, pp. 69–73, 1990.
5. D.J. Slate and L.R. Atkin, Chess 4.5 – The Northwestern University Chess Program, in *Chess Skill in Man and Machine*, Springer-Verlag, 82–118, 1977.
6. A. Reinfeld, An Improvement on the Scout Tree Search Algorithm, *ICCA Journal*, Vol. 6, No. 4, pp. 4-14, 1983.
7. C. Donninger, Null Move and Deep Search: Selective-Search Heuristics for Obtuse Chess Programs. *ICCA Journal*, Vol. 16, No. 3, pp. 137–143, 1993.

Appendix: Championship Games

The following games can be played at http://arimaa.com/arimaa/gameroom. Follow the links to Computer Championship and World Championship.

The second best computer player was BOT_OCCAM, and the championship games are 5038 and 5039. In both games BOT_BOMB was able to pull the enemy Camel to its side of the board and get a very strong position in the opening, and games were over at move 30 and 33.

The human championship games lasted much longer, 38 to 96 moves, with a 56 move average. BOT_BOMB is able to defend its strong pieces in the opening, but

Fig. 4. After move 7

Fig. 5. After move 17

Fig. 6. After move 29

Omar was usually able to pull weaker pieces to his side of the board, trapping them or immobilizing other pieces. BOT_BOMB gave higher value to threatening Omar's strong pieces than to defending the cats, but the threats rarely led to any lasting advantage.

The fourth challenge match game, #5247: Move 7 (Figure 4), Omar (Gold) has pulled a Horse onto his trap, and immobilized BOT_BOMB's Elephant. Bomb thinks this is ok, since it takes more gold pieces to keep the Horse on the trap. But the gold Elephant can move away at any time, so BOT_BOMB's Camel can always be attacked, but Omar's Camel is free to move.

Omar launched an attack on the lower left trap, capturing a Dog, and leading to the position of Figure 5. Omar will capture the other Dog now. But Omar lets the Knight get off the upper right trap, and BOT_BOMB manages to immobilize Omar's Elephant at the lower right trap on move 29 (Figure 6).

Fig. 7. After move 36

Fig. 8. After move 51

Omar gives up the Horse to attack the lower left trap, but BOT_BOMB is able to immobilize Omar's Elephant defending the Camel at move 36 (Figure 7).

In spite of this disadvantage, Omar is able to trade a piece and capture several Rabbits, leading to this position at move 51 (Figure 8). BOT_BOMB was never able to get its Camel out to attack the upper right trap.

Now BOT_BOMB does not have enough pieces left to prevent the goal at move 67.

Blockage Detection in Pawn Endings

Omid David Tabibi[1], Ariel Felner[1], and Nathan S. Netanyahu[1,2]

[1] Department of Computer Science,
Bar-Ilan University, Ramat-Gan, Israel
{davoudo, felner, nathan}@cs.biu.ac.il
[2] Center for Automation Research,
University of Maryland, College Park, MD, USA
nathan@cfar.umd.edu

Abstract. In this article we introduce a blockage-detection method, which manages to detect a large set of blockage positions in pawn endgames, with practically no additional overhead. By examining different elements of the pawn structure, the method checks whether the Pawns form a blockage which prevents the King from penetrating into the opponent's camp. It then checks several criteria to find out whether the blockage is permanent.

1 Introduction

In the early days of computer chess, many were pessimistic about the prospects of creating grandmaster-level chess-playing programs via the conventional search methods which incorporate chess knowledge only in the evaluation function applied to the leaf nodes. As De Groot [3] mentioned, there are positions in which computers have trouble to arrive at the correct evaluation by exhaustive search, yet humans recognize the correct evaluation at a glance.

For many years, complex pawn endgames remained the toughest stage of the game, as many combinations require very deep searches, infeasible for the computers to conduct. Several efforts to improve the pawn endgame play of computers have been performed. We mention PEASANT [8], the co-ordinate squares approach [2], and the chunking method [1]. But since then, most of such positions have succumbed to the sheer processing power of computers, in particular by using transposition tables. Nowadays, tournament-playing programs search to depths of tens of plies in pawn endgames, and thus, manage to solve many positions once thought as never solvable by computers. For example, Newborn [8] estimated that the position shown in Figure 1 (Fine #70, [4]) would require 25,000 hours for the computers to solve. Yet today it takes less than a second for most programs to search sufficiently deep to find the correct move 1. Kb1.

However, there are still position types which cannot be evaluated realistically even by the deepest searches. Positions in which the element of *blockage* is present, are a prominent example of such positions. In these positions there is a fortress of Pawns which blocks one side from penetrating into the other's camp. Assuming there is no blockage-detection knowledge present in the program, only a 100-ply search will declare the position a draw by the fifty-move rule.

Fig. 1. Fine #70:White to move, 1. Kb1 wins.

In this article, we focus on pawn endgames featuring the blockage motif. In these position types, the Pawns divide the board into two separate disconnected areas. That is, the Pawns are blocked, and the King is not able to attack the opponent's Pawns in order to clear the way for its own Pawns. So, a draw will be the outcome of the game. Computers using conventional search methods will not be able to detect such draws, because no matter how deep they search, ultimately their assessment of the positions is based on the static evaluation which lacks the knowledge of blockage detection.

Although such positions are not very frequent in practice, whenever they arise, they badly expose this deficiency of computers. For example, while the position shown in Figure 2 is clearly a draw, all top tournament-playing programs tested evaluate the position as a clear win for White (see the Appendix). This was also the case in the first game of DEEP FRITZ vs. Vladimir Kramnik in their Bahrain match [7]. Throughout the game, DEEP FRITZ maintained an edge over Kramnik, until it found a clear advantage in one of the variations and opted for it. Figure 3 shows the position which DEEP FRITZ saw in its search, and evaluated it as a clear advantage for White. However, by a simple observation a human player recognizes that the position arrived at forms a blockage, resulting in no more than a draw (which was indeed the outcome of the game). If DEEP FRITZ had had the knowledge to detect the true outcome in that blockage position, it would have opted for another variation with greater winning chances.

In this contribution we introduce a blockage-detection method which manages to detect a large set of such blockage positions in pawn endgames, with practically no additional overhead. The article is organized as follows. In Section

Fig. 2. Despite the material advantage for White, the position is a blockage

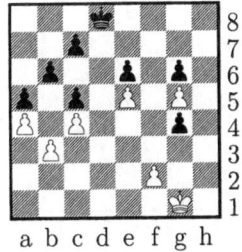

Fig. 3. DEEP FRITZ vs. Kramnik, Game 1, after White's 28th move

2 we define blockage, and present a method to detect the presence of blockage on the board. In Section 3 we provide additional criteria to check whether the formed blockage is permanent or can be broken by dynamic Pawns. In Section 4 we discuss some practical issues of blockage detection in search trees. Section 5 contains concluding remarks.

2 Blockage Detection

For the sake of simplicity, throughout this paper we will assume that we are interested in proving that White's upper bound is a draw, i.e., white cannot penetrate the blockage (assuming that Black is playing best defense). The process applies similarly for Black.

2.1 Applying Blockage Detection Within the Search

Figure 4 shows the simplest form of blockage: the pawns form a fence which divides the board into two separate parts, so that White's King cannot attack any black Pawns, and there are no mobile Pawns on the board. So despite White's material advantage, the game is a draw. A human can easily notice the blockage present. However, the computer's static evaluation will deem the position as better for White, due to the material advantage.

Looking at the position, a human sees that neither side has a mobile Pawn, and that the white King has no path to any black Pawn. Thus, White cannot achieve more than a draw, so White's upper bound is a draw (i.e., $\beta = DRAW$). Analogously, the black King does not have any path to a white Pawn, so Black's upper bound is also a draw. Therefore, the outcome is a draw. This thought process could be modelled so that a computer could follow it: if the upper bound of the side to move is greater than the draw score ($\beta > DRAW$), apply blockage detection. In case of blockage, the upper bound will be set to draw ($\beta = DRAW$). In case the lower bound is also greater than or equal to the draw score ($\alpha \geq DRAW$), we cut off the search at the current node immediately (as $\alpha \geq \beta$) (see Figure 5). Blockage detection can also be applied within the search using interior-node recognizers [10, 6].

Fig. 4. A simple blockage

```
int search(alpha, beta, depth) {
    // blockage detection
    if (beta > DRAW) {
        if (blockage()) {
            if (alpha >= DRAW)
                return DRAW;
            beta = DRAW;
        }
    }
    // continue regular search
    ...
}
```

Fig. 5. Use of blockage detection

Thus, blockage detection can be used to prune the search or to narrow the search window. For example, if the program reaches the position shown in Figure 4, it will usually evaluate it as advantageous for White. After applying the blockage detection, it will realize that the true upper bound is a draw, and so will avoid stepping into this position if it can find another truly advantageous position in its search tree.

2.2 Marking the Squares

In blockage positions, the most prominent feature is the presence of *fixed Pawns* (Pawns that cannot move forward or capture a piece) which cannot be attacked. These Pawns might form a fence, dividing the board into two disconnected regions (i.e., there is no path from one region to the other). Thus, the fixed Pawns in the position should be identified at the first stage of blockage detection.

More formally, we define a *fixed Pawn* for White as a Pawn which is blocked by a black Pawn, or by another white fixed Pawn, and cannot capture any piece. A fixed Pawn cannot move unless the black Pawn which blocks it is captured. We also define a *dynamic Pawn* as a non-fixed Pawn. For example, in Figure 6 white Pawns on a4, c4, e4, and e5 are fixed Pawns. Note that even though the g2 Pawn

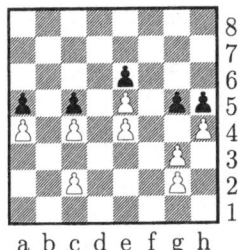

Fig. 6. White's fixed Pawns on a4, c4, e4, and e5

Fig. 7. The marked squares are: a4, c3, c4, e3, h3 (White's fixed Pawns), and b4, d4, d3, f3, g3, g6 (attacked by black Pawns)

cannot currently move, it is not considered fixed according to our definition, and is therefore dynamic.

In order to determine whether a blockage is present, we mark the squares on the board which the white King cannot step into, and check whether they form a fence (to be defined below). These squares can belong to one of the following two classes:

- Squares of White's fixed Pawns, such as a4, c3, c4, e3, h3 in Figure 7.
- Squares attacked by black Pawns, such as b4, d4, d3, f3, g3, g6 in Figure 7.

Now we have 64 squares, some of which are marked. Next, we check whether the marked squares form a fence. If they do, we have a blockage. Thus, the blockage detection will work as follows.

1. Detect fixed Pawns for White and mark them.
2. Mark the squares attacked by black Pawns.
3. Apply the fence-detection procedure to determine whether the marked squares form a fence.
4. Verify that White cannot penetrate the fence.

2.3 Fence Detection on Marked Squares

We now present a procedure that determines whether the marked squares form a fence. We define a *fence* as a 4-type path [9], i.e., a line of squares, such that each square has a neighbouring square either from east, north, west, or south. For blockage to take place, a necessary condition is that the fence divides the board into two separate regions, such that the Kings do not reside in the same region.

The presence of a fence can be detected by means of connected component labelling (see, e.g., [9, 5]). We can view the board as an 8x8 pixel image, where the pixels corresponding to marked squares are taken out of consideration. Connected component labelling will be applied by scanning all the remaining pixels

```
// marked squares have the value "true" and unmarked squares the value "false"
bool marked[8][8];
// squares which are identified as part of the fence, will be set to "true"
bool fence[8][8];
// visited squares will be set to "true"
bool processed[8][8];

// returns "true" if a fence is detected
bool blockage() {
    // find the topmost marked square on the A-file
    for (i = RANK_7; i > RANK_1; i--)
        if (marked[FILE_A][i] && forms_fence(FILE_A, i))
            return true;
    return false;
}

// recursively check whether a fence exists from the current square.
// function arguments are the file and rank of the current square.
bool forms_fence(f, r) {
    processed[f][r]= true;
    // the rightmost file (H-file)
    if (f == FILE_H) {
        fence[f][r]= true;
        return true;
    }
    // check neighbors in the order north, east, south
    // north
    if (marked[f][r+1] && !processed[f][r+1] && forms_fence(f, r+1)) {
        fence[f][r]= true;
        return true;
    }
    // east
    if (marked[f+1][r] && !processed[f+1][r] && forms_fence(f+1, r)) {
        fence[f][r]= true;
        return true;
    }
    // south
    if (marked[f][r-1] && !processed[f][r-1] && forms_fence(f, r-1)) {
        fence[f][r]= true;
        return true;
    }
    return false;
}
```

Fig. 8. DFS fence detection

from top to bottom and left to right, and assigning the label of each pixel to the east and south neighbouring pixels. If a pixel has no top or left neighbour, then it is assigned a new label. By the end of the process we check whether the pixels corresponding to the squares of the two Kings have different labels. If so, then a fence is present. Otherwise, the two Kings are in connected regions and so there is no fence present.

However, there exists a more efficient way of detecting the fence by activating a simple depth-first search procedure on the marked squares. We start from the leftmost marked square (which has to be on the a-file). If there are several such marked squares on that file, we choose the topmost square, as we are interested in finding the topmost fence line. Then we consider all its marked 4-neighbours and choose one with the priority of north, east, south. We continue this procedure until reaching the rightmost h-file. If a square in question does not have any neighbouring marked squares, we backtrack to the previous marked square. In case we have backtracked to the leftmost file, and there are no marked squares left on that file, this implies that there is no fence present. Figure 8 illustrates this procedure. Note that this procedure can be implemented more efficiently using bitboards, but as the experimental results in Section 4 show, the overhead of the implementation below is already very small.

After applying the above algorithm, in case the presence of a fence is detected, each square (on file f, rank r) that belongs to the fence is marked as fence[f][r]= true.

In order to ensure that White cannot penetrate the fence, the white King must be below the fence line, and the black King must have at least one vacant square to step into (i.e., Black is not in zugzwang). Now, if White has no non-fixed (dynamic) Pawns, and Black has no Pawns below the fence line (that might be captured by the white King), White has no way of penetrating the fence, and a blockage is present. Therefore, the upper bound for White is a draw.

If not all white Pawns are fixed, we have to verify that the dynamic Pawns cannot break the blockage. This is discussed in the next section.

3 Blockages with Dynamic Pawns

In most blockage positions, in addition to fixed Pawns, the side to move has also dynamic Pawns. Since as mentioned in the previous section, we are interested only in determining whether one side's upper bound is a draw, we only consider the dynamic Pawns of that side (here, we describe the detection methods for White). Assuming that Black is not in zugzwang, Black's dynamic Pawns cannot possibly increase White's upper bound. Additionally, black Pawns below the fence might get captured by the white King, and so Black cannot rely upon them. Thus, all black Pawns below the fence line are removed from the board henceforward (and the white Pawns in front of them are considered dynamic). Note that no white Pawn should be able to capture a black Pawn, as this will change the pawn structure.

Now we will check each dynamic Pawn of White to determine whether it can break the blockage. We divide these dynamic Pawns into three categories: dynamic Pawns above the fence line, dynamic Pawns on the fence, and dynamic Pawns below the fence line.

3.1 Dynamic Pawns Above the Fence Line

A dynamic white Pawn above the fence line might be an unstoppable passed Pawn (see Figure 9), in which case the result will not be a draw but a win for White. In order to prevent this, we require that the black King be in front of this dynamic Pawn (on any square in front of the Pawn on that file); see Figure 10. Requiring that the King be in front of the Pawn deals with all possible cases of white Pawns beyond the fence line. For example, in Figure 11, the black King cannot be on the files B and F at the same time, and so the position will not be deemed a blockage.

3.2 Dynamic Pawns on the Fence Line

Each square in the fence line is either occupied by a fixed white Pawn or is attacked by a black Pawn (see Section 2). Therefore, if a dynamic white Pawn is on the fence line, that square is attacked by a black Pawn (see Figure 12). In order to prove that this dynamic Pawn cannot remove the blockage, we have to show that this dynamic Pawn can neither penetrate the fence by capturing the black Pawn, nor turn into a passed Pawn by moving forward.

To satisfy both these points, the presence of a black Pawn two squares in front of this dynamic Pawn is required. For example, in Figure 12 Black has a Pawn two squares in front of White's f4 Pawn (on f6). This ensures that the blockage will not be removed in any of the following two ways: if the f4 Pawn captures a black Pawn by either 1. fxg5 or 1. fxe5, Black will recapture by 1. ... fxg5 or 1. ... fxe5 respectively, and the fence line will remain intact; and if the Pawn moves forward by 1. f5, it will turn into a fixed Pawn blocked by Black's f6 Pawn (and the new fence line will pass through e5-f5-g5 in that section).

Additionally, this dynamic Pawn should be White's only Pawn on that file, so that after an exchange White is not left with another Pawn in that file which can move forward and break the blockage; and the square in front of it should not be attacked by a black Pawn, so that moving this Pawn forward will turn it into a fixed Pawn. Figures 13 and 14 illustrate two positions in which a lack of either of these two requirements results in the removal of the blockage.

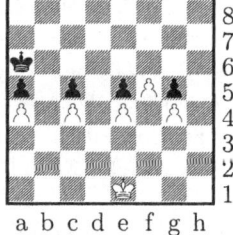

Fig. 9. The fence line is a4-b4-c4-d4-e4-f4-g4-h4. The f5-Pawn is above the fence line and cannot be stopped by the black King.

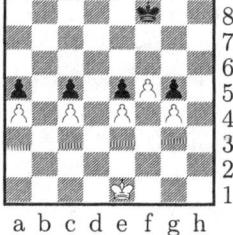

Fig. 10. The fence line is a4-b4-c4-d4-e4-f4-g4-h4. The f5-Pawn is stopped by the black King which is in front of the Pawn.

Blockage Detection in Pawn Endings   195

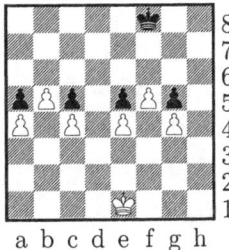

Fig. 11. The fence line is a4-b4-c4-d4-e4-f4-g4-h4. Black King cannot stop both Pawns.

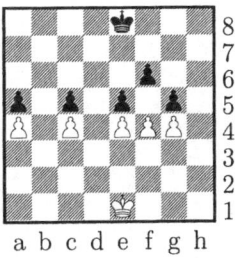

Fig. 12. The fence line is a4-b4-c4-d4-e4-f4-g4-h4. White's f4-Pawn is on the fence line.

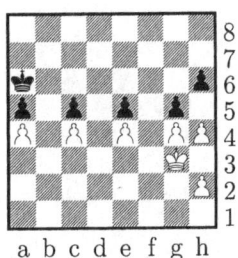

Fig. 13. Since the h4-Pawn is not White's only Pawn on the H-file, White can break the blockage by 1. hxg5 hxg5 2. h4.

Fig. 14. Since one square in front of the h4-Pawn (h5) is attacked by a black Pawn, White can break the blockage by 1. h5 gxh5 2. gxh5.

3.3 Dynamic Pawns Below the Fence Line

A dynamic Pawn below the fence line can move forward until it either:

- gets blocked by a white Pawn (e2 Pawn in Figure 15 gets blocked by the e4 Pawn),
- engages a black Pawn below the fence line (b2 Pawn in Figure 15 engages the black c4 Pawn after reaching b3), or
- reaches the fence line (g2 Pawn in Figure 15 reaches g4).

The latter case (dynamic Pawn on the fence line) was already discussed in the previous subsection (note that also when a Pawn passes the fence line by doing a 2-square move from the second rank, the position still remains blocked, provided that the conditions described in the previous subsection are met). In the first case, it is clear that when the Pawn is blocked by a friendly Pawn before reaching the fence line, it cannot break the blockage. In the second case, when engaging a black Pawn, the pawn structure can change and so the blockage might be removed.

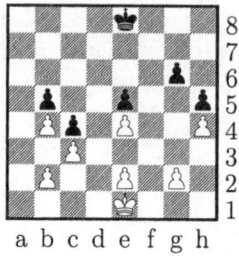

Fig. 15. Different types of dynamic Pawns at b2, e2, and g2. The fence line is a4-b4-c4-d4-e4-f4-g4-h4.

Thus, we start scanning the squares in front of the Pawn, until we either reach a white Pawn, or a square attacked by a black Pawn. In case we stop at a white Pawn (e.g., the e2 Pawn in Figure 15 which is stopped by the Pawn on e4), the Pawn will not break the blockage. In case we stop at a square attacked by a black Pawn, and this square is below the fence line (e.g., the b2 Pawn in Figure 15 which stops at b3), the pawn structure might change as a result of a pawn capture, and so we do not deem the position a blockage (the blockage detection immediately returns "false"). Otherwise, the Pawn reaches the fence line (covered in Subsection 3.2).

3.4 An Exception

A frequent pattern that appears in many blocked positions is illustrated in Figure 16. While not all the criteria mentioned in the previous subsections are satisfied, the position is still a blockage. White's f2 Pawn can reach the fence line (at f4) which is attacked by a black Pawn. However, there is no black Pawn on f6 to defend the fence (as required in Subsection 3.2; see Figure 12). In this pawn structure the blockage will not be broken if the black King is on the dynamic Pawn's file, and there is exactly one white Pawn on each of the neighbouring files, when these Pawns are (1) fixed Pawns, (2) on the fence line, and (3) both on the same rank, which is below the sixth rank. In Figure 16 all these conditions hold true:

- black King is on the f2 Pawn's file, i.e., the f-file,
- exactly one white Pawn exists on each of the neighbouring files (e4 and g4 Pawns on files e and g),
- both the e4 Pawn and g4 Pawns are fixed Pawns,
- both the e4 Pawn and g4 Pawns are on the fence line,
- both the e4 Pawn and g4 Pawns are on the fourth rank, which is below the sixth rank.

If all these requirements hold true, even though White might be able to break the blockage by moving the dynamic Pawn to the fence line, its upper bound will still remain a draw, since Black will capture the Pawn and will have a supported passed Pawn (e.g., Black's f4 Pawn after gxf4 in our example), and White's new passed Pawn (on g4) will be stopped by the black King.

4 Analysis

In this section we discuss some practical issues of blockage detection. We have implemented the blockage-detection method described in Sections 2 and 3 in the FALCON engine. FALCON is a fast yet sophisticated grandmaster-strength chess program. Blockage detection was incorporated as described in Figure 5. This was done in all the nodes.

Adding the blockage-detection procedure at each node of the search tree might seem costly. However, in practice it incurs almost no overhead. The blockage detection is called only if there are no pieces on the board apart from King and Pawns. Additionally, the already visited pawn structures are stored in a pawn hash table. Thus, the costly DFS fence detection is performed once for each pawn structure, and retrieved later from the hash table.

In order to measure the performance of the blockage-detection method we used a database of 3,118 positions where each side had at least six Pawns on the board and no additional pieces, and the game ended in a draw (see Appendix). It is clear that when in addition to Pawns there are other pieces on the board, or when there are only a few Pawns on the board, blockage detection will either not be triggered at all, or will take a negligible time. So, in order to measure the speed overhead in worst case, we used positions where each side has at least six Pawns on the board, since in these positions there is a higher potential for

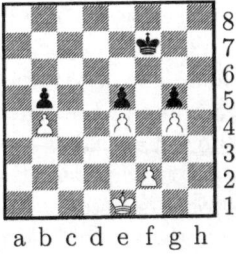

Fig. 16. The fence line is a4-b4-c4-d4-e4-f4-g4-h4. Although White's f2-Pawn can reach f4, and a black Pawn is not present at f6, White still cannot break the blockage.

Table 1. Total number of nodes searched, blockages detected, positions returning the draw score, nodes per second speed rate, and the speed overhead of blockage detection. Number of positions: 3,118. Analysis time: 10 seconds per position.

Blockage Det.	Nodes	Blockages	Draws	Speed (NPS)	Overhead
On	7,086,119,946	446,824	895	227,264	3%
Off	7,323,190,952	0	569	234,868	-

blockage detection. We used two identical versions of FALCON, one employing blockage detection, and the other having it turned off. We let each version analyse each position for 10 seconds. Table 1 provides the results. The results show that blockage detection has only a 3% speed overhead in this extreme worst case. We can further see that 446,824 times in our search, the blockage detection returned "true". From the total of 3,118 positions, the search on 895 of them returned the draw score (0.00) when using blockage detection, while only 569 returned that score without blockage detection being employed.

The results show that applying blockage-detection has practically no additional overhead. Whenever there are other pieces in addition to Pawns on the board, blockage detection is not triggered at all. In pawn endgames, blockage detection is applied to each pawn structure only once, since the result is hashed, so the cost is negligible. And even when there are many Pawns on the board and they can form many pawn formations in our search, the overhead of blockage detection will still be at most about 3%, as our results indicate. Even there, the small cost is negligible in comparison to the benefit coming from blockage detection. As our results show, the version using blockage detection managed to detect 326 draws more than the version without blockage detection.

The blockage-detection knowledge contributes to the search in many ways. When the program evaluates a position totally unrealistically, this can wreak havoc to the whole search tree, leading the program into playing a blunder. Figure 17 is an example of a position in which, owing to blockage-detection knowledge, Black manages to avoid a loss, and instead draws the game. The only correct move for Black is 1. ... Rxc4, which results in a blockage. While FALCON chooses this move instantly and displays the draw score together with the principal variation of 1. ... Rxc4 2. dxc4 g6, all top tournament-playing programs tested choose other moves which result in a loss for Black (see the Appendix). After 1. ... Rxc4 2. dxc4 g6 the position is a blockage: the fence line is a4-b4-c4-d4-d5-e5-f5-g5-h5, and none of White's dynamic Pawns (b2, e2, and h4) can break the blockage (according to the criteria mentioned in the previous section). Thus, White's upper bound is a draw.

In other positions, the lack of blockage-detection knowledge might cause the program to trade its advantage for a "greater advantage", which is in fact a mere draw. This was the case in the first game of DEEP FRITZ vs. Kramnik in their

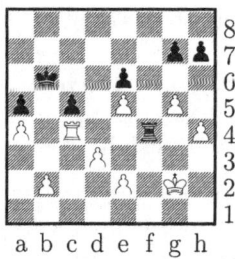

Fig. 17. Black to move.1. ... Rxc4 results in a draw, all other moves lose.

Fig. 18. DEEP FRITZ vs. Kramnik. Here White played 25. h4 resulting in a draw after 25. ... hxg4 26. Bg5 Bxg5 27. hxg5.

Bahrain match. After the 24th move the position illustrated in Figure 18 arose on the board. Here DEEP FRITZ played 25. h4, which after 25. ... hxg4 26. Bg5 Bxg5 27. hxg5 resulted in the position illustrated in Figure 3. After White has captured Black's g4 Pawn the position is evaluated as advantageous for White, but in fact it is a blockage, and so a draw. All the programs tested again fail to see the draw, except FALCON which includes the blockage-detection knowledge. If DEEP FRITZ had had the knowledge to detect the blockage, it would have chosen another variation with greater winning chances.

In addition to helping avoid blunders, blockage detection can also serve as a safe pruning method for pawn endgames. Since we apply blockage detection in all the nodes, whenever a blockage occurs in the search tree, we know it immediately and so can usually cut off the search at once. Thus, the program will spend its time searching other moves more deeply, avoiding a waste of resources for blockage positions.

5 Conclusion

In this article we have introduced a blockage-detection method which manages to detect a large subset of blockage positions. Our method provides just as much knowledge needed for the search to detect the blockage. For example, if in Figure 3 we place the black King on a6 and the white King on g4 (see Figure 19), in our search we will see that the black King manages to reach the f-file just in

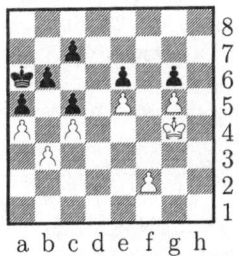

Fig. 19. Black King manages to reach the f-file in time to stop an f4-f5 penetration

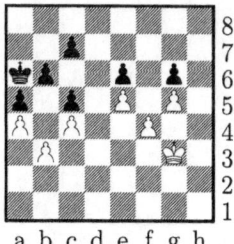

Fig. 20. 1. f5 wins.

time to stop a white penetration of the fence with f4-f5. But if we further move the white King to g3 and the white Pawn to f4 (see Figure 20), then White will win by 1. f5. That will be noticed in the search, as the black King will not reach the f-file in time to satisfy the conditions mentioned in subsection 3.4.

Lack of blockage-detection knowledge has always been a blind spot of even top tournament-playing programs. Our presented blockage-detection method enables the programs to evaluate such positions realistically, and avoid the blunders which result from a lack of blockage-detection knowledge. This method can be incorporated in all the nodes of the search tree at a negligible cost. This enables the program to detect the blockage instantly, and narrow the search window or cut off the search at the node immediately.

More criteria can be added to the blockage-detection method, resulting in detection of more blockage positions. However, a position wrongly labelled as a blockage might direct the search into an incorrect direction, and so the additional criteria must prove to be correct in all the positions.

Acknowledgements

We would like to thank Dieter Buerssner for providing the database of games for our empirical studies. We would further like to thank the anonymous referees for their valuable remarks and suggestions.

References

1. Berliner, H. and Campbell, M. (1984). Using chunking to solve chess pawn endgames. *Artificial Intelligence*, Vol. 23, No. 1, pp. 97–120. ISSN 0004-3702.
2. Church, K.W. (1979). Co-ordinate squares: A solution to many pawn endgames. *IJCAI 1979*, pp. 140–154.
3. De Groot, A.D. (1965). *Thought and Choice in Chess*. Mouton Publishers, The Hague. ISBN 90-279-7914-6.
4. Fine, R. (1941). *Basic Chess Endings*. Random House, 2003, ISBN 0-812-93493-8.
5. Haralick, R.M. and Shapiro, L.G. (1992). *Computer and Robot Vision*. Addison-Wesley, Cambridge, MA. 1992, ISBN 0-201-10877-1.
6. Heinz, E.A. (1998). Efficient interior-node recognition. *ICCA Journal*, Vol. 21, No. 3, pp. 156-167.

7. Müller, K. (2002). The clash of the titans: Kramnik – FRITZ BAHRAIN. *ICGA Journal*, Vol. 25, No. 4, pp. 233–238.
8. Newborn, M. (1977). PEASANT: An endgame program for kings and pawns. *Chess Skill in Man and Machine*, (Ed. P.W. Frey), pp. 119–130. Springer-Verlag, New York, N.Y., 2nd ed. 1983, ISBN 0-387-90790-4/3-540-90790-4.
9. Rosenfeld, A. and Kak, A. (1982). *Digital Picture Processing*. Morgan Kaufmann, 2nd ed. 1982, ISBN 0-125-97301-0.
10. Slate, D.J. (1984). Interior-node score bounds in a brute-force chess program. *ICCA Journal*, Vol. 7, No. 4, pp. 184–192.

Appendix

All the programs were run using ChessBase FRITZ 8 GUI, with 64MB of hash table, running on a 733 MHz Pentium III system with 256MB RAM, using the Windows XP operating system. The positions used in the Table-1 experiment were extracted by searching a database for games where at some point both sides had at least six Pawns on the board and no additional pieces, and the game ended in a draw. From each of the 3,118 games found, we chose the first position where both sides had at least six Pawns on the board and no additional pieces.

Table 2. Analysis of Figure-2 position. Analysis Time: 30 minutes. FALCON instantly detects the blockage at a depth of 2 plies, and displays the draw score of 0.00. Scores are from White's point of view.

	JUNIOR 8	FRITZ 8	SHREDDER 7.04	CHESS TIGER 15	HIARCS 8	FALCON
Score	+3.42	+2.91	+5.54	+2.36	+3.82	0.00
Depth	41	28	31	30	25	2

Table 3. Analysis of Figure-3 position. Analysis Time: 30 minutes. FALCON instantly detects the blockage at a depth of 2 plies, and displays the draw score of 0.00. Scores are from White's point of view.

	JUNIOR 8	FRITZ 8	SHREDDER 7.04	CHESS TIGER 15	HIARCS 8	FALCON
Score	+0.35	+0.84	+0.43	+0.66	+0.73	0.00
Depth	41	30	33	29	25	2

Table 4. Analysis of Figure-17 position. Analysis Time: 60 minutes. FALCON instantly detects the correct move leading to the blockage at a depth of 3 plies, and displays the draw score of 0.00. Scores are from White's point of view.

	JUNIOR 8	FRITZ 8	SHREDDER 7.04	CHESS TIGER 15	HIARCS 8	FALCON
Move	1...Rf5	1...Rf7	1...Rf5	1...Rf7	1...Rf7	1...Rxc4
Score	+2.42	+2.06	+3.64	+2.46	+2.53	0.00
Depth	24	19	22	21	18	3

Dao: A Benchmark Game

H. (Jeroen) H.L.M. Donkers, H. Jaap van den Herik, and Jos W.H.M. Uiterwijk

Department of Computer Science, Institute for Knowledge and Agent Technology,
Universiteit Maastricht, Maastricht, The Netherlands
{donkers, herik, uiterwijk}@cs.unimaas.nl

Abstract. Dao is an attractive game to play, although it is solvable in a few seconds on a computer. The game is so small that the complete game graph can be kept in internal memory. At the same time, the number of nodes in the game graph of Dao is large enough to allow interesting analyses. In the game spectrum, Dao resides between on the one hand trivial games such as Tic-Tac-Toe and Do-Guti and on the other hand games, such as Connect-Four and Awari that are solved but of which the game graph cannot be kept in memory. In this paper we provide many detailed properties of Dao and its solution. Our conclusion is that a game like Dao can be used as a benchmark of search enhancements. As an illustration we provide an example concerning the size of transposition tables in α-β search.

1 Introduction

In research on machine learning, a set of standard benchmark databases from a repository such as UCI [3] is frequently used to compare different machine-learning algorithms. Each database represents a typical difficulty that can be encountered in real tasks. In the area of computer game-playing, the idea of benchmarking is less standardized. There are sets of test positions which are used for the assessment of new algorithms in a specific game and there are standardized methods for comparing the overall effects of the various search enhancements [2]. But the ultimate test for a new idea is its result in a machine-machine contest or in a man-machine tournament. However, tournaments are time-consuming and require fully functional game-playing programs. Furthermore, these programs have so many factors influencing each other that they are too complex to isolate the effect of a single enhancement. Therefore, computer-game researchers usually set up their own controlled experiments in order to analyse the effect of their ideas. Unfortunately, it is not always easy to compare the results of these experiments, since different games are used, or different sets of training data.

This paper presents a small game, called Dao, that we propose as a benchmark game. It is so small that the game graph can be stored in memory and the game can be strongly solved [1] in just a few seconds. Yet, the game is complex enough to be non-trivial. The availability of the game graph and the complete solution makes it possible to investigate the working of an algorithm in more detail than is normally feasible. As an example we investigate the effect of the transposition-table size in α-β search: it appears that the number of nodes in the game graph visited during search does not change much with the size. It is the number of visits per node that changes drastically with the growth of a transposition table.

In Section 2 we first present an even smaller game, Do-Guti; this game will illustrate some important concepts. The rules of Dao are explained in Section 3. Section 4 provides combinatorial properties of the game and Section 5 analyses random games. The solution of the game is given in Section 6. An example experiment is described in Section 7. In Section 8 conclusions are given.

2 Do-Guti

We start with a small game, called Do-Guti [10]. This traditional game can be viewed as a rudimental version of Dao and can be used to illustrate some of the basic ideas dealt within this paper. Do-Guti (*'two-game'*) is played in Punjab, India. Like Tic-Tac-Toe, the game is a two-player, zero-sum game with perfect information. It is fun to play the game up to the moment that both players know the 'trick'. The goal of the game simply

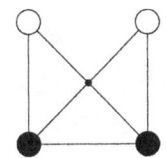

Fig. 1. Opening of Do-Guti

is to blockade the opponent. Do-Guti is played on a board with a network of 5 lines and 5 points. The points lie on the places where the lines meet (or cross) each other (see Figure 1). One player has two white stones, the other player has two black stones. The players move in turn, passes are not allowed. In order to move, a player selects one of the own stones and shifts it along a line to a neighbouring free point. The start configuration is given in Figure 1.

It is not difficult to compute the number of possible configurations of Do-Guti: $\binom{5}{2} \times \binom{3}{2} = 30$. The board of Do-Guti only has a vertical symmetry axis; based on this symmetry, the 30 configurations can be divided into 16 equivalence classes (see Figure 2). When we aim to solve the game with the help of the game graph, it is possible to draw a graph of 16 nodes and use retrograde analysis on it. However, if we include colour switches (i.e., black stones become white and white stones black) in the equivalence relation, the 30 configurations can be grouped into just 9 different equivalence classes (see Figure 3). In both game graphs, every class is represented by one of the configurations in it. In the sequel we will not distinguish between equivalent configurations and use the word 'position' when we in fact mean an equivalence class of configurations. In the first game graph (Figure 2), all equivalence classes are represented by configurations having White to move (WTM). Therefore, the arrows in this graph have two implications: (1) they indicate the move played by White and (2) they imply colour exchanging and side switching (which is not part of the game). In the game graph of Figure 3, colour symmetry is included; in some configurations both players can move. It means that we need to label the arrows in the graph indicating whether moves are made by Black or White and whether colours are switched from one to another position.

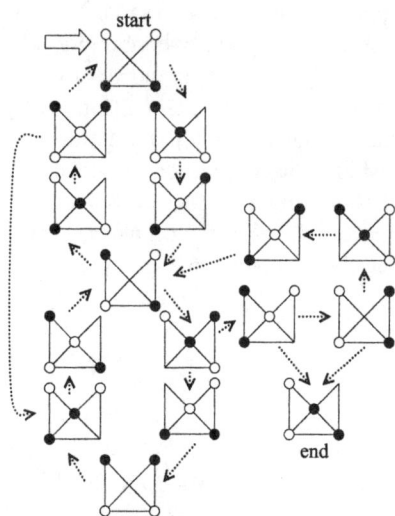
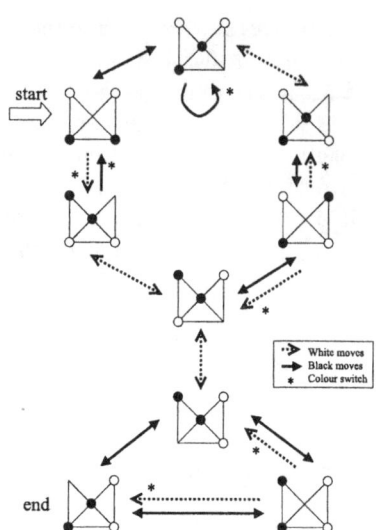

Fig. 2. A graph of 16 equivalence classes (with White to move)

Fig. 3. A graph of 9 equivalence classes (colour symmetry is included)

The game-graph representation of Figure 3 immediately unveils that Do-Guti is a drawn game, irrespective of which player starts. Only when one of the players makes a mistake by moving to the lower triangle of nodes, the game can be forced to an end. However, both players can force the game to stay within the upper six nodes. Not making the mistake and, at the same time, knowing how to deal with it when the opponent does, constitutes the 'trick' of Do-Guti.

Although it is possible to follow the move sequences in the graph of Figure 3, the graph of Figure 2 makes it easier to deduce that the shortest game from start to a decisive end takes 6 moves, irrespective of who starts the game. The graph of Figure 2 is the type that will be used in the next game, Dao.

3 Dao Game Rules

In 1999 Jeff Pickering and Ben van Buskirk [7] designed the game of Dao. It is played on a 4 × 4 board (see Figure 4). (The board has only 16 points. The crossings in the middle of the nine small squares have no meaning.) Each player has four pieces on the board. In the commercial version of the game [11], these pieces are glass spheres or Buddha-like statuettes. However, we will use plain white pieces for the first player (called White) and black pieces for the second player (called Black).

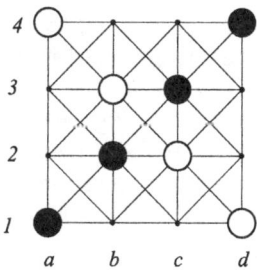

Fig. 4. Opening of Dao

The players move in turn as follows: one piece of their own colour is selected and shifted horizontally, vertically, or diagonally (i.e., in one of eight directions, following the lines indicated in Figure 4). A piece is shifted as far as possible, but it cannot jump over any piece or occupy an already occupied point. We indicate moves by an origin-destination notation: for example, a4-a2 indicates moving the upper left white piece downwards in the opening position. There are no captures in Dao and passes are not allowed. The goal of the game is to be the first to reach one of the following five winning configurations:

I: all your pieces form one vertical line,
II: all your pieces form one horizontal line,
III: your pieces form a 2 × 2 square anywhere on the board,
IV: your pieces occupy the four corners of the board,
V: one of your pieces is enclosed by three opponent pieces in one of the corners of the board.

The last winning configuration differs from the other four configurations since it depends on an opponent's move. The outcome of a Dao game is either a loss or a win; a tie is not defined. There is no rule to prevent move repetition.

4 Dao Properties

The shortest possible game of Dao takes 5 ply: [1: b3-a3, b2-b4; 2: c2-a2, a1-b2; 3: d1-a1]. Of course, this game is based on a serious mistake by Black. If both players play optimally, a Dao game will never end, as we will see in Section 6. In the following subsections we will discuss five properties of the Dao game graph: the number of configurations (4.1), the game-graph statistics (4.2), the in-degree and out-degree (4.3), the eccentricity (4.4), and the cycles (4.5).

4.1 Number of Configurations

The computation of the number of possible Dao board configurations (White to move) is straightforward: 4 white pieces choose 4 squares out of 16; 4 black pieces choose 4 squares out of the remaining 12, resulting in $\binom{16}{4} \times \binom{12}{4} = 900,900$. This number does not account for board symmetries (one horizontal, one vertical, and two diagonal flips, plus three rotations). The configurations are partitioned into 113,028 symmetry-equivalence classes with 7.97 configurations per class on average. If colour-switch would also be taken along as a symmetry, the number of equivalence classes becomes 56,757 (15.87 configurations per class on average). In the sequel we only consider the 113,028 symmetry-equivalence classes, to which we will refer as 'positions'. We constructed a Dao game graph (see Subsection 6.1 for details) analogous to the Do-Guti game graph in Figure 2. Figure 5 shows a part of the game graph consisting of the opening position and its direct neighbours.

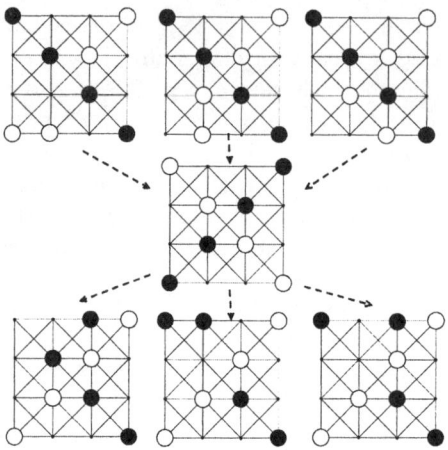

Fig. 5. Part of the Dao game graph: the opening position and its neighbours. All positions are WTM.

4.2 Game-Graph Statistics

From the game graph of Dao the following statistics can be derived. Within the 113,028 positions there are 4,277 terminal positions and 108,751 non-terminal positions. The terminal positions consist of 2,154 positions that constitute a win for White and 2,123 positions that constitute a loss for White. Not all terminal positions are reachable during play. For instance, the position in which the a-file contains all four white stones and the b-file contains all black stones is not reachable since the game is already terminated when the a-file is completed. In total 1,024 of the won positions and 1,135 of the lost positions are not reachable. This makes the number of reachable terminal positions: 1,130 won and 988 lost ones. All reachable terminal positions are reachable from the opening position.

There are only 2 non-terminal positions that are not reachable (i.e., their nodes have in-degree 0). These positions are represented in Figure 6. By inspection of these positions, it is obvious that they are not reachable: none of the black stones could have reached its current spot, at least not starting from a non-terminal position. This is due to the rules that a piece is shifted as far as possible (both positions in Figure 6) and that a position is won if four pieces form a square (right position). All reachable non-terminal positions are reachable from the opening position (and the opening position is reachable from all non-terminal positions). This means that it is possible to create a sequence of moves that visits all 108,749 non-terminal, reachable positions. (This is different from games such as Tic-Tac-Toe, Checkers, or Kalah in which the opening position cannot be reached anymore.) From now on we disregard all non-reachable positions and concentrate on the 110,867 reachable positions.

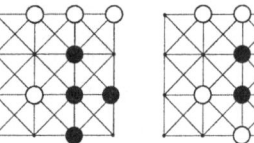

Fig. 6. The two non-reachable non-terminal Dao positions (WTM)

Table 1. Distribution of in-degree and out-degree in the Dao game graph

d	in-degr.	out-degr.	d	in-degr.	out-degr.	d	in-degr.	out-degr.
0	0	2,118	9	12,243	11,263	18	1,451	83
1	41	0	10	13,759	15,482	19	763	13
2	112	18	11	13,663	18,305	20	307	0
3	380	97	12	12,704	17,962	21	110	0
4	1,002	302	13	10,480	14,635	22	35	0
5	2,273	780	14	8,207	9,798	23	12	0
6	3,931	1,798	15	6,005	5,141	24	9	0
7	6,970	3,612	16	4,104	2,096	25	0	0
8	9,746	6,803	17	2,559	561	26	1	0

4.3 In-Degree and Out-Degree

Table 1 gives the in-degree and out-degree distribution in the (reachable part of the) game graph; d stands for degree, it ranges from 1 to 26. The in-degree is the number of arrows leading into a position, the out-degree is the number of arrows leading away from a position. The in-degree and out-degree of a position can differ. The 2,161 nodes with in-degree zero are excluded from the graph in this count since these nodes relate only to non-reachable positons. The average in-degree of a position is 10.99, and, necessarily, the average out-degree of a position is 10.99 too. However, the spread of in-degrees is larger than that of out-degrees throughout the graph. The average in-degree for terminal positions alone is 8.58. The average branching factor of the game tree might be expected to be larger than the average out-degree in the game graph since moves can lead to different but equivalent configurations. However, the average branching factor of the game tree in Dao happens to be 10.83.

Particularly interesting is the fact that there are 41 nodes with in-degree 1. The edges leading to these nodes could be edges that separate parts of the graph similar to the separating edge in the graph of Do-Guti (see Figure 3). However, a graph scan reveals that this is not the case in Dao: none of these edges separates the game graph. Another interesting observation is that there are no nodes with out-degree 1, which means that there are no positions with forced moves.

4.4 Eccentricity

Another metric that characterizes the Dao game graph is the distribution of the shortest distance from the start position to all nodes in the graph. The longest of these shortest distances is called the *eccentricity* of the start position. The distances can be measured in two ways: (1) by taking the minimal distance between nodes in the plain game graph (only White to move and colour switches at the edges), and (2) by transforming the game graph into a *bipartite graph* of separate positions with White to move (WTM) and with Black to move (BTM) and taking the minimal distance in this bipartite graph. The average shortest distance is 7.99 in the first case and 8.49 in the bipartite case. The longest of

Table 2. Distribution of the shortest distance from the opening position in the Dao game graph. Counts for BTM positions in the bipartite graph are marked with an *. The fourth column gives the number of positions that are reachable during play within the distance.

dist.	Plain	Bipartite	Reachable
1	3	*3	3
2	35	35	38
3	253	*288	291
4	1,800	2,056	2,091
5	6,429	*8,230	8,521
6	23,792	30,221	32,313
7	39,531	*63,323	71,844
8	35,636	75,167	107,480
9	3,313	*38,949	110,793
10	74	3,387	110,867
11	0	*74	110,867

all shortest distances are 10 and 11, respectively (see Table 2). The counts of the BTM and of the WTM positions in the bipartite graph both add up to 110.867. This means that from the start position (WTM), every position can be reached with Black to move at that position as well as with White to move. Table 2 also provides the number of positions that are reachable during play within the given number of plies. The entries of this column are almost equal to the cumulative sum of the entries in column 2. However, the opening position itself can be reached again after 4 ply, which is included in column 4 but not in the columns 2 and 3.

4.5 Cycles

The number and size of cycles that exist in game graphs determine the amount of transpositions that appear in a search tree. The more and the smaller the cycles, the more transpositions will occur during search. We recorded for every reachable non-terminal position in the Dao game graph the *smallest* directed cycle in which that position takes part. Note that positions represent equivalence classes of configurations: when a position is re-entered after following a cycle, the actual configuration might have changed.

There are no *self-loops* (cycles of length 1), but there are 1,888 positions on directed cycles of length 2. These cycles form twin-positions between which the players can alternate infinitely (after your move and the opponent's move, you are back in the same – or an equivalent – position). Since there are no positions with out-degree 1, these cycles are no traps from which the opponent cannot escape. The cycles are not connected, so their number is 944. Of the remaining positions, 3,322 lie on directed cycles of length 3, and all other 103,539 reachable non-terminal positions lie on directed cycles of length 4. (In passing we note that the opening position of Do-Guti, see Figure 2, lies on a directed cycle of length 6.) When returning in a position after following a cycle of length 3, sides are switched. So, there are two cycles (six ply) needed to get back to the original position, unless the node also lies on a cycle of length 4.

The fact that the smallest cycle for each position is at most 4 ply long means that transpositions in Dao are abundant. Any search tree of 4 ply depth or more will encounter many of them, so a transposition table is expected to be quite effective in Dao.

5 Random Games

In addition to the numerical and combinatorial approaches of the game graph in the previous section, we played a series of random games. We hoped that the distribution of the length of the random games revealed some additional information on the structure of the game graph. A random game is basically a random walk on a bipartite graph [9], starting at the opening position and ending when a terminal position is reached. However, we introduced a small difference by selecting at every position one of the legal moves with uniform probability. This means that in the game graph, edges that represent more than one move obtain a higher probability than edges representing a single move.

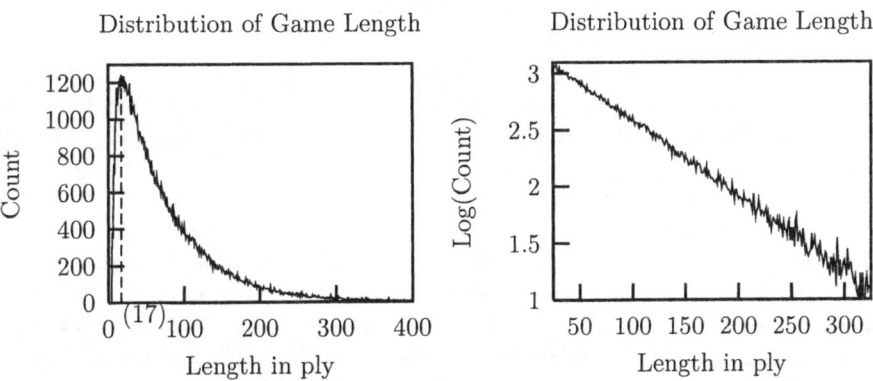

Fig. 7. Distribution of the length of randomly played Dao games. The left figure is a normal plot, the right figure is a log-plot of the same data.

Figure 7 shows the result of 100,000 random games. Obviously no game shorter than 5 ply occurred. The longest game encountered took 800 ply (not in the figure). The maximum of the histogram is at 17 ply; the average game length is 74.8 ply. The log-plot in Figure 7 suggests that the tail of the distribution (starting at ply 25) is negative exponential. A linear regression on the log-plot provides an exponent of -6.66×10^{-3}. This means that from ply 25 on, there is a constant probability that a randomly played game comes to an end. It indicates that a stationary distribution over the graph is reached fast, which may be caused by the small diameter (11) of the graph.

6 Solving Dao

The final part of our Dao investigations is strongly solving [1] the game. By strongly solving we mean determining for every node in the graph whether the player to move can force a win, whether the other player can force a win (i.e., forcing the player to move into a lost position), or whether none of the players can force a win (i.e., a drawn position). The size of the game graph (113,028 nodes) makes it possible to keep it in a PC's internal memory, next to a series of bookkeeping arrays. It enables us also to find the strong solution of the game in a very short period of time. With the current means, the game graph is built and the complete solution is arrived at within three seconds (on a AMD Athlon 1.3 Ghz, 256 Mb PC). Below we first describe how the game graph is constructed (6.1) and then discuss the solving method and its results (6.2).

6.1 Building the Game Graph

Building the game graph was done in three steps. The first step was developing a ranking function by using standard combinatorics, that maps Dao configurations one-to-one to non-negative integers smaller than 900,900. For this mapping, we represented each configuration by two sets of four numbers, indicating (in increasing order) the points that were occupied by the four white pieces and the four black pieces, respectively. Then we used a combination of two colex orderings for increasing functions [15], one for the white pieces and one for the black pieces, to produce the unique rank number. Simultaneously, we developed an unranking function that translates each rank number back to a Dao configuration.

The second step was the determination of the equivalence classes (which we equate with positions); they are the nodes of the game graph. To determine the classes we created and initialized an array SymClasses[] of 900,900 entries and filled it as follows. We enumerated all configurations and determined per configuration the set of equivalent configurations that forms the equivalence class. To achieve this we applied the seven symmetry transformations mentioned in Subsection 4.1. We checked for every configuration in the class whether that configuration was already labelled. If so, all configurations received that same label in SymClasses[], else a new class-label was generated and the configurations were labelled with that new label. This process resulted in 113,028 equivalence classes.

The array SymClasses[] acts as a function that maps configurations to equivalence classes. We also created an inverse function that mapped an equivalence class to one of the configurations in it. This function was realized by an array SymRefs[] containing the rank of a reference configuration in each of the 113,028 equivalence classes.

The third step was the construction of the edge list. Since the average degree of the graph is very low in relation to the number of nodes (0.02 %), the edge list is the most efficient data structure for the graph. We decided to represent the edge list not by linked lists, but by two fixed arrays of 26 (incoming) and 19 (outgoing) entries for all of the 113,028 classes. For every class, we took its reference configuration, generated all legal moves, and determined the equivalence class of each resulting configuration. The class numbers were stored in two appropriate edge lists.

6.2 Retrograde Analysis

After building the game graph, (strongly) solving the game was straightforward. We applied retrograde analysis [13,8,12] to the game graph, which essentially is an application of dynamic programming. We generated an array SolLabel[] and initialized the labels of the terminal positions. The next step was to enumerate all unlabelled positions that had a labelled successor position and to inspect the rest of its successor positions. If all successor positions were labelled as a win, the position was labelled as loss (since a colour switch takes place at each edge). If one of the successor positions was labelled as a loss, the position was labelled a win.

The labelling was performed in two stages during each step: (1) a temporary label was assigned to each newly labelled position, and (2) after all assignments, the temporary labels were converted into definitive labels. The process was repeated until no new positions could be labelled anymore. Moreover, the process guaranteed the detection of the shortest distance to win.

The procedure started with labelling the 2,154 won terminal positions and the 2,123 lost terminal positions. After 11 rounds, the procedure ended with 21,052 positions labelled as forced wins for the player to move and 2,860 as losses (see Table 3). The opening position of Dao was left unlabelled. This implied that a Dao game cannot be forced to an end (either a win or a loss), when both sides play optimally, and therefore will continue forever, again assuming optimal play. One could say that Dao is a drawn game, although that outcome is not defined in the rules. Like Do-Guti, the game can only be won if one of the players makes a mistake. Obviously, the 'trick' of Dao is much more concealed than the 'trick' of Do-Guti and therefore, the game is still challenging for human beings, despite the drawn solution.

Table 3. Number of wins and losses added during the subsequent steps of the retrograde analysis. Step 0 represents the labelling of the reachable terminal positions.

Step	Wins added	Losses added
0	1,130	988
1	9,950	0
2	0	1,407
3	8,087	0
4	0	357
5	1,480	0
6	0	86
7	324	0
8	0	20
9	75	0
10	0	2
11	6	0
total	21,052	2,860

The fact that no losses are encountered in step 1 indicates that there is no position in which the player to move, irrespective of the move, loses directly. One can lose the game only in a single move if one makes a move that results in configuration V (cf. Section 3). The player cannot be forced to play such a move: there is always an alternative move available. The absence of losses-in-1 explains the zeros in Table 3.

Table 3 also shows that there are 6 positions with a maximal depth (11) of forced win and 2 positions with a maximal depth (10) of forced loss. Figure 8 shows two examples of each. A sequence of optimal moves, starting from position B is: [1: c4-c3, d3-c4; 2: b4-a4,

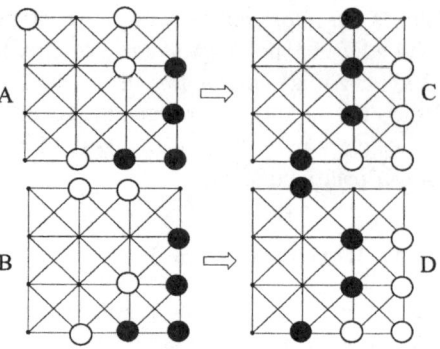

Fig. 8. Two examples of WTM Dao positions with the maximal distance of 11 ply to a forced win (A and B) and two WTM examples with a maximal distance of 10 ply to a forced loss (C and D). The arrows indicate optimal moves from A leading to C and from B leading to D.

c4-a2; 3: b1-b4, a2-a3; 4: a4-b3, a3-a1; 5: c2-d3, c1-b1; 6: d3-c4]. It is remarkable that the length of the longest forced win is equal to the eccentricity of the opening position (11). It is not clear whether there is a relation between both.

7 Experiment: The Effect of Transposition-Table Size

The usefulness of the Dao game is in its benchmarking character. For the investigation of game-tree search methods, we performed an example experiment concerning the effect of the transposition-table size. The aim of the experiment was to find out how the size of the transposition table relates to the portion of the game graph that is investigated. From practice it is clear that a transposition table can be used to reduce the search tree. Breuker [5,4] performed a series of experiments in which he quantified this effect for a number of replacement schemes.

Before we discuss the effect of a transposition table, we provide some figures of plain Minimax search on Dao, starting from the opening position. During the search we recorded the number of nodes visited in the game graph and counted how often each node was visited. We recorded four types of events: (1) the visit, (2) whether the visit took place with White next to move, or (3) with Black next to move, and (4) whether the position was evaluated during the visit. Furthermore, we counted (5) the number of leaf nodes in the search tree. Table 4 gives the results of these counts. The table makes it clear that in Minimax search, nodes in the game graph are visited many times without using the knowledge from previous visits. We use a similar type of recording in the experiments below.

Table 4. Minimax search in Dao. The first column contains the search depth d in ply. The next four pairs of columns give the number of nodes in the game graph visited and the average number of visits per node. The numbers are given for four events: (1) visit, (2) visit with WTM next, (3) visit with BTM next, and (4) evaluation. The last column (5) presents the number of leaf nodes in the search tree.

d	1. All Visits Nodes	Avg.#	2. WTM Nodes	Avg.#	3. BTM Nodes	Avg.#	4. Evaluations Nodes	Avg.#	5. Leaf Nodes
1	3	4.0	0	0	3	4.0	3	4.0	12
2	38	4.0	35	4.0	3	4.0	35	4.0	140
3	294	6.3	35	4.0	291	5.8	288	5.9	1,688
4	2,121	10.2	2,092	9.6	291	5.8	2,092	9.5	19,888
5	8,570	28.9	2,092	9.6	8,521	26.7	8,521	26.5	225,664
6	32,562	85.0	32,313	78.6	8,521	26.7	32,374	77.8	2,520,140
7	72,166	424.1	32,313	78.6	71,844	390.6	72,076	386.3	27,845,896
8	108,861	3,077.0	107,480	2,855.5	71,844	390.6	108,789	2,799.6	304,568,204

7.1 Experimental Set-Up

In the experiments, we performed an 8-ply α-β search on the opening position of Dao. The search was enhanced with iterative deepening and a transposition table. The transposition table was used to store the game score, a flag, and the best move of a position encountered. We used the replacement scheme called *deep*: positions that are searched more deeply replace positions that have been searched less deeply [4]. The search algorithm was not provided with move ordering, killer move, history heuristic, or any other enhancement except the transposition table. The transposition table was also used to put the best move, according to the table, in front at each position.

Before we could apply α-β search in a way that resembles practice, we needed a reasonable static heuristic evaluation function for Dao. This evaluation function was constructed *ad hoc* as follows: we took the number of winning configurations I – IV (cf. Section 3) that had three pieces in common with the position, and multiplied it with 100. Then we added 200 if the position was a win. Finally, we added a uniformly distributed pseudo-random variable with a range of [−50, 50]. We pre-computed the evaluation function for all positions and stored them in an array. For the transposition tables we had to generate a hash code for all positions. We generated one pseudo-random long integer for every position and stored all of them in an array; they are used during the search. (Most practical hash functions, such as the Zobrist approach, do not take into account the symmetry equivalence classes of the positions.) An additional fixed random number was used to indicate the player to move. The same evaluation function and hash code were used in all experiments below.

7.2 Results

The Tables 5 and 6, and the Figures 9 and 12 show the results for transposition-table sizes of 0 to 2^{21} entries. The size is indicated by the number of bits per entry,

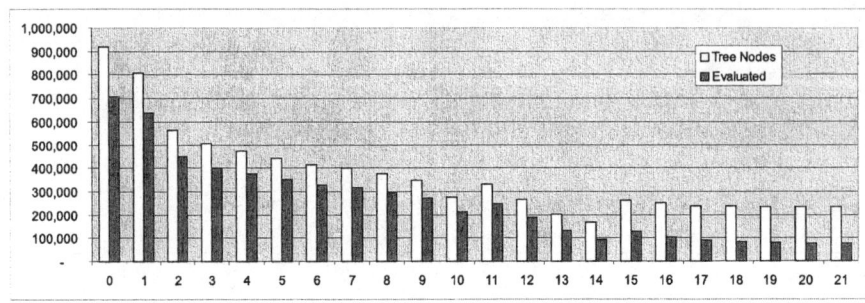

Fig. 9. Number of nodes in the search tree (number of graph nodes times the visits per node) and the number of evaluated leaf nodes (see Table 5). On the x-axis, the size of the transposition table in bits is represented.

Table 5. Effect of the transposition table on search. The first column contains the size of the transposition table in bits (size $= 2^{bits}$). The other columns have the same meaning as in Table 4. The last column only counts the leaf nodes of which the values are computed, not the ones that have been retrieved from the transposition table.

Bits	1. All Visits Nodes Avg.#		2. WTM Nodes Avg.#		3. BTM Nodes Avg.#		4. Evaluations Nodes Avg.#		5. Leaf Nodes
0	71,215	12.90	66,855	10.99	36,634	5.01	66,828	10.59	707,547
1	66,668	12.13	59,507	8.89	39,465	7.08	64,291	9.92	637,700
2	52,512	10.77	45,512	8.14	30,102	6.48	49,899	8.99	448,578
3	51,900	9.74	45,060	7.42	29,126	5.87	49,318	7.66	377,864
5	52,459	8.41	45,388	6.25	29,492	5.34	49,825	7.04	350,719
6	53,589	7.76	46,691	5.85	29,366	4.86	50,869	6.47	329,181
7	53,612	7.49	46,771	5.66	29,275	4.67	50,833	6.22	316,384
8	53,823	7.00	46,963	5.31	29,378	4.33	50,976	5.79	295,178
9	54,024	6.47	47,101	4.96	29,421	3.94	51,059	5.32	271,837
10	51,878	5.31	44,526	4.01	28,436	3.41	49,128	4.31	211,906
11	56,010	5.93	49,217	4.56	30,685	3.52	52,995	4.65	246,217
12	52,628	5.05	45,579	3.79	29,289	3.18	50,038	3.77	188,593
13	47,696	4.24	39,935	3.08	27,420	2.88	45,327	2.89	130,846
14	43,045	3.90	35,095	2.80	25,566	2.73	40,858	2.31	94,491
15	57,192	4.58	50,301	3.40	31,656	2.88	53,609	2.41	129,294
16	56,986	4.43	50,266	3.26	31,170	2.83	53,417	1.99	106,065
17	56,044	4.24	48,960	3.08	30,967	2.80	52,504	1.70	89,517
18	55,980	4.20	48,745	3.06	31,030	2.79	52,374	1.58	82,776
19	55,865	4.18	48,665	3.03	30,971	2.78	52,272	1.51	78,859
20	55,963	4.17	48,747	3.03	31,013	2.77	52,352	1.48	77,244
21	55,910	4.17	48,702	3.02	30,983	2.77	52,302	1.46	76,344

which means that the size is 2^{bits}. With size 0 (i.e., no transposition table), no iterative deepening was used. Table 5 shows the effect of the transposition table on the search by indicating how often nodes in the game graph are visited and how many leaves of the search tree are evaluated. In Figure 9, we present the size of the search tree and the number of nodes in the tree that have been evaluated. Table 6 shows the usage of the transposition table. The first two columns indicate the size ($= 2^{bits}$) of the transposition table. The next four columns indicate the total number of look-ups in the table, the number of times the look-up resulted in a change of the search window, in a cut-off (hit), and the number of times the look-up had no effect. Collisions (i.e., different positions with the same hash value) are not included in the look-up counts. The last three columns indicate the number of filled entries (at the end of the search), its proportion of the table size, and its relation to the number of visited nodes. In Figure 12, some of the results of Table 6 are presented in relative proportions.

From these results we derive five applications for Dao as a benchmark game. The first application is the comparison between Minimax and α-β. It appears that even in the case of no transposition table, α-β is quite efficient in comparison with Minimax search. α-β needs only 0.23 per cent of the number of evaluations that Minimax needs. In terms of the number of nodes visited the difference is still significant but less profound: α-β visits 71,215 nodes of the 108,861 nodes that Minimax visits at the same search depth (according to Table 4). This means that α-β still needs 70 per cent of the information in the reachable region of the game graph to determine the (heuristic) game value of the start position. This is a rather large number in comparison to the massive pruning that takes place in the search tree.

The second application is the comparison between different sizes of the transposition table with respect to the number of nodes visited. When a transposition table is used, and thus some move ordering is provided, the number of visited nodes is lower (around 55,000). However, the number of visited nodes does not change much with the size of the transposition table. The number of nodes visited at size 2 bits is even smaller than the number at 21 bits.

The third application is the comparison of the number of visits per node: in Minimax, the average number of visits per node is more than 3,000 at search depth 8, whereas in α-β, the average number of visits per node is around 12 without a transposition table. The number of visits per node decreases even more and becomes less than 5 when the size of the transposition table grows larger than 12 bits.

The fourth application is the comparison of the number of leaf evaluations. In the experimental setting, leaf nodes are stored in the transposition table as well. On top of the reduction by 430 with respect to Minimax, the transposition tables add a further reduction of a factor 10. As is visible in Figure 9, roughly 80% of the nodes in the search tree is evaluated for table sizes up to 13 bits. For table sizes 15 and larger, the percentage of evaluated nodes is only 30%. The local maxima that appear at size 11 and 15 in Figure 9 can be explained by accidental configurations in the game tree and transposition table. Figure 10 may explain this. Assume that position A and B in the figure share the b least significant bits of their hash codes, but differ at bit $b+1$. Assume further that we use a transposition table of size b and the replacement scheme *deep* [5]. The information that is stored in the table after the visit of the right-most position A (value 9) will be overwritten by information on position B, due to the replacement scheme and the common hash entry.

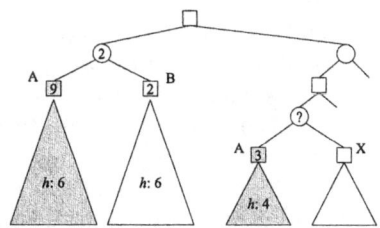

Fig. 10. An example tree that shows how a larger transposition table can lead to less pruning

When position A is visited again in the tree (a true transposition), no information is available in the transposition table, so the subtree under the node is searched, resulting this time in value 3. This value is used as β-value in the subtree under position X in order to prune.

Now assume that the size of the transposition table is increased to $b+1$ bits. If the tree is searched in this case, both the information on position A and on position B will be stored in the transposition table. When the second occurrence of A is encountered, a value of 9 is found in the table and the subtree under A is not searched. This means that in the subtree under X value 9 is used as β-value, which is higher than 3 and therefore might lead to less pruning.

Similar effects can occur with other replacement schemes although the necessary configurations might differ from the example in Figure 10. The effect disappears when averaging over multiple experiments with different hash codes. Figure 11 shows the average over 100 samples. In each sample, a different hash function was generated and used for all table sizes.

A particular interesting application of Dao as a benchmark game is the fifth application, viz. the comparison of the usage of the transposition table. The usage numbers in Table 6 indicate that there appears to be a discontinuity between sizes 14 and 15. At sizes of 14 and larger, the transposition tables have unused space. The number of look-ups that lead to a change in the α-β window is maximal at size 15 and decreases slowly for larger sizes. However, the relative proportions as presented in Figure 12 do not show the same discontinuity.

Based on the results of the five benchmark applications, we suggest that in this setting the size of a transposition table should be selected at that point where the table just starts to have some unused space. At the same point, the number

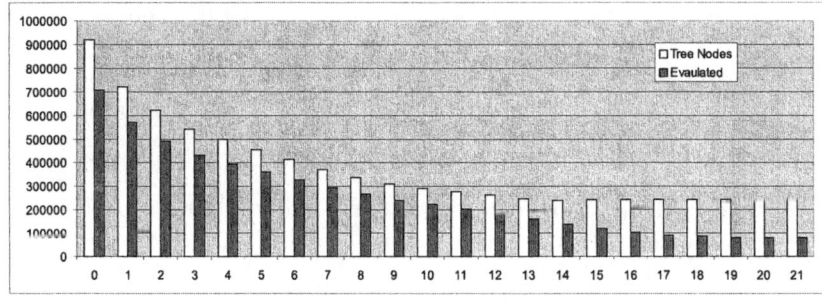

Fig. 11. Number of nodes in the search tree (number of graph nodes times the visits per node) and the number of evaluated leaf nodes, averaged over 100 samples, each generated with a different random hash function. On the x-axis, the size of the transposition table in bits is represented.

Table 6. Usage of the transposition table. The first two columns indicate the size (= 2^{bits}) of the transposition table. The next four columns indicate the total number of look-ups in the table, the number of times the look-up resulted in a change of the search window, in a cut-off (hit) and the number of times the look-up had no effect. The last three columns indicate the number of filled entries (at the end of the search), its proportion of the table size, and its relation to the number of visited nodes.

Bits	Size	Look-up	Change	Hit	Nothing	Filled	(prop.)	Nodes
0	0	0	0	0	0	0	100%	0.00%
1	2	377	8	363	6	2	100%	0.00%
2	4	463	54	401	8	4	100%	0.01%
3	8	842	105	720	17	8	100%	0.01%
4	16	1,428	225	1,177	26	16	100%	0.02%
5	32	2,205	432	1,729	44	32	100%	0.04%
6	64	3,203	734	2,385	84	64	100%	0.08%
7	128	5,118	1,263	3,710	145	128	100%	0.17%
8	256	7,495	1,961	5,274	260	256	100%	0.34%
9	512	10,241	2,661	7,147	433	512	100%	0.67%
10	1,024	14,346	3,607	9,951	788	1,024	100%	1.40%
11	2,048	27,705	6,883	19,400	1,422	2,048	100%	2.56%
12	4,096	36,248	7,668	26,099	2,481	4,096	100%	5.47%
13	8,192	45,368	7,496	33,665	4,207	8,191	100%	12.16%
14	16,384	57,521	7,399	43,254	6,868	15,960	97%	26.31%
15	32,768	110,488	14,041	84,567	11,880	29,986	92%	36.59%
16	65,536	130,651	14,031	100,515	16,105	46,206	71%	56.74%
17	131,072	137,284	13,089	105,264	18,931	59,335	45%	74.24%
18	262,144	144,720	13,271	110,552	20,897	68,044	26%	85.29%
19	524,288	148,071	13,170	112,951	21,950	73,063	14%	91.75%
20	1,048,576	150,367	13,245	114,620	22,502	75,786	7%	95.02%
21	2,097,152	151,220	13,248	115,199	22,773	77,140	4%	96.81%

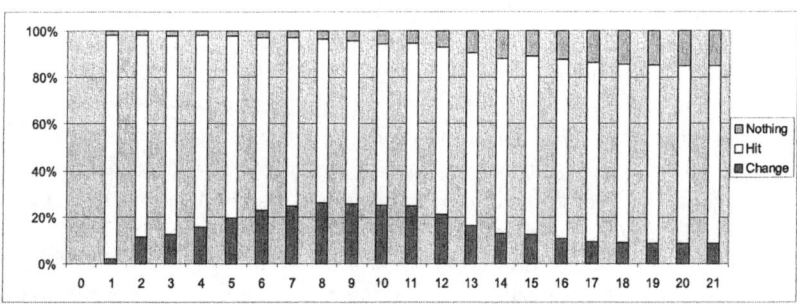

Fig. 12. Relative distribution of the three outcomes of a transposition-table look-up (see Table 6). On the x-axis, the size of the transposition table in bits is represented.

of evaluations in relation to the search-tree size drops. Any further increment of the size seems not to have a significant effect. Both criteria (size and starting point of unused space) are measurable during the search itself, they do not need knowledge of the game graph.

8 Conclusions

In this paper we presented two small games that are played by humans: the traditional game Do-Guti and the modern game Dao. The first game is interesting to *explain* some of the concepts of games since its game graph fits into a single illustration. The second game is interesting as a *benchmark game* since its game graph fits into internal memory of a computer but is large enough to allow aggregated analysis.

We showed that both games are drawn games. In the case of Do-Guti, this is directly visible from the game graph. In the case of Dao, it is proven by a retrograde analysis. In passing we note that the game graph of Dao seems to be governed by the number 11: it is the average in-degree and out-degree of the nodes, it is the eccentricity of the start position, and it is the longest optimal win. Whether this fact has significant meaning for other games is subject of further research.

The example experiment in this paper showed the advantage of having a complete game graph of a non-trivial game available for benchmarking. It was possible to pinpoint exactly how much of the actual game graph was used during the search. The α-β search needed only 70 per cent of the information in the game graph without transposition table, and 50 per cent with a table. The size of the table did not matter that much as long as it had at least 4 entries. The experiment hinted at a possible method to determine an optimal size for the transposition table. Of course, the hypothesis should be tested on more domains before it is applicable.

We just tested the influence of a single factor: the size of a transposition table. It is obvious that the game of Dao can be used to test many other factors. From these observations we may conclude that Dao can splendidly serve as a benchmark for all kinds of search algorithms and their enhancements. It can also act as a benchmark for machine-learning techniques in games. In order to stimulate its use, we provide the Java source code at our website [6].

The particular properties of Dao (it is a fully connected, very loopy, drawn game) make it a prototype for some domains but unsuited for others. For instance, the game is not suitable for the investigation of deep searches or large branching factors. Therefore, it is worthwhile to find games of the same overall size with different properties. This could lead to a set of benchmark games, similar to the benchmark databases that are popular in the field of machine learning (e.g., the UCI repository [3]). Examples of such games would be some of the Chess endgames [13,12], small instances of Kalah [8], or some of the many solved games (see the overview in [14]).

Acknowledgement

We thank the audience at the Computer and Games Conference 2004 for the constructive remarks making it possible to improve this paper.

References

1. L.V. Allis. *Searching for Solutions in Games and Artificial Intelligence.* PhD thesis, Rijksuniversiteit Limburg, Maastricht, The Netherlands, 1994.
2. D.F. Beal and M.C. Smith. Quantification of search-extension benefits. *ICCA Journal*, 18(4):205–218, 1995.
3. C.L. Blake and C.J. Merz. UCI repository of machine learning databases. http://www.ics.uci.edu/~mlearn/mlrepository.html, 1998.
4. D.M. Breuker. *Memory versus Search in Games.* PhD thesis, Universiteit Maastricht, Maastricht, The Netherlands, 1998.
5. D.M. Breuker, J.W.H.M. Uiterwijk, and H.J. van den Herik. Replacement schemes for transposition tables. *ICCA Journal*, 17(4):183–193, 1994.
6. H.H.L.M. Donkers. Dao page. http://www.cs.unimaas.nl/ ~donkers/games/dao., 2003.
7. K. Handscomb. Dao (game review). *Abstract Games Magazine*, 6:7, 2001.
8. G. Irving, H.H.L.M. Donkers, and J.W.H.M. Uiterwijk. Solving Kalah. *ICCA Journal*, 23(3):139–148, 2000.
9. L. Lovász. Random walks on graphs: A survey. *Combinatorics: Paul Erdös is Eighty*, 2:1–46, 1993.
10. H.J.R. Murray. *A History of Board Games other than Chess.* Oxford University Press, Oxford, UK, 1952.
11. PlayDAO.com. Official Dao home page. http://www.playdao.com., 1999.
12. K. Thompson. Retrograde analysis of certain endgames. *ICCA Journal*, 9(3):131–139, 1986.
13. H.J. van den Herik and I.S. Herschberg. The construction of an omniscient endgame database. *ICCA Journal*, 8(3):141–149, 1985.
14. H.J. van den Herik, J.W.H.M. Uiterwijk, and J. van Rijswijck. Games solved: Now and in the future. *Artificial Intelligence Journal*, 134(1-2):277–311, 2002.
15. S.G. Williamson. *Combinatorics for Computer Science.* Computer Science Press, Rockville, MD, 1985.

Incremental Transpositions

Bernard Helmstetter and Tristan Cazenave

Laboratoire d'Intelligence Artificielle,
Université Paris 8, Saint-Denis, France
{bh, cazenave}@ai.univ-paris8.fr

Abstract. We introduce a distinction, in single-agent problems, between transpositions that are due to permutations of commutative moves and transpositions that are not. We show a simple modification of a depth-first search algorithm which can detect the transpositions of the first class without the use of a transposition table. It works by maintaining, for each node, a list of moves that are known to lead to transpositions. This algorithm is applied to two one-player games: a solitary card game called *Gaps*, and a game called *Morpion Solitaire*. We analyze, for each domain, how often transpositions are due to commutative moves. In one variant of Gaps, the algorithm enables to search more efficiently with a small transposition table. In Morpion Solitaire, a transposition table is not even needed. The best known sequence for this game is proved optimal for more than one hundred moves.

1 Introduction

In games, transpositions often arise because the same moves appear in two sequences but in different orders. For example, when two moves a and b are commutative, the sequences (a, b) and (b, a) lead to the same position. This is not the only possibility; for instance, if m and m^{-1} are inverse moves, then the sequence (m, m^{-1}) leads to the same position as the null sequence.

We state that we do not need a transposition table to detect transpositions of the first class. Instead, they can be detected by maintaining incrementally, for each node of the game tree, a list of the moves that are known to lead to transpositions.

We present an algorithm that implements this idea, called the incremental transpositions algorithm. When a search is made with a small transposition table, comparatively to the real size of the tree, this may result in a drastic increase of the number of nodes searched. We show that our algorithm can help attenuate this phenomenon.

The incremental transpositions algorithm is presented in Section 2. Section 3 presents its application to a solitary card game called *Gaps*; we analyze two variants of this game. In Section 4, we introduce a game called *Morpion Solitaire*, which is a perfect application domain since all transpositions can be detected without a transposition table. Section 5 contains our conclusion.

2 The Incremental Transpositions Algorithm

We present a modification of a depth-first search algorithm that recognizes all transpositions that are due to permutations of commutative moves (in 2.1), and we show that it is complete (in 2.2). Finally, we discuss the conditions for applicability of the algorithm (in 2.3).

2.1 Algorithm

The main function of the algorithm is called dfs_it (depth-first search, incremental transposition). The function takes as argument a set of moves T which are known to lead to transpositions, that is to say positions that have already been completely searched. The function is to be called at the root position with $T = \emptyset$.

function dfs_it(T) {
  for each legal move m, $m \notin T$,
    $T' = \{\ t \in T,\ t$ is legal after m, m is legal after t,
         t and m are commutative$\}$;
  do_move(m);
  dfs_it(T');
  undo_move(m);
  $T = T \cup \{m\}$;
}

The set T' is built so that it is restricted to moves that will be legal after the current move m; therefore, for all recursive calls to dfs_it, the argument set T will be a subset of the set of legal moves. Consequently, this algorithm needs only very little memory.

2.2 Proof of Completeness

We state that the incremental transpositions search algorithm is complete. We have to show that, assuming the set T given as parameter to the function dfs_it contains only moves which lead to positions that have already been completely searched, the same property holds for the recursive calls of dfs_it on the child nodes.

Let m be a move at the current node a and $t \in T$, such that t is legal after m, m is legal if played after t, and the moves t and m are commutative. This means that we have the commutative diagram shown in Figure 1. At node a, we know that move t

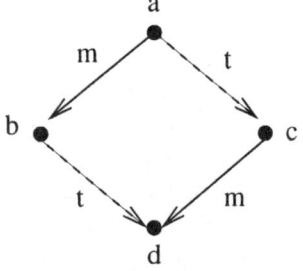

Fig. 1. Transposition

leads to a node, c, that has already been completely searched. Therefore node d, which can be reached from node b with move t, has also been completely searched. Since node d can also be reached from node c with move m, we conclude that move t at node b leads to a transposition.

2.3 Applicability of the Algorithm

The incremental transpositions algorithm is applicable whenever the game contains commutative moves. It is therefore broadly applicable. However, since it does not detect all transpositions, it should be viewed as a complement to a transposition table rather than a substitution to it. This being said, the question to assess is what can be gained from the algorithm.

First, what is the proportion of transpositions that are due to permutations of commutative moves? This question is very domain-dependent. For example, there are absolutely no commutative moves in the sliding-tile puzzle. In Rubik's cube, there are few commutative moves (only moves on opposite faces are commutative), so the algorithm would not be very useful. The two domains that we will study, Gaps and Morpion Solitaire, feature a large number of commutative moves; in the second domain, we will show that all transpositions are due to permutations of commutative moves. Moreover, we have also been able to apply the algorithm with some success in Sokoban.

Secondly, the incremental transpositions algorithm is only interesting for large searches, when a transposition table alone is not sufficient. In chess, experiments by D. Breuker [1] show that a transposition table can detect most transpositions even when the size of the transposition table is much less than the size of the search space. One of his experiments shows that, for a search that would take optimally about 100×10^6 nodes, the situation is already almost optimal with a transposition table of size 1M positions, and we lose less than a factor two when using a transposition table of size 8K. This situation, however, does not appear to be general. Our experiments with Gaps will show that the number of nodes searched explodes when the transposition table is smaller than the size of the search space.

Finally, our algorithm only works for a depth-first search in one player games. It would be difficult to apply it to two-player games, because it is not sufficient to know that a move leads to a transposition, it is also necessary to know the value of the position after the move. This value has to be stored in memory.

3 Gaps

In 3.1 and 3.2 we describe the rules of two variants (named *basic* and *common* of the game called *Gaps* (sometimes also called *Montana* or *Superpuzz*). In 3.3 we give examples of non-incremental transpositions for both variants. Subsequently, in 3.4 we give experimental results of the incremental transpositions algorithm for the basic variant. We compare three possibilities: transposition table alone, incremental transpositions alone, and both at the same time.

3.1 Rules

We start describing the variant we call the *basic* variant. The game is played with a 52-card deck. The cards are placed in 4 rows of 13 cards each. The 4 Aces are removed, resulting in 4 gaps in the position; then they are placed in a new column at the left in a fixed order (e.g., 1st row Spade, 2nd Heart, 3rd Diamond, 4th Club). The goal is to create ordered sequences of the same suit, from Ace to King, in each row. A move consists in moving a card to a gap, thus moving the gap where that card was. A gap can be filled only with the successor of the card on the left (that is, the card of the same colour and one higher in value), provided that there is no gap on the left and that the card on the left is not a King, in which case we can place no card in that gap. Figure 2 shows an initial position with only 4 cards per suit, before and after moving the Aces, and the possible moves.

Fig. 2. An initial position with 4x4 cards, before and after moving the Aces. This position can be won.

We will be concerned with another variant of the game which we call the *common* one, since it is the one most often played. In this variant, the Aces are not placed in a column on the left of the first column but are definitely removed. Instead, it is allowed to place any Two in a gap if it is on the first column. As we will see, this rule is the cause of many transpositions that are not due to permutations of moves.

3.2 Previous Work

T. Shintani has worked on a game he calls *Superpuzz*; it is actually another name of Gaps, precisely what we call the common variant [6]. He has analyzed the structure of the strongly connected components in the search graph. This is complementary to our research in this paper since, as we will see, the moves inside the strongly connected components are the cause of many transpositions that are not due to permutations of moves.

In [2], the game is shown to be decomposable in relatively independent subgames, and the dependencies between them are analyzed. An algorithm called *blocks-search* has been designed, which works by grouping several positions in one block and making a search on the blocks rather than on the positions alone. As a side effect (it was not the primary objective of the algorithm), it was observed that blocks-search could detect more transpositions when the size of the search space exceeds the size of the transposition table. The incremental transpositions algorithm presented here is simpler; it does not perform a simplification of the search space as blocks-search does, but it does a better job viz. by detecting the transpositions. Up to now, attempts to combine both algorithms have not been successful.

3.3 Examples of Non-incremental Transpositions

Below we provide two examples of transpositions that are not due to permutations of moves. Although there are other possibilities, it gives a rough idea of the structure of the transpositions in the two variants of Gaps.

The first example, shown in Figure 3, occurs in both variants. The sequences m_1, m_2 and m_3, m_1, m_4 give the same position. It may be possible to make the algorithm more efficient by adding a rule to detect automatically this pattern, and thus increase the number of transpositions detected without a transposition table.

The second example, shown in Figure 4 (a), is specific to the common variant. The position after moves m_1 and m_2 is the same as the position after move m_3. The position shown in Figure 4 (b), with many gaps in the first column and long ordered sequences on the right, features many similar transpositions. Therefore, the incremental transpositions algorithm is not well suited to these kinds of positions. Here again, it would probably be possible to add rules to detect those patterns, such as preventing the same card from moving twice in a row. Also, the methods developed by T. Shintani [6] seem to be particularly efficient precisely in these kinds of positions.

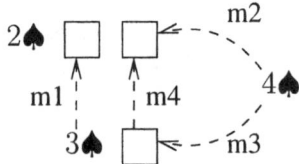

Fig. 3. Example of transposition occurring in both variants

Transpositions of the kind of the second example appear more often than those of the kind of the first example. Therefore, the proportion of transpositions that are not due to permutations of moves is larger for the common variant than for the basic one. The next section shows experimental results for the basic variant.

3.4 Experimental Results for the Basic Variant

The experimental results have been made for the basic variant with 52 cards. We are performing a complete search and we do not stop after a winning sequence has been found. For all the random positions tested it is possible that the game could be searched completely with a depth-first search and a transposition table. The transposition table has been implemented with a hash table of a fixed number of entries, and one position per entry. When collisions occur, the old position is replaced by the new. More complex replacement schemes with two positions per entry may be slightly more efficient [1].

First, we analyze how much we lose by using incremental transpositions alone, rather than combining it with a transposition table. We compare the number

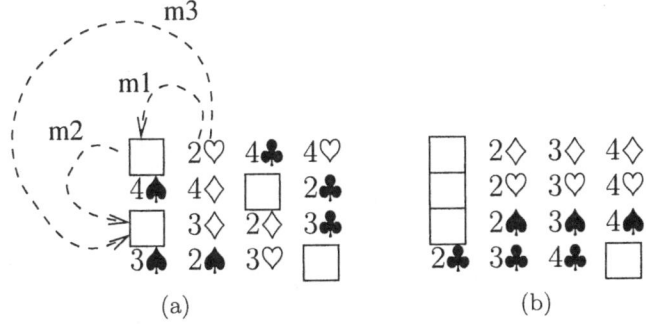

Fig. 4. Examples of transpositions occurring only in the common variant

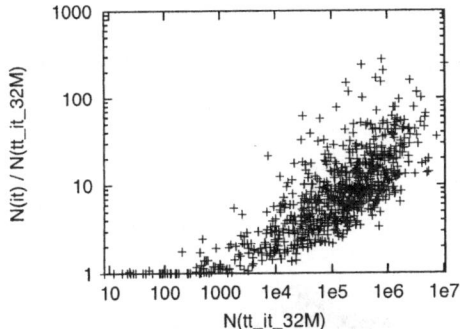

Fig. 5. Comparison between transposition table (32M entries) combined with incremental transpositions, and incremental transpositions alone

of nodes searched using a transposition table and incremental transpositions, $N_{tt\_it\_32M}$, with the number of nodes searched using incremental transpositions alone, N_{it}. Figure 5 shows the ratio $N_{it}/N_{tt\_it\_32M}$ depending on $N_{tt\_it\_32M}$, for 1000 random initial positions. Here the size of the transposition table is $2^{25} = 32M$ entries, more than the number of positions searched, so $N_{tt\_it\_32M}$ is very close to the real size of the search space when all transpositions are detected. The good point of this experiment is that we manage to make a complete search without a transposition table, although the cost in number of nodes relative to the optimal number of nodes is between a factor 1 and 1000.

Secondly, we show the usefulness of combining incremental transpositions with a transposition table of limited size. Here we work with a transposition table of a reduced size: $2^{20} = 1M$ entries. We compare the number $N_{tt\_it\_1M}$, with the number of nodes searched when using a transposition table alone, N_{tt_1M}. Figure 6 shows the ratio $N_{tt\_1M}/N_{tt\_it\_1M}$ depending on $N_{tt\_it\_1M}$, for 3200 random initial positions. This shows how useful the incremental transpositions algorithm can be, in conjunction with a transposition table. When the size of the search space exceeds the size of the transposition table, the number of positions searched when using only a transposition table explodes.

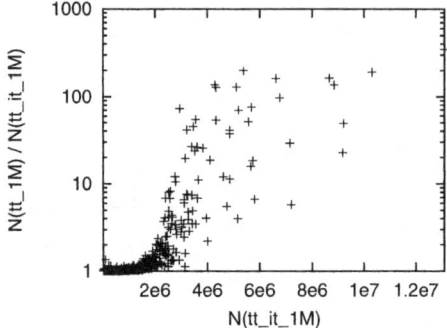

Fig. 6. Comparison between transposition table (1M entries) combined with incremental transpositions, and transposition table alone

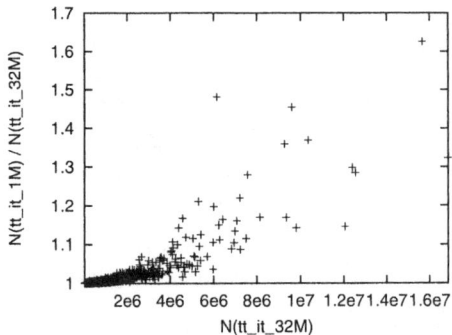

Fig. 7. Transposition table combined with incremental transpositions, with transposition table size 32M and 1M

In Figure 6 we also see the large increase of nodes searched when using a limited size transposition table. This is a loss compared to $N_{tt\_it\_1M}$, so *a fortiori* it is a loss compared to the optimal size. This increase is particularly sensitive for problems of size more than 2×10^6 nodes.

Our third experiment shows that this phenomenon is much weaker when the transposition table is combined with the incremental transpositions algorithm. We compare the number of nodes searched for two different sizes of the transposition table: 1M and 32M. We call these two numbers $N_{tt\_it\_1M}$ and $N_{tt\_it\_32M}$. Figure 7 shows the ratio $N_{tt\_it\_1M}/N_{tt\_it\_32M}$, depending on $N_{tt\_it\_32M}$, for 5000 random initial positions. The figure shows that, when the transposition table is backed up with the incremental transpositions algorithm, it is much less affected by the size of the search space.

The experiments have been run on a Pentium 3GHz with 1GB memory. Our implementation of the depth-first search, without a transposition table and incremental transpositions, can search 20.4×10^6 positions/s. This drops down to 9.5×10^6 positions/s with incremental transpositions, and 1.7×10^6 with a transposition table. With both, the search is actually faster than with a transposition table alone: 3.4×10^6 positions/s. The reason is that incremental transpositions are tested before accessing the transposition table; our implementation of the incremental transpositions algorithm is very well optimized and can detect transpositions faster than the transposition-table lookup.

4 Morpion Solitaire

Below we introduce a relatively new game called Morpion Solitaire. We show that, when doing a complete depth-first search of the game tree, using the incremental transpositions algorithm is sufficient to detect all transpositions, and therefore a transposition table is useless. Although we do not beat the current record, we are able to prove its optimality for more than one hundred moves.

4.1 Rules

Morpion Solitaire is a one player deterministic game, with a fixed starting position. General information about the game can be found in [5]. It is played using a paper with a grid layout, ideally infinite, and a pencil. The initial position, with 28 dots, is shown in Figure 8 (a). A dot can be added if it makes an alignment of five dots, horizontally, vertically, or diagonally, with the dots already drawn. The move consists in adding the dot and the segment of line, of length four, that goes through the five dots. Additionally, an alignment cannot be drawn if it shares more than one dot along the same axis as any previous line segment. Figure 8 (b) shows an example of a legal move. The goal is to maximize the number of moves.

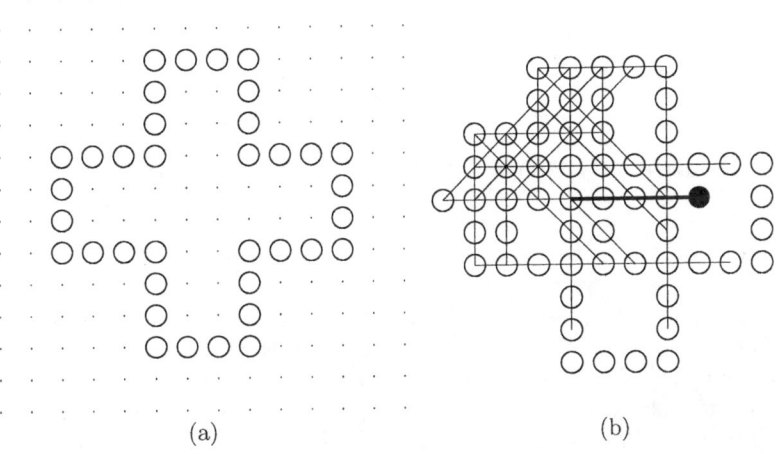

Fig. 8. Positions at the beginning and after a few moves

4.2 Bound on the Length of the Game

It has been proved that the length of the game is bounded. We explain the argument briefly. Each dot has 8 possibilities to be used for drawing an alignment, one for each direction. Some of them may have already been used. For each dot, the sum of the number of free possibilities is an invariant during the game. Indeed, when we make a move, this number is increased by 8 because we add a new dot, and decreased by 8 because of the segment of line drawn: the 3 dots inside the alignment lose 2 possibilities each , and the 2 dots at the extremities lose 1 each. The more dots are drawn, the more there are dots "on the boundary" (dots that are not completely surrounded by other dots). Those have free possibilities, so the length of the game is bounded. Achim Flammenkamp has proved a bound of 324 moves [5].

4.3 Finding Long Sequences: Previous Work

The current best published record for the game is 170; it has been obtained by hand, by Charles-Henri Bruneau, in the 1970s [5].

Hugues Juillé has applied a method which he calls Evolving Non Determinism to the game [3,4]. This algorithm works by making a population of prefix sequences evolve; the best ones are selected depending on the results of random games starting with the prefixes; the prefix sequences are gradually made longer until the end of the game is reached. The best result obtained by Juillé with this algorithm is 122. Pascal Zimmer has later improved the algorithm and found a sequence of length 147 [5].

Although this is not precisely the subject of this paper, we have developed a program to find good sequences. This has been done independently with Juillé's work, but the method is similar, since it also makes heavy use of statistics on random games. Our current record is 136.

In the following, our goal will be to determine whether the record, of length 170, is optimal in the endgame, starting from as as many moves as possible before the end.

4.4 All Transpositions Are Due to Permutations of Moves

We show that, in Morpion Solitaire, all transpositions are due to permutations of commutative moves. First we remark that, whenever two moves a and b are legal as well as the sequences (a, b) and (b, a), they are always commutative. Therefore, all we need to show is that, given a position, we can find the set of moves that have been made to reach it. This is done in two steps.

First, we group all the segments of a line drawn in segments of length four, each being the effect of only one move. We can do it separately for all the segments lying on the same infinite line, and in this case it is easy: we just have to group them four by four in order.

Secondly, we find the exact position where a dot has been added on each of the segments of a line just found. We start from the initial position, and, whenever a move is possible that corresponds to one of the segments of length four found, we make it. We lose nothing by making such a move as soon as it can be done. On the one hand, no other move will ever be possible using this segment of length four even after other moves have been made; on the other hand, making this move now or later does not change the possibilities for the other moves.

We have thus shown theoretically that the transpositions are due to permutations of moves; this is confirmed experimentally by the fact that, whether we make a depth-first search with incremental transpositions and a transposition table or incremental transpositions alone, the same number of nodes are searched.

4.5 Experimental Results

We have performed some experiments to test whether the best known sequence, of length 170, is optimal for the last moves. We have also made a similar search for the best sequence we have found in our own experiments, of length 136.

Since we do not need a transposition table, we have been able to parallelize the search. This search has run for about two months on the computers of our laboratory. Our program can search 3.7×10^6 positions/s on a Pentium 3GHz.

Figure 9 (a) shows the evolution of the total number of nodes searched depending of the depth of the starting position. Starting at move 61, a total of 3.97×10^{13} nodes have been searched, and the record has not been beaten. Figure 10 shows the positions after move 61 and after the last move for the best known sequence.

Figure 9 (b) shows the same work for the best sequence found in our own experiments (Subsection 4.3). We have put less effort on this one, but we have still been able to search it completely starting at move 42. This sequence has not been improved either.

The fact that the sequences considered in this section are optimal for so many moves is surprising, especially for the record of length 170 which has been obtained by hand. In our opinion, this indicates that the endgame is relatively

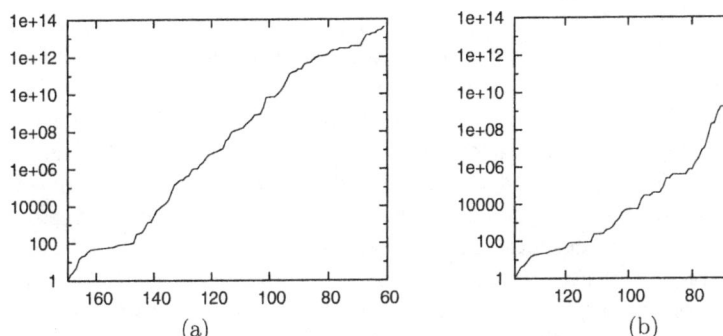

Fig. 9. number of positions searched depending on the starting move (a) in the best known sequence of length 170, (b) in a sequence of length 136

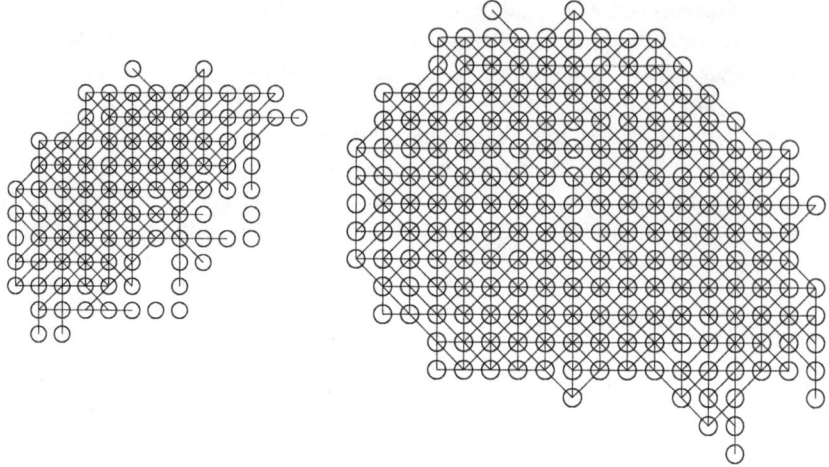

Fig. 10. Positions after move 61 and after move 170

easy. With the help of the methods described in Subsection 4.3, both human players and computers are able to play well.

5 Conclusion

We have introduced a distinction between transpositions that are due to permutations of moves and transpositions that are not. The first category of transpositions can be easily detected using a method we have called incremental transpositions. Unlike a transposition table, this method costs very little memory. It does not seem, however, to be applicable in two-player games.

The proportion of transpositions that are due to permutation of moves, and therefore have an impact on the efficiency of the algorithm, depends on the domain. In the basic variant of Gaps, the algorithm can be used in conjunction with a transposition table to detect more transpositions. Results are particularly good when the search space is larger than the size of the transposition table. The method would be less efficient in the common variant. Finally, we have shown a perfect application domain for the algorithm, Morpion Solitaire, where the incremental transpositions algorithm makes a transposition table useless.

In the domain of Morpion Solitaire, our parallel search program has proved the optimality of the best known sequence for more than one hundred moves before the end. We know that methods using statistics on random games are efficient in finding the best sequences in the endgame. Therefore, as further research, we plan to make use of statistics on random games to make cuts in the search, so that we can test the optimality for more moves before the end, and still be quite sure of the result.

References

1. Dennis M. Breuker. *Memory versus Search in Games*. PhD thesis, Maastricht University, 1998.
2. Bernard Helmstetter and Tristan Cazenave. Searching with analysis of dependencies in a solitaire card game. In H.J. van den Herik, H. Iida, and E.A. Heinz, editors, *Advances in Computer Games 10: Many Games, Many Challenges*, pages 343–360. Kluwer, 2003.
3. Hugues Juillé. Incremental co-evolution of organisms: A new approach for optimization and discovery of strategies. In *European Conference on Artificial Life*, pages 246–260, 1995.
4. Hugues Juillé. *Methods for Statistical Inference: Extending the Evolutionary Computation Paradigm*. PhD thesis, Brandeis University, 1999.
5. Jean-Charles Meyrignac. Morpion solitaire progress. http://euler.free.fr/morpion.htm.
6. Toshio Shintani. Consideration about state transition in Superpuzz. In *Information Processing Society of Japan Meeting*, pages 41–48, Shonan, Japan, 2000. (In Japanese).

Kayles on the Way to the Stars

Rudolf Fleischer[1] and Gerhard Trippen[2],*

[1] Shanghai Key Laboratory of Intelligent Information Processing,
Department of Computer Science and Engineering,
Fudan University, Shanghai, China
fleischer@acm.org
[2] Department of Computer Science,
The Hong Kong University of Science and Technology, Hong Kong
trippen@cs.ust.hk

Abstract. We present several new results on the impartial two-person game Kayles. The original version is played on a row of pins ("kayles"). We investigate variants of the game played on graphs. We solve a previously stated open problem in proving that determining the value of a game position needs only polynomial time in a star of bounded degree, and therefore finding the winning move - if one exists - can be done in linear time based on the data calculated before.

1 Introduction

Studying combinatorial games may show new interesting algorithmic challenges and may lead to precise problems in complexity theory. Games as models for a diverse set of practical problems attract researchers in mathematics and computer science. Combinatorial games are intrinsically beautiful with interesting features and a theory whose study also provides entertainment on its own.

Many relations between graph theory and combinatorial game theory exist. Like many other games that are being played on graphs, the versions of Kayles we are going to investigate are played on graphs too.

Kayles is a combinatorial game played by two persons who move alternately. Both players can make any move for all game positions. Therefore the game is impartial. On the contrary for example chess is partizan: White can only move white pieces, and Black can only move black pieces. All information about a position is known to both players at any time. There are no chance moves, i.e., no randomization generated by a dice for example is involved.

When playing Kayles on a graph choosing one vertex will remove this vertex and all its neighbors from the graph. Thus the game played on a finite graph is obviously of bounded play. In every move at least one vertex will be taken (this might disconnect the graph in several components), and therefore the game must terminate in time linear in the size of the initial position.

---

* The work described in this paper was partially supported by a grant from the Research Grants Council of the Hong Kong Special Administrative Region, China (Project No. HKUST6010/01E).

We only consider games of normal type, which means a player loses if he[1] is unable to make a move, opposed to the misere version where the last player to move loses.

All these features of the game make it possible to analyze it with the Sprague-Grundy theory [5,1]. Grundy has shown that there exists a function $\mathcal{G} : \mathcal{P} \to \mathcal{Z}$ from the set of possible game positions $\mathcal{P}$ into the set of integers $\mathcal{Z}$ with the following properties.

For any terminal position P we have $\mathcal{G}(P) = 0$. For any other position P, $\mathcal{G}(P)$ can always be calculated (inductively, starting from the terminal positions). Hereby a position P will take as its value the smallest non-negative integer, called the minimal excluded value (mex), different from all values of $\mathcal{G}(Q_i)$, where position Q_i can be reached from position P by one single move.

The value of a disjunctive combination of games is the *Nim-sum* of the values of the components. The Nim-sum is obtained as the sum of the binary representations of the values added together using the XOR (eXclusive OR) operation, denoted by the commonly used symbol $\oplus$.

If $\mathcal{G}(P)$ is a positive integer, then the game position P can be won by the first player; otherwise the second player can win.

This paper is organized as follows. In Section 2, we give background information to the game we are analyzing. Also different variants will be mentioned. In Section 3, we describe an algorithm how to find a winning move, and show that it always can be found in polynomial time. Determining that there is no winning move in a particular position can be done in polynomial time, too. In Section 4 and Section 5, we present how differences in the underlying sequences affect the sequences of the star game. We close with conclusions and some open problems in Section 6.

2 Background

Below we provide some history on Kayels (2.1) and discuss Kayles on Graphs (2.2).

2.1 History of Kayles

The game Kayles was introduced by Dudeney and independently also by Sam Loyd, who originally called it 'Rip Van Winkle's Game'. It was supposed to be played by skillful players who could either knock down exactly one or two adjacent pins ("kayles") out of a row of pins [3].

We could also think of starting with one heap (or several heaps) of beans, and give the following description of the rules. Each player, when it is his turn to move, may take 1 or 2 beans from a heap, and, if he likes, split what is left of that heap into two smaller heaps [1]. This game is well studied and the value of every game position can be determined in constant time [1].

---

[1] In this contribution 'he' is used when both 'he' and 'she' are possible.

2.2 Kayles on Graphs

Node-Kayles, a variant of the original Kayles, is played on a graph $G = (V, E)$ with n nodes in the node set V and m edges in the edge set E.

In this paper we will mainly consider the variant in which a move consists of selecting a node v and thereafter removing it and all its neighbors $N(v)$ from the graph. In case of a directed graph, only those vertices $w \in V$ will be called neighbors that can be reached using a directed edge $e = (v, w)$ from node v to node w. In addition, we will look at a variant that is closer to the original Kayles: either one single node or a node together with an arbitrary neighbor can be taken away.

For all variants of Kayles considered we will only investigate normal play, i.e., the first player unable to move loses. Schaefer proved that the problem to determine which player has a winning strategy for Node-Kayles played on arbitrary graphs is PSPACE-complete [7]. Bodlaender and Kratsch investigated the game on special classes of graphs, which include graphs with a bounded asteroidal number, cocomparability graphs, circular arc graphs, and cographs, and showed that the problem is polynomial time solvable in these cases [2].

In this paper we will describe how to find the value of a game position in polynomial time for stars. This solves an open problem in [2]. A *star* is an acyclic connected graph with one distinguished node, the *center* of the star. The center may have any degree Δ, all other nodes have degree at most 2, i.e., they lie on paths emanating from the center, so-called rays.

We will denote a star of degree Δ by $S_{l_1, l_2, ..., l_\Delta}$, where the l_i are natural numbers and stand for the length of ray i. Naturally rays of zero-length can be neglected. A degenerated star consisting of only one vertex has degree 0. Degenerated stars of degree one or two form a simple path only. In case of degree one the path length is $l_1 + 1$; for the degree-two case we have a path of length $l_1 + l_2 + 1$.

Kano considered stars where rays always have length 1 [6]. Furthermore in his game not vertices will be removed but any number of edges whereby these edges must belong to the same star. Kano gave some results on double-stars and on forks. A double-star is a graph obtained from two stars by joining their two centers by a new edge, and a fork is defined to be a graph which is obtained from a star and a path by joining the center of the star to one of the end vertices of the path by a new edge.

Node-Kayles has the same characteristics as Dawson's Chess where two phalanxes of Pawns are facing each other just one row apart. The Pawns step forward and capture diagonally as usual in chess. But capture is obligatory so queening is not possible. The winner is the last player to move. (In the original version by Dawson the last player to move loses [1].)

Both games have the same octal encoding .137. Octal encoding is used to describe different variants of take-and-break games, see Table 1 [5,1]. In that sense, .137 means: one single vertex can only be removed if it has no outgoing edges (anymore). Two vertices can only be removed if they form one connected component or if after their removal the remaining part of that connected com-

Table 1. Interpretation of Code Digits

Value of digit d_k	Condition for removal of k beans from a single heap.
0	Not permitted.
1	If the beans removed are the whole heap.
2	Only if some beans remain and are left as a single heap.
3	As 1 or 2.
4	Only if some beans remain and are left as exactly two non-empty heaps.
5	As 1 or 4 (i.e., not 2).
6	As 2 or 4 (i.e., not 1).
7	Always permitted (i.e., 1 or 2 or 4).

ponent is still connected. For three vertices we have more possibilities. We are allowed to remove them if the remaining vertices are left in at most two connected components.

The Grundy values for Node-Kayles have the periodicity 34. The 34 values for $n \equiv 0, 1, 2, \ldots, 33 \pmod{34}$ are
$(8, 1, 1, 2, 0, 3, 1, 1, 0, 3, 3, 2, 2, 4, 4, 5, 5, 9, 3, 3, 0, 1, 1, 3, 0, 2, 1, 1, 0, 4, 5, 3, 7, 4)$,
with the following seven exceptions.

- $\mathcal{G}(n) = 0$ when $n = 0, 14, 34$;
- $\mathcal{G}(n) = 2$ when $n = 16, 17, 31, 51$.

3 A Step into the No Man's Land

Fraenkel mentioned that there lies a huge no man's land between the polynomial time 0.137 and the PSPACE-hard Node-Kayles [4]. To explore a part of that no man's land we take a look at the stars first. Node-Kayles played on stars will be called Star-Kayles for short.

We can characterize the possible moves in Star-Kayles as follows.

1. Choosing the center vertex of the star will remove $\Delta + 1$ vertices, and split the star in up to Δ simple paths. It seems difficult to denote this game in the octal notation, or any other analogous polynomial notation.
2. Removing a vertex adjacent to the star center will split the star into either $\Delta - 1$ or Δ simple paths. Thereby either three or only two vertices will be removed (if the corresponding ray had only length 1).
3. Choosing the second next vertex to the center of a ray will leave a star of degree $\Delta - 1$ and (a possibly empty) simple path.
4. For any other vertex we will get a star of the same degree with one ray shortened accordingly to the position of the chosen vertex as well as a simple path (possibly empty again).

As the smallest representative for non-degenerated stars we choose stars of degree 3. This saves time and space for computations, but the following arguments can easily be extended to higher degrees. S_{l_1, l_2, l_3} denotes a star with rays of length l_1, l_2, and l_3, respectively.

3.1 A Star Is Born

Let us begin with the simplest star. The center appears first and then three rays will be sent out from the center. First we want to find the Grundy values for the class of stars $S_{l_1,1,1}$.

For $l_1 = 1$ the whole star can be removed in one move by choosing the center vertex (case (1) above). Therefore this situation is a winning position for the first player. Choosing any other of the vertices (2) will always have the same outcome, namely two single vertices. Adding up their values will of course result in a zero-game. No other configuration can be achieved and therefore $\mathcal{G}(S_{1,1,1}) = 1$.

For $l_1 = 2$ we have more possibilities:
Removing the center vertex (1) will leave one single vertex, Grundy value 1.
Choosing the vertex of ray r_2 or r_3 (2) will leave a simple path of length 2 and a single vertex. Adding up these two components will result in a zero-game.
We will also get the value 0 when choosing the first vertex of r_1 (2), because only two single vertices remain.
The last choice (3) will remove ray r_1 completely, and therefore leave a degenerated star behind with one center vertex and two rays of length 1 each. This is actually a simple path of length 3. Its Grundy value is 2.
Now we can determine the mex of these values, which is 3.

We will continue to determine the Grundy values of the stars $S_{l_1,1,1}$ in this way. Since the first move on a larger star may split it into a smaller one, it is best to compute the Grundy values iteratively for increasing values of l_1.

We analyze all four cases of star-decompositions mentioned above occurring for the stars $S_{l_1,1,1}$.

1. Removing the center will just leave a simple path of length $l_1 - 1$, whose value is $\mathcal{G}(l_1 - 1)$.
2. Choosing the first vertex of one of the rays r_2 or r_3 leaves a simple path of length l_1 and a single vertex. The value of this position is $\mathcal{G}(l_1) \oplus \mathcal{G}(1) = \mathcal{G}(l_1) \oplus 1$.
   Choosing the first vertex of r_1 leaves two single vertices and a simple path of length $l_1 - 2$, i.e., a position of value $\mathcal{G}(l_1 - 2) \oplus \mathcal{G}(1) \oplus \mathcal{G}(1) = \mathcal{G}(l_1 - 2) \oplus 0 = \mathcal{G}(l_1 - 2)$.
   To include into this more general description the case $S_{1,1,1}$ we assume that $\mathcal{G}(k) = 0$ for $k \leq 0$.
3. Choosing the second vertex of ray r_1 leaves two simple paths with length $l_1 - 3$ and 3 with value $\mathcal{G}(l_1 - 3) \oplus \mathcal{G}(3) = \mathcal{G}(l_1 - 3) \oplus 2$.
4. When choosing the very last vertex of ray r_1 we get a position of value $\mathcal{G}(S_{l_1-2,1,1})$. For all other vertices (along r_1) the set of all possible values is

$$\bigcup_{i=2}^{l_1-1} (\mathcal{G}(S_{l_1-i,1,1}) \oplus \mathcal{G}(i-3)).$$

To write all possible cases in a more compact way let us define the Grundy values for stars with non-positive length of ray r_1:

- $\mathcal{G}(S_{0,1,1}) = \mathcal{G}(3) = 2$;
- $\mathcal{G}(S_{-1,1,1}) = \mathcal{G}(1) \oplus \mathcal{G}(1) = 0$;
- $\mathcal{G}(S_{-2,1,1}) = \mathcal{G}(0) \oplus \mathcal{G}(0) = 0$.

Removing the vertex of one of the rays r_2 or r_3 (case 2) leads to a game value of $\mathcal{G}(l_1) \oplus 1$ (actually it could also be denoted by $\mathcal{G}(S_{l_1,-1,1})$). In all other cases, we get game value

$$\bigcup_{i=1}^{l_1+1} (\mathcal{G}(l_1 - i) \oplus \mathcal{G}(S_{i-3,1,1})).$$

Next we prove that the class of stars $S_{l_1,1,1}$ is ultimately periodic. The proof is very similar to the proof of the ultimate periodicity of the game 0.137 itself given in [1]. The fundamental idea is that the calculations of Nim-sequences is made easier by using a Grundy scale. Figure 1 shows such a scale being used for the computations of our star classes (a more detailed description follows below.) In general, successive values are written on squared paper and the arrowed entry is computed as the mex of all (or sometimes some accordingly to the game chosen) preceding entries. Then the scale is moved on one place. A very nice tool to do calculations of such kind (and many more) is the Combinatorial Game Suite by Aaron Siegel [8].

In our case we can also use Grundy scales. However, on the first scale we only write the Grundy values for 0.137. We can do this for a few numbers at the beginning first, and then just append more if needed. We fix this scale on the table. Our second scale will carry the Grundy values for our stars in reverse order. It will be shifted step by step to the right. The Grundy value for the star we are calculating will always be written into the first empty cell to the left of the already computed values. For an illustration see Figure 1. The arrow indicates the next value to be determined.

With this picture in mind we can easily transfer the proof given for 0.137 to our class of stars. We keep on calculating game values until two complete periods p lie between the last irregularity of 0.137, which is already known to be $i_1 = 51$,

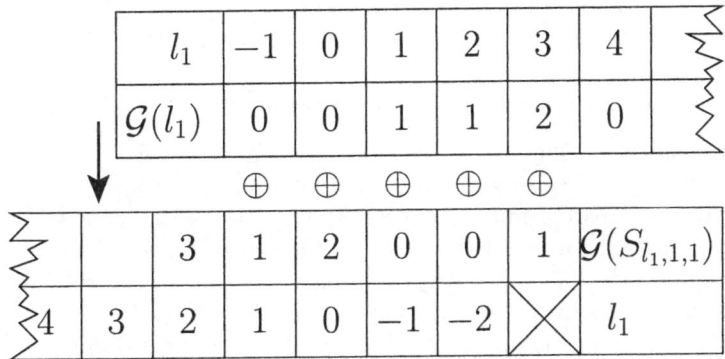

Fig. 1. Calculating $\mathcal{G}(S_{l_1,1,1})$ in Star-Kayles

and the last observed irregularity of our star class, which we call i_2. As at most three nodes can be cut out of a ray we calculate $t = 3$ additional values. So the last value that needs to be computed to verify that the period persists is

$$\mathcal{G}(i_1 + i_2 + 2p + t).$$

This gives the following ultimately periodic sequence for $l_1 \equiv 0, 1, 2, \ldots, 33 \pmod{34}$: (2, 9, 3, 15, 14, 1, 9, 4, 4, 14, 5, 13, 4, 0, 8, 1, 2, 4, 8, 5, 13, 2, 4, 8, 5, 9, 4, 12, 8, 6, 9, 9, 0, 8); Table 2 shows the first 374 game values of that class; $S_{310,1,1}$ is the last irregular value.

Table 2. Game values for the star class $S_{l_1,1,1}$. (A table entry gives the game value for l_1 = row-header + column-header.).

l_1	1	2	3	4	5	6	7	8	9		11		13		15		17		19		21		23		25		27		29		31		33	
0	1	3	0	0	1	1	4	0	5	1	1	0	0	3	1	2	0	0	1	2	2	3	3	5	2	4	3	3	2	2	1	0	0	2
34	1	3	0	0	1	7	4	4	6	5	7	0	0	3	1	2	0	0	1	2	2	4	3	5	6	4	7	3	6	2	1	0	0	2
68	1	3	0	8	1	9	4	2	8	5	9	4	0	3	1	2	0	0	1	2	2	4	8	5	6	4	7	8	6	9	1	0	0	2
102	1	3	0	8	1	9	4	4	8	5	9	4	0	8	1	2	0	0	1	2	2	4	8	5	9	4	12	8	6	9	9	0	0	2
136	1	3	0	8	1	9	4	4	14	5	13	4	0	8	1	2	0	0	1	2	2	4	8	5	9	4	12	8	6	9	9	0	0	2
170	1	3	0	8	1	9	4	4	14	5	13	4	0	8	1	2	4	0	1	2	2	4	8	5	9	4	12	8	6	9	9	0	8	2
204	9	3	0	8	1	9	4	4	14	5	13	4	0	8	1	2	4	0	1	2	2	4	8	5	9	4	12	8	6	9	9	0	8	2
238	9	3	0	8	1	9	4	4	14	5	13	4	0	8	1	2	4	0	1	2	2	4	8	5	9	4	12	8	6	9	9	0	8	2
272	9	3	0	8	1	9	4	4	14	5	13	4	0	8	1	2	4	0	5	2	2	4	8	5	9	4	12	8	6	9	9	0	8	2
306	9	3	15	8	1	9	4	4	14	5	13	4	0	8	1	2	4	8	5	13	2	4	8	5	9	4	12	8	6	9	9	0	8	2
340	9	3	15	14	1	9	4	4	14	5	13	4	0	8	1	2	4	8	5	13	2	4	8	5	9	4	12	8	6	9	9	0	8	2

3.2 The Magic Number of the Game .137 Is 34

When we just gave the proof of the ultimate periodicity of the class $S_{l_1,1,1}$ we did this with the implicit understanding that Star-Kayles has the same period as the basic game. This is indeed the fact and not too surprising as the only difference in this variant to the original game occurs in splitting the star by either removing the center or one of its neighbors. All other moves only break rays which is similar to breaking a row in the original game.

3.3 Dynamic Programming and Memoization

Having calculated all values of the class $S_{l_1,1,1}$ as far as necessary we take a look at the class $S_{l_1,2,1}$. The only difference is that we need to include two further values before we can determine the mex of the set, $\mathcal{G}(l_1+1+1)$ and $\mathcal{G}(l_1) \oplus \mathcal{G}(2)$. These positions arise from either removing the second vertex of r_2 or the first and only vertex of r_3. Furthermore, choosing the center vertex will now leave a game position consisting of two simple paths with value $\mathcal{G}(l_1 - 1) \oplus \mathcal{G}(1)$.

The situation becomes different when we start to examine $S_{l_1,3,1}$. For the first time we need to consider values of stars that do not lie in the same class. Choosing the last vertex of r_2 will leave us with a star of the class $S_{l_1,1,1}$. Here memoization comes into play. Having stored all the values of our former computations we simply look up the desired value.

To show which former values are needed to calculate the value of any star S_{l_1,l_2,l_3} we shall generalize the star decomposition.

1. Choose the center vertex:
   $\mathcal{G}(l_1 - 1) \oplus \mathcal{G}(l_2 - 1) \oplus \mathcal{G}(l_3 - 1)$.
2. Choose the first vertex of any ray:
   For each $i \in \{1,2,3\}$, and $i \neq j \neq k \neq i \in \{1,2,3\}$
   $\mathcal{G}(l_i - 2) \oplus \mathcal{G}(l_j) \oplus \mathcal{G}(l_k)$.
3. Choose the second vertex of any ray:
   For each $i \in \{1,2,3\}$, and $i \neq j \neq k \neq i \in \{1,2,3\}$
   $\mathcal{G}(l_i - 3) \oplus \mathcal{G}(l_j + l_k + 1)$.
4. Choose any other vertex:
   For each $i \in \{1,2,3\}$, and $i \neq j \neq k \neq i \in \{1,2,3\}$
   $$\bigcup_{l=4}^{l_i+1} (\mathcal{G}(l_i - l) \oplus \mathcal{G}(S_{l-3,l_j,l_k})).$$

After defining

- $\mathcal{G}(S_{0,l_2,l_3}) = \mathcal{G}(l_2 + l_3 + 1) = 2$,
- $\mathcal{G}(S_{-1,l_2,l_3}) = \mathcal{G}(l_2) \oplus \mathcal{G}(l_3)$, and
- $\mathcal{G}(S_{-2,l_2,l_3}) = \mathcal{G}(l_2 - 1) \oplus \mathcal{G}(l_3 - 1)$,

we can again state a more compact formula for (1) – (4):

For each $i \in \{1,2,3\}$, and $i \neq j \neq k \neq i \in \{1,2,3\}$ $\displaystyle\bigcup_{l=2}^{l_i+1}(\mathcal{G}(l_i - l) \oplus \mathcal{G}(S_{l-3,l_j,l_k}))$.

We remark that the term for case (1) will appear three times in the above union of all values. Taking the mex of the union of all those sets will give us the game value $\mathcal{G}(S_{l_1,l_2,l_3})$ for any star S_{l_1,l_2,l_3}. Based on this approach we can use dynamic programming to find the Grundy value of any star. Given that the values of all stars smaller than the considered star have already been calculated it takes only $O(n)$ time to find its value, whereby n is the size of the star.

Observation
While using memoization we only need to store the values of stars $S_{l_1,l_2,...,l_\Delta}$, where $l_1 \geq l_2 \geq \ldots \geq l_\Delta$. This comes from the fact that we are only interested in the combinatorial structure of the graph, and therefore all permutations of rays are isomorphic in our viewpoint.

3.4 Leaving the Orbit

While looking at all the data and comparing ultimately periodic sequences of different star classes we first thought that there might be one last exceptional star, after which all other stars can be evaluated by looking at ultimately periodic sequences. However, we had to realize that we could not find that last irregularity. Instead, we observed a constant growth of the last irregular value with the growth of the ray lengths. Therefore we will now investigate the relation between the last irregular value and the ray lengths of the stars.

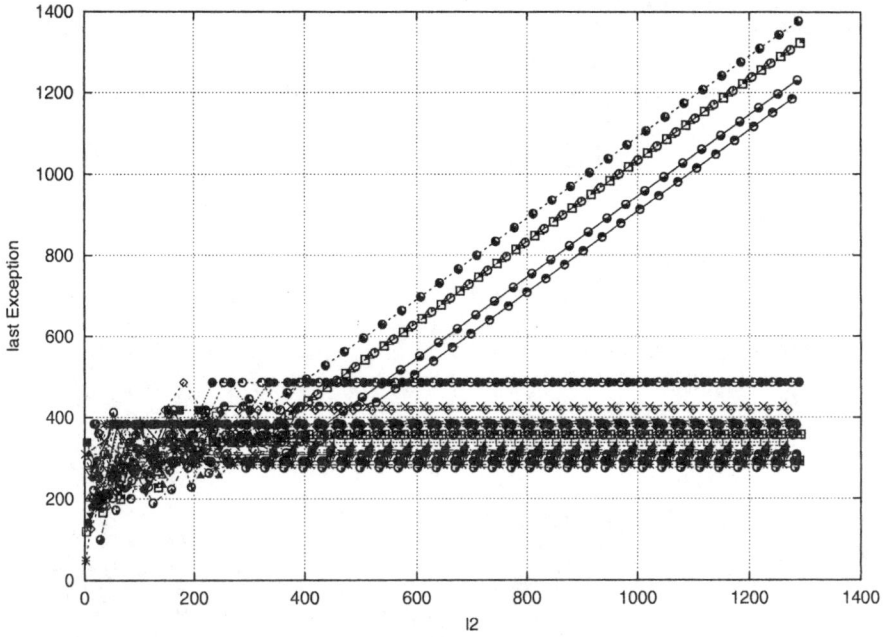

Fig. 2. Last irregular values of the star class $S_{l_1,l_2,1}$

As an example we look at all classes with a fixed ray length $l_3 = 1$ and arbitrary ray length for r_1 and r_2 (see Figure 2). The diagram shows all last irregular values (vertical axis) of the sequences as a function of the length l_2 of the second ray (horizontal axis). A detailed look at all values is not necessary. The aim is to point out that nearly all of these classes become ultimately periodic, while for a few classes the last irregularity will not stop to grow.

All star classes can be grouped in 34 equivalence classes by taking the modulo of the ray length l_2. Table 3 shows all ultimately periodic sequences of star classes with the third ray fixed to length 1, and the second ray has length 1 (mod 34). For greater values of l_2 there is no change anymore in the ultimately periodic sequence.

Most of these equivalences actually become ultimately periodic. However, the star classes $S_{l_1,l_2,1}$ with $l_2 \equiv 7, 15, 18, 27, 29, 32 \pmod{34}$ do not show this behavior. Instead, the last irregularity will grow with the length l_2 of the second ray (as depicted in Figure 2). To give an explanation we prove an easy lemma first.

Table 3. Ultimately periodic sequences of star classes for $l_3 = 1$ and $l_2 \equiv 1 \pmod{34}$

l_2	ultimately periodic sequence
1	9 3 15 14 1 9 4 4 14 5 13 4 0 8 12 4 8 5 13 2 4 8 5 9 4 12 8 6 9 9 0 8 2
35	1 14 0 11 18 1 9 0 17 14 14 19 4 18 6 2 15 0 8 18 2 16 3 21 1 24 23 3 13 2 10 19 0 18
69	1 14 0 8 1 1 10 0 24 14 26 23 0 3 1 2 20 0 22 2 2 3 3 16 25 17 27 3 28 2 1 0 0 23
103	1 14 0 11 1 1 31 0 25 10 28 23 0 25 1 2 24 0 18 2 2 3 3 32 1 23 27 3 21 6 14 0 0 18
137	1 33 0 27 32 1 33 0 24 40 33 33 0 18 1 2 21 0 32 2 2 3 3 32 1 23 33 3 15 6 30 0 0 23
171	1 40 0 36 39 1 40 0 36 37 33 34 0 35 1 2 38 0 32 2 2 3 3 36 1 34 38 3 15 6 33 0 0 29
205	1 40 0 34 18 1 35 0 25 10 28 40 0 34 1 2 24 0 18 2 2 3 3 40 1 24 23 3 26 6 14 0 0 25
239	1 37 0 36 41 1 42 0 25 10 28 44 0 35 1 2 20 16 47 2 2 3 3 21 1 28 23 3 17 6 14 0 0 43
273	1 36 0 47 41 1 42 0 25 10 28 47 0 34 1 2 20 16 46 2 2 3 3 21 1 28 23 3 17 6 14 0 0 34
307	1 35 0 34 41 1 35 0 25 10 28 44 0 34 1 2 20 16 48 2 2 3 3 21 1 28 23 3 17 6 14 0 0 34
341	1 35 0 34 47 1 35 0 25 10 28 48 0 34 1 2 20 16 49 2 2 3 3 21 1 28 23 3 17 6 14 0 0 34
375	1 35 0 34 47 1 35 0 25 10 28 28 0 34 1 2 20 16 46 2 2 3 3 21 1 28 23 3 17 6 14 0 0 34
409	1 35 0 34 47 1 35 0 25 10 28 28 0 34 1 2 20 16 46 2 2 3 3 21 1 28 23 3 17 6 14 0 0 34
443	1 35 0 34 47 1 35 0 25 10 28 28 0 34 1 2 20 16 46 2 2 3 3 21 1 28 23 3 17 6 14 0 0 34
477	1 35 0 34 47 1 35 0 25 10 28 28 0 34 1 2 20 16 46 2 2 3 3 21 1 28 23 3 17 6 14 0 0 34
511	1 35 0 34 47 1 35 0 25 10 28 28 0 34 1 2 20 16 46 2 2 3 3 21 1 28 23 3 17 6 14 0 0 34
545	1 35 0 34 47 1 35 0 25 10 28 28 0 34 1 2 20 16 46 2 2 3 3 21 1 28 23 3 17 6 14 0 0 34
579	1 35 0 34 47 1 35 0 25 10 28 28 0 34 1 2 20 16 46 2 2 3 3 21 1 28 23 3 17 6 14 0 0 34

Lemma 1. *Let l_2 and l_3 be fixed lengths of rays r_2 and r_3, and w. l. o. g. $l_2 \geq l_3$. For all l_2 greater than all last irregularities of all classes involved in the calculation of the Grundy-values of stars with ray r_1 of length i from 1 up to $l_2 - 1$ we get $\mathcal{G}(S_{i,l_2,l_3}) = \mathcal{G}(S_{i,l_2+34,l_3})$.*

Proof. The values $\mathcal{G}(S_{i,l_2,l_3})$ and $\mathcal{G}(S_{i,l_2+34,l_3})$ for all $i = 1, \ldots, l_2 - 1$ will be looked up as $\mathcal{G}(S_{l_2,i,l_3})$ and $\mathcal{G}(S_{l_2+34,i,l_3})$. As l_2 is greater than the last irregularity of this class and this class also has period 34 these two values must be equal. □

We observe that after passing this last irregularity the sequence will also become ultimately periodic. But for special classes the last irregularity keeps on growing with the lengths of the rays. Therefore we can never find an ultimate, huge star whose sequence builds the ultimately periodic sequence for the sequences of earlier stars. Furthermore, our calculations show that the classes that reveal the feature of a growing last exception are not the same for different values of l_3; not even values for values of l_3 that lie in the same equivalence class modulo 34.

3.5 The Milky Way

As we have seen so far, we can calculate the value of a single star using memoization in polynomial time. As already mentioned in the introduction, a game position composed by a disjunctive combination of several components can be evaluated as the Nim-sum of them. Therefore a game position consisting of many stars is simply the Nim-sum of the Grundy values of all these stars. Hence, computing the value of a game position of several stars needs time polynomial in their sizes. If the value of the position is zero, then there is no wining move. For any positive value we know that there exists a winning move. Every star exhibits only a linear number of decompositions. Therefore, the actual winning move can be found in linear time after the values of the decompositions have been calculated.

4 If the World Were Regular

In this section, we try to find out why Node-Kayles exhibits irregularities in its periodicity of 34, whether this is just an artifact from the underlying Kayles or whether there are also other reasons. To this purpose, let us assume in this section that Kayles was periodic without any irregularities. Then, how would Star-Kayles behave?

If we repeat the same investigations as before, the sequences of the star classes are looking quite regular. As an example we give Table 4, which shows all different ultimately periodic sequences of star classes with the third ray fixed to length 1, and the second ray has length 1 (mod 34).

Table 4. Ultimately periodic sequences of star classes for $l_3 = 1$ and $l_2 \equiv 1$ (mod 34) based on a regular sequence

l_2	ultimately periodic sequence
1	9 3 8 8 1 9 4 2 8 5 9 4 0 3 1 2 4 8 5 10 2 3 3 5 5 4 4 8 2 2 9 0 8 2
35	1 16 0 0 1 1 16 0 8 9 1 10 0 3 12 2 0 0 11 2 10 14 3 12 2 9 10 3 2 17 1 0 0 2
69	1 17 0 0 1 1 10 0 16 9 1 18 0 3 16 2 0 0 13 2 22 3 3 19 2 9 24 3 2 2 1 0 0 2
103	1 17 0 0 1 1 32 0 21 9 1 20 0 3 16 2 0 0 13 2 10 3 3 31 2 32 28 3 2 2 1 0 0 2
137	1 17 0 0 1 1 32 0 34 9 1 20 0 3 16 2 0 0 13 2 10 3 3 26 2 17 28 3 2 2 1 0 0 2

Already for $l_2 = 137$ the ultimately periodic sequence has reached the point where it will not change anymore; for all greater values of l_2 with $l_2 \equiv 1$ (mod 34) we get the same ultimately periodic sequence. The same holds for all other equivalence classes of l_2(mod 34). However, the considerations above were only for a fixed length 1 of the third ray.

To build a basis for all games on degree-3 stars we need to look at different lengths of r_3 as well. From 2 to 20 we get the same regular behavior. But then the unexpected happens: if $l_3 = 21$, then there is an irregular value in the class $l_2 \equiv 22$ (mod 34) which is growing with increasing length l_2. There is one very remarkable point about this: those stars have their last irregular value when l_1 is equal to l_2.

5 Back to the Roots

After observing the stars in the sky for a while we start to feel the infinity of the universe: the last irregular value will move farther and farther away as the

Table 5. Changes in the length of the period of sequences of star classes for $l_3 = 1$. (A table entry gives the period multiplier for l_2 = row-header + column-header.).

l_2	1	2	3	4	5	6	7	8	9	10	11	12
0	1	1	1	1	1	1	1	1	1	1	1	1
12	1	1	1	1	1	1	1	1	1	1	1	14
24	1	1	1	1	1	1	1	1	1	1	1	1
36	14	1	14	1	1	1	14	14	1	14	1	1
48	1	5	1	1	1	1	5	5	1	10	14	1
60	1	10	1	1	14	1	10	1	1	10	10	1
72	5	1	1	1	10	1	1	1	10	1	1	10
84	14	1	1	1	1	1	1	5	1	14	1	1
96	1	1	1	70	10	1	1	5	1	5	1	10
108	10	1	10	1	70	1	10	1	1	1	14	1
120	1	1	14	1	1	5	1	1	70	70	70	5
132	1	1	1	1	14	10	1	1	5	70	1	1
144	1	1	1	1	70	10	10	1	1	70	70	10
156	10	1	1	70	1	14	10	1	70	70	70	1
168	10	1	1	70	70	10	10	1	70	70	70	10
180	10	1	1	14	14	10	10	1	70	70	70	1
192	1	1	1	14	1	10	1	1	14	70	14	1
204	10	70	1	14	1	10	1	1	70	70	14	1
216	1	70	1	14	70	10	1	1	70	70	70	1
228	1	70	1	70	1	1	1	1	14	70	14	1
240	1	70	1	14	1	1	1	1	70	70	14	1
252	1	1	1	14	1	1	1	1	14	70	14	1
264	1	1	1	14	1	1	1	1	14	70	14	1
276	1	1	1	14	1	1	1	1	14	70	70	1
288	1	1	1	14	1	1	1	1	14	70	70	1
300	1	1	1	14	1	1	1	1	14	70	14	1
312	1	1	1	14	1	1	1	1	14	70	14	1
324	1	1	1	14	1	1	1	1	14	70	14	1

rays of the stars get longer and longer. This makes us think whether we should come back to earth to take a look at the roots: Pin-Kayles.

In Pin-Kayles a player can knock out one or two adjacent pins only. Thereby the remaining pins might also be split into several groups. In the original game we have only one row of pins, and therefore at most two groups emerge, but in our star-shaped setting we can get up to d groups if we remove the center pin.

The Grundy values for Pin-Kayles have the periodicity 12, and show the following sequence for $n \equiv 0, 1, 2, \ldots, 11 \pmod{12}$: $(4, 1, 2, 8, 1, 4, 7, 2, 1, 8, 2, 7)$, with the following 14 exceptions.

- $\mathcal{G}(n) = 0$ when $n = 0$;
- $\mathcal{G}(n) = 3$ when $n = 3, 6, 18, 39$;
- $\mathcal{G}(n) = 4$ when $n = 9, 21, 57$;
- $\mathcal{G}(n) = 5$ when $n = 28$;
- $\mathcal{G}(n) = 6$ when $n = 11, 22, 34, 70$;
- $\mathcal{G}(n) = 7$ when $n = 15$.

5.1 An Alteration Spell

The magic number of Pin-Kayles is 12, but as we start to play the game on stars we quickly encounter a class of stars that shows a different period. Stars with $l_2 = 24$ and $l_3 = 1$ have a 14-times lengthened period. Every 168 numbers we will get exactly the same numbers again. Only one out of the original 12 positions causes this new period. The next candidate that shows the same behavior and same period has a ray length of $l_2 = 37$. Here, two different positions display irregularities. Three further multipliers of 12 can be found which are 5, 10, and 70.

Table 5 lists the changes in periods as multipliers of 12 for an increasing ray length l_2 and a fixed ray length $l_3 = 1$. The last three rows show the same pattern, and also later rows do not present any changes anymore.

In contrast to the other two Node-Kayles versions we could not find yet a class that does not stop shifting the position of its last irregular value.

Another fact seems notable. The last irregular value of the underlying Pin-Kayles sequence appears at position 70. This coincides with the greatest multiplier of the period we have found so far.

6 The Forest Lies Ahead

We have shown that the problem of determining game values for stars can be done in polynomial time using dynamic programming and memoization. We have seen that Node-Kayles played on stars cannot lead to an ultimate periodic sequence that can be used to describe all stars bigger than a certain size. Instead we presented star classes whose last irregular value will continue to move farther and farther away with growing length of the rays.

Pin-Kayles played on stars seems more promising but further studies are necessary to prove or disprove the existence of such an ultimate periodic sequence.

Our investigations have given a solution for stars, a special kind of trees. It still remains an open problem to show how difficult it is to calculate a game position for arbitrary trees. If one can compute this, solving the problem for a forest is as simple as adding the Grundy values for the single trees together in the usual manner.

References

1. Elwyn R. Berlekamp, John H. Conway, and Richard K. Guy. *Winning Ways for Your Mathematical Plays.* A K Peters, Ltd., 2nd edition, 2001.
2. Hans L. Bodlaender and Dieter Kratsch. Kayles and Nimbers. In *J. Algorithms*, volume 43 (1), pages 106–119, 2002.
3. John H. Conway. *On Numbers and Games.* A K Peters, Ltd., 2nd edition, 2001.
4. Aviezri. S. Fraenkel. Recent results and questions in combinatorial game complexities, Expanded version of invited lecture at Ninth Australasian Workshop on Combinatorial Algorithms, Perth, Western Australia, 27—30 July 1998. In *Proc. AWOCA98 (C. Iliopoulos, ed.)*, pages 124–146, 1998.

5. Richard K. Guy and Cedric A. B. Smith. The G-values of various games. In *Proc. Cambridge Philos. Soc.*, volume 52, pages 514–526, 1956.
6. Mikio Kano. Edge-removing games of star type. In *Discrete Mathematics*, volume 151, pages 113–119, 1996.
7. Thomas J. Schaefer. On the complexity of some two-person perfect-information games. In *Journal of Computer and System Sciences*, volume 16, pages 185–225, 1978.
8. Aaron Siegel. *Combinatorial Game Suite*. http://cgsuite.sourcefourge.net/, 2003.

Searching over Metapositions in Kriegspiel

Andrea Bolognesi[1] and Paolo Ciancarini[2]

[1] Dipartimento di Scienze Matematiche e Informatiche "Roberto Magari",
University of Siena, Italy
abologne@cs.unibo.it
[2] Dipartimento di Scienze dell'Informazione,
University of Bologna, Italy
cianca@cs.unibo.it

Abstract. Kriegspiel is a chess variant similar to war games, in which players have to deal with uncertainty. Kriegspiel increases the difficulty typical of chess by hiding from each player his opponent's moves[1]. Although it is a two-person game it needs a referee, whose task consists in accepting the legal moves and rejecting the illegal ones, with respect to the real situation. Neither player knows the whole history of moves and each player has to guess the state of the game on the basis of messages received from the referee. A player's try may result in the statement *legal* or *illegal*, and a legal move may prove to be a *capture* or a *check*.

This paper describes the rationale of a program to play basic endgames of Kriegspiel, where a player has only the King left. These endings have been theoretically studied with rule-based mechanisms, whereas very little research has been done on a game-tree-based approach.

We show how the branching factor of a game tree can be reduced in order to employ an evaluation function and a search algorithm. We then deal with game situations which are dependent on a stochastic element and we show how we solve them during the tree visit.

1 Introduction

Kriegspiel is a chess variant similar to wargames, in which players have to deal with uncertainty. All chess rules are valid, but the players are not informed of their opponent's moves. Although it is a two-person game, it needs a referee, whose task consists in accepting the legal moves and rejecting the illegal ones, with respect to the real situation. As the game progresses, each player tries to guess the position of his opponent's pieces by trying moves to which the referee can stay *silent*, if the move is legal, or can state *"illegal move"*, if the move is illegal. Then the player has to try again. If the move is legal and it gives check or captures a piece, the referee says *"check"* or *"capture"*, respectively. In order to speed up the game if a Pawn can capture an opponent's piece this is also announced. For instance, here is a simple game (we omit illegal tries):

---

[1] In this contribution 'he' is used where 'he' or 'she' are both possible.

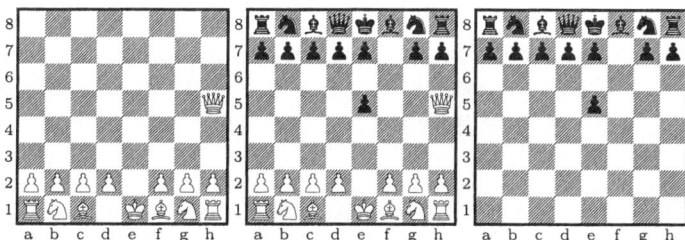

Fig. 1. Three views on one position

1. e4 f6 – the referee stays silent (both moves are legal);
2. e5 – the referee announces: "Black has a Pawn try";
2. ..fe5 – the referee says "Pawn captured on e5";
3. ♕h5 – the referee announces: "Check on short diagonal" (see Figure 1);
3. ..g6 – the referee stays silent;
4. ♗e2 – the referee announces: "Black has a Pawn try";
4. ..gh5 – the referee says "piece captured on h5";
5. ♗xh5 – the referee announces: "Checkmate".

The three diagrams in Figure 1 (left for White, center for referee, right for Black) shown after move three are a typical way of displaying the partial knowledge that players have about the current state of a Kriegspiel game: a player does not know what his opponent has done up to his turn to move. Therefore Kriegspiel is considered a game of imperfect information. Incidentally, we call the leftmost and rightmost boards *reference boards* for White and Black, respectively.

The design of a Kriegspiel-playing program is an interesting problem, because we can adapt most techniques already developed for computer chess. However, it remains a problem to adapt the game-tree search and the evaluation function typical of chess-playing programs to Kriegspiel, because each player is uncertain about the position of his opponent's pieces. In theory it would be possible to build a huge game tree taking into account all possible positions compatible with past announcements from the referee. In practice this is an impossible task, because the complete game tree of Kriegspiel is much larger than in the case of chess.

In this article we propose a way to reduce the game tree which leads to a representation through metapositions instead of normal chess positions. In order to simplify our task, in the next sections we will consider simple endings, i.e., Black having only his King and White having a King and a Rook (in the rook ending or ♔♖♚), a King and a Queen (♔♕♚), a King and two Bishops (♔♗♗♚), and a King and a Pawn (♔♙♚).

These endings are simple but all are quite difficult to play with uncertainty about the position of the opponent King. We have developed algorithmic solutions, which solve a given ending in all positions, however we will have to distinguish between positions without a stochastic element, that are states where it is possible to find the best move to play among the possible ones deterministically, and positions with a stochastic element, that are states from which a player may reach several equivalent metapositions through different moves. We will deal with the latter by randomly choosing one of the equivalent moves.

In Section 3 we begin to describe how we represent uncertainty using metapositions and we show the adjustments done on the game tree in order to have a deterministic search, then we deal with the case of search including moves randomly chosen. In Section 4 we propose the evaluation function for basic endgames and we describe the search algorithm which use the evaluation function. In this section we deal with several endings (including ♔♖♚, ♔♕♚, ♔♗♗♚, ♔♙♚) and we show some examples of games played by the program.

2 Related Works

Although it is a fascinating game, played by several hundred people every day on the Internet Chess Club, only a small number of papers have studied some aspects of Kriegspiel or Kriegspiel-like games. Below we provide some instances of related work.

Boyce proposed a procedure to solve the ♔♖♚ ending, that we have implemented to be able to evaluate our algorithm [2]. Ferguson analysed the endings ♔♗♘♚ ([5]) and ♔♗♗♚ ([6]), respectively. Ciancarini, Dalla Libera, and Maran ([4]) described a rule-based program to play the ♔♙♚ ending according to some principles of game theory. Sakuta and Iida in ([7]) described a program to solve Kriegspiel-like problems in Shogi (Japanese Chess). Bud and others ([3]) described an approach to the design of a computer player for a sub-game of Kriegspiel, called Invisible Chess. Finally, our paper [1] describes preliminary research on ♔♖♚ endings in Kriegspiel. The present paper expands and generalizes that work to other endings.

3 Metapositions

3.1 Game Situation Without Stochastic Element

Figure 2 shows an example of a position during a game of ♔♖♚ ending. Assume it is White's turn to move. At the first ply, the game tree whose root is the position of Figure 2, has a branching factor of 11 moves, corresponding to White's possible moves, plus Black's possible moves, of which there are 5 in 7 cases; 6 in 2 different cases and, finally, 3 in the last 2 cases.

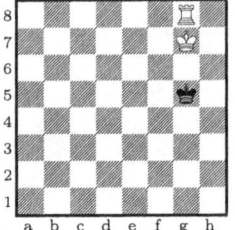

Fig. 2. Example of ♔♖♚ ending position

Black's choices compose the information sets for White, who does not know where Black has moved his pieces. Thus there are 32 information sets for White, in all 53 nodes; with the tree visit we have to handle numerical growth, which is not easily handled with brute force. Consider the tree depicted in Figure 3 obtained from the miniature in diagram 2 with move ♖f8: Black's possible moves are ♖h5, ♖h4 or ♖g4. White now faces two information sets of 2 and 1 elements respectively. The former includes 21 possible moves[2], while the latter includes 19 moves. Thus, the game tree branches out with $2 \cdot 21 + 19 = 61$ further nodes.

In Figure 3 x, y, and z denote the value of positions computed by an evaluation function we will deal with in Section 4.

In general, whether there is an information set with more than an element, it is not possible to employ a search algorithm on the game tree to find a optimal solution. If Black had a priority over his choices, which lead to the same information set, White could perform a tree visit to deduce which move of Black is the most dangerous. The problem arises when in the information set with more than one element we have to deal with alternative positions having the same probability.

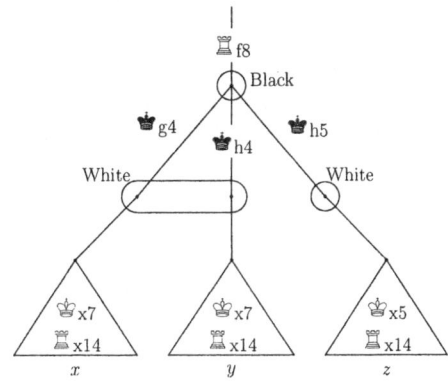

Fig. 3. White has 1 information set with more than 1 element

A possible solution consists of joining the information sets into a single position which describes them all. By collecting those states reached with same likelihood we can simultaneously represent them, without the constraint of choosing one of them. In this way moves with different priorities can be safely represented without the risk of losing information. Thus, we adopt the notion of *metaposition* [7], which is a special position denoting a set of positions. In addition to the referee's board, each player updates his own board, which is annotated with all of the opponent's possible positions. We recall that we refer to the metaposition representing the knowledge of a player with the term *reference board*.

Now we join those moves made by Black which lead to the same information set, obtaining identical metapositions with similar uncertainty about the White's pieces positions. The example in Figure 3 is reduced as in Figure 4. Moves ♛g4 ∧ ♛h4, which led to the same information set, lead now to a unique metaposition.

According to the definition, the tree in Figure 4 represents a *game of perfect information*. In the example above the unification of moves leads to sub-game trees whose evaluation is given by the minimum value of the

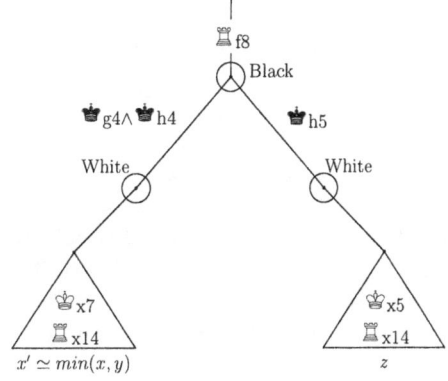

Fig. 4. White has 2 information sets with 1 element

---

[2] 7 for the King and 14 for the Rook.

original branches as evaluated before the join. The formula $x' = min(x, y)$, where x' is the new evaluation of the metaposition, requires the calculation of both x and y. Thus, the problem is not dissimilar to the previous one.

An improvement to the tree visit can be made considering the evaluation function not only with its recursive role, returning a value for a position at a particular depth, but also with a static meaning, in order to give a value during the tree visit and to distinguish the promising one among several branches.

Because of the complexity of a procedure which distinguishes from among Black's moves which lead to different information sets for White, that is moves that lead to metapositions or to simple positions, we define the game tree in a simpler but equivalent way.

We define the notion of *metamove*, which is a move that the black King can perform and that transforms a metaposition into another one. A metamove allows White to update his reference board expanding all the possible moves for Black. The metamove includes not only the moves that lead to the same information set, but comprises all the possible moves that Black can play from that particular metaposition. In some sense, we are transforming a basic Kriegspiel endgame into a new game where White has to confront several black Kings. In doing this we loose the property of information sets which claims that from each node into the set there are the same moves. We introduce the concept of *pseudomove* to indicate the moves by White on a metaposition. Pseudomoves are moves whose legality is not known to White.

Black's moves joined for the previous example are ♚h5, ♚h4 and ♚g4. Figure 5 shows the new game tree with metapositions.

The set P of pseudomoves has a cardinality equal to that of the union of possible legal moves by White for each position.

$$|P| = |\cup \text{(legal moves)}| \quad (1)$$

For the example in Figure 5 White always considers 21 pseudomoves, even if he is in the case with 19 legal moves, because he cannot distinguish between the two situations.

In order to have the new tree equivalent to the former, we introduce the information given by the referee. In fact, what characterizes the pseudomoves is the referee's answer, which can be *silent* (S), *check* (C) or *illegal* (I). Thus, the game tree has a branching factor equal to $3 \cdot i$, where i is the number of metapositions.

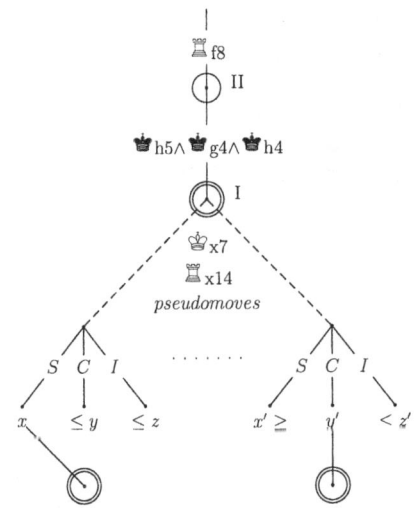

Fig. 5. The game tree with metapositions

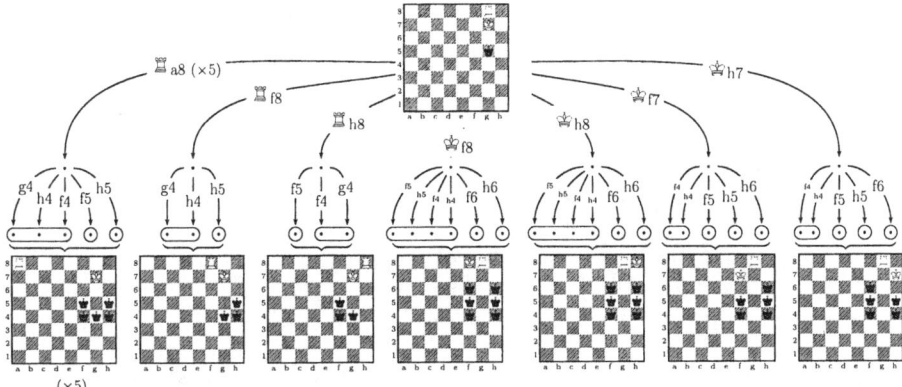

Fig. 6. Example of game tree

Starting from position depicted in Figure 2, there are 11 moves for White which, after considering Black's move, lead to 11 metapositions, as showed in Figure 6.[3] If we separately considered moves made by Black for each move by White we would have obtained 33 information sets and 53 positions in total. If we consider only metapositions, rather then all possible positions, we obtain just 11 metapositions. Thus, the game tree with the referee's answers has a branching factor equal to $3 \cdot 11 = 33$ nodes.

In Figure 5, metapositions are depicted with a double circle and pseudomoves are depicted with a dotted line. During the search visit on the game tree, we use the heuristic that chooses the worse referee's answer among the three. Thus, in the previous example the branching becomes equal to 11 nodes. In Figure 5 we indicate with x, x', y, y', z, z' the vote given statically by the evaluation function to the metapositions reached after playing each pseudomove.

3.2 Game Situation with Stochastic Element

Below we deal with game situations where players have to consider probability. For example, a situation with a stochastic element may happen when players have to choose between moves that lead to symmetric positions and therefore they have to draw moves by lot.

Assume we have the position where ♔c5, △c6 and ♚c7, depicted in Figure 7. In this case, the unification of all Black's moves leads to a unique information set, and White's pseudomoves can be actually considered legal.

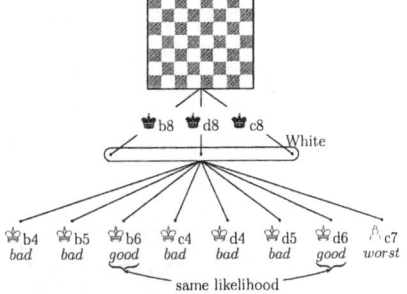

Fig. 7. Moves have same likelihood

---

[3] The move ♖a8 from Figure 2 is similar to ♖b8 or ♖c8 or ♖d8 ♖e8, so in Figure 6 we depicted only the first one and we indicated the whole number of moves (×5).

We define an evaluation function which allows us to classify the choices for White. This function is based on the following rules, in decreasing order of importance.

1. it never risks the capture of the Pawn;
2. it favours the advancing of the Pawn;
3. it pushes the Pawn to the seventh row if the white King is on the seventh row;
4. it keeps the white King and Pawn adjacent;
5. among those moves that lead the King on the row below the Pawn, it favours the move which brings the King on the same column of the Pawn.

Using these rules as an evaluation function, the move ♗c7 is considered the worst move, then it is discarded; with ♛b8, ♛c8 or ♛d8, after ♗c7 the Pawn would risk to be captured. Also the moves ♔b4, ♔c4 and ♔d4 are discarded, because they move the King away from the Pawn; moves ♔b5 and ♔d5 are better but not really good, because they do not push the Pawn. Finally, moves ♔b6 and ♔d6 are equivalent and best, since they push the Pawn and let the Pawn stay adjacent to its King.

The equivalence between ♔b6 and ♔d6 is inevitable. Figure 8 shows the symmetries between the two metapositions reached with these moves. Thus it is not possible to have a numerical value which correctly represents the grade for a metaposition and which is not correct for the symmetric one. In other words, in the game tree we would face two different nodes with same evaluation.

During the tree search, we use a random number generator to assign randomly a bonus with likelihood 1/2. In this case we use the term *aleatory metaposition*. We remark that, with an aleatory metaposition, each visit of its game tree becomes aleatory. A negative consequence of using aleatory metapositions is that we cannot employ techniques to accelerate the search, such as hash tables or Zobrist keys, since we would lose the stochastic nature of tree search.

In our example, let ♔b6 be the move randomly chosen, so White's reference board is the one on the left in Figure 8. The White's pseudomoves ♔a7, ♔a6 and ♔a5 are discarded because they move the King away from the Pawn; the pseudomoves ♔b5 and ♔c5 are discarded, since they do not help to advance the Pawn. If ♔c7 is illegal, then White chooses the remaining ♔b7, followed by ♗c7 if the referee's answer is silent. Otherwise, if ♔b7 proves to be illegal, White plays ♔c5, owing to the fifth rule, since the Pawn is on the c column.

 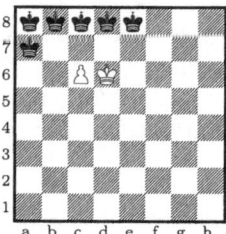

Fig. 8. Metapositions reached with ♔b6 and ♔d6

4 The Evaluation Function

The evaluation function contains the rules which synthesize the notion of progress leading the player towards victory. It is a linear weighted sum of features such as

$$\text{EVAL}(m) = w_1 f_1(m) + w_2 f_2(m) + ... + w_n f_n(m) \qquad (2)$$

where, for a given metaposition m, w_n indicates the weight assigned to a particular subfunction f_n. For example, a weight might be $w_1 = -1$ and $f_1(s)$ may indicate the number of black Kings.

The EVAL function is different according to each single ending, but it has some invariant properties: it avoids playing those moves that lead to stalemate and it immediately returns the move which gives checkmate directly, if it exists.

In the following subsections we briefly describe the search algorithm used for some basic Kriegspiel endings (4.1), then we examine the evaluation of metapositions more in depth (4.2 to 4.5).

4.1 The Search Algorithm

As we have seen in Subsection 3.1, we consider that each node of the game tree consists of a metaposition. For example, assume that the white reference board is the one depicted in Figure 9 and that it is White's turn to move.

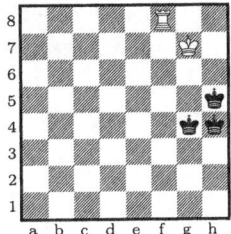

The search algorithm proceeds by generating all the pseudomoves and, for each metaposition reached, it creates three new metapositions according to the three possible answers from the referee. Then it chooses the one with the smallest value as given by the evaluation function. In the example we have 21 pseudomoves which lead to 63 metapositions, but after filtering the information from the referee we obtain again 21 nodes.

Fig. 9. Rook ending metaposition

Then, if the search algorithm has reached the desired search depth it simply returns the evaluation for the best node, that is the max value, otherwise it applies the metamove on each nodes, it decrements the depth of search and it recursively calls itself obtaining a value from the subtree.

Finally, it retracts the pseudomove played and adds to the metaposition's value the vote which is returned by the recursive call. Then it updates the max on that particular search depth.

When the algorithm ends visiting the tree, it returns the best pseudomove to play. Since the same candidate pseudomove may be proposed in two different sequential turns to move, creating a loop and thus not progressing, the algorithm avoids choosing the pseudomoves which appear in the history of recently played moves.

4.2 The Rook Ending (♔♖♚)

The evaluation function for this ending considers the following $n = 6$ different features:

1. it avoids jeopardizing the Rook: $w_1 = -1000$ and f_1 is a boolean function which is true if the white Rook is under attack;
2. it brings the two Kings closer: $w_2 = -1$ and f_2 returns the distance (number of squares) between the two Kings;
3. it reduces the number of black Kings on the quadrants of the board as seen from the Rook and it favors having the black Kings grouped together in as few quadrants as possible: $w_3 = -1$ and $f_3 = c \sum_{i=1}^{4} q_i$ where $c \in \{1, 2, 3, 4\}$ is a constant which counts the quadrants that contains a black King and q_i counts the number of possible black Kings on i^{th} quadrant;
4. it avoids the black King to go between white Rook and white King: $w_5 = -500$ and f_5 is a boolean function which returns true if the black King is inside the rectangle formed by white King and white Rook on two opposite corners;
5. it keeps the white pieces close to each other: $w_5 = +1$ and f_5 is a boolean function which returns true if the Rook is adjacent to the King;
6. it pushes the black King toward the corner of the board: $w_6 = +1$ and $f_6 = \sum_{i=0}^{63} v[i]$, where v is a numerical 64-element vector, shown in Figure 10, that returns a grade for each squares which possibly holds the black King or returns 0 otherwise.

$$\begin{pmatrix} 1 & 1 & 0 & 0 & 0 & 0 & 1 & 1 \\ 1 & 0 & 0 & 0 & 0 & 0 & 0 & 1 \\ 0 & 0 & -2 & -4 & -4 & -2 & 0 & 0 \\ 0 & 0 & -4 & -4 & -4 & -4 & 0 & 0 \\ 0 & 0 & -4 & -4 & -4 & -4 & 0 & 0 \\ 0 & 0 & -2 & -4 & -4 & -2 & 0 & 0 \\ 1 & 0 & 0 & 0 & 0 & 0 & 0 & 1 \\ 1 & 1 & 0 & 0 & 0 & 0 & 1 & 1 \end{pmatrix}$$

Fig. 10. The simple numerical matrix $v[]$

Here we show some example games. We consider that the program plays White against Black whose strategy consists of centralizing himself on the board. Starting from the metaposition depicted in Figure 11, where ♔ is on b6 and ♖ is on c1, the game continues as follows.

Considering ♚ on b8:
1. ♔c7 I, ♚a6; ♚a8
2. ♖c8 #.

Considering ♚ on a8:
1. ♔c7; ♚a7
2. ♖a1 #

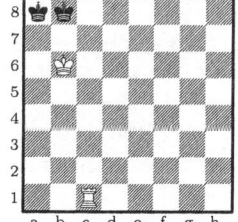

Fig. 11. Rook ending example

Starting from the metaposition depicted in Figure 12, where ♔ is on d6 and ♖ is on d7, the game continues as follows.

Assume ♚ on c8:	Assume ♚ is on b8:	Assume ♚ on a8:
1. ♔c7 I, ♔c6; ♚b8	1. ♔c7 I, ♔c6; ♚c8	1. ♔c7; ♚a7
2. ♔c7 I, ♖d6; ♚c8	2. ♔c7 I, ♖d6; ♚b8	2. ♖d5; ♚a6
3. ♔d7 I, ♔b6; ♚b8	3. ♔d7; ♚b7	3. ♔b6 I, ♔c6; ♚a7
4. ♖d8 #	4. ♔c7 I, ♔c6 I, ♖e6; ♚a7	4. ♖c5; ♚a6
	5. ♔c7; ♚a8	5. ♔c7; ♚a7
	6. ♔b6; ♚b8	6. ♖a5 #
	7. ♖e8 #	

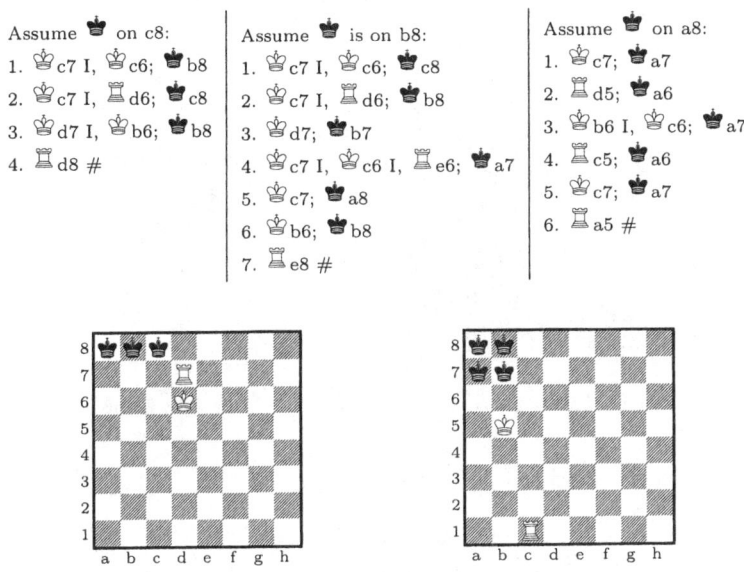

Fig. 12. Rook ending examples

Starting from the metaposition depicted in Figure 12 (right), where ♔ is on b5 and ♖ is on c1, the game continues as follows.

Assume ♚ on b7:	Assume ♚ is on a7:	Assume ♚ on a8:	Assume ♚ on b8:
1. ♔b6 I, ♖c4; ♚b8	1. ♔b6 I, ♖c4; ♚b7	1. ♔b6; ♚b8	1. ♔b6; ♚a8
2. ♖c5; ♚b7	2. ♖c5; ♚b8	2. ♔c7 I, ♔a6; ♚a8	2. ♔c7; ♚a7
3. ♖c4; ♚b8	3. ♖c4; ♚b7	3. ♖c8#	3. ♖a1#
4. ♖c6; ♚b7	4. ♖c6; ♚b8		
5. ♔c5; ♚b8	5. ♔c5; ♚b7		
6. ♔b6; ♚a8	6. ♔b6I, ♔b5; ♚b8		
7. ♔c7; ♚a7	7. ♔b6; ♚a8		
8. ♔c8; ♚a8	8. ♔c7; ♚a7		
9. ♖a6#	9. ♔c8; ♚a8		
	10. ♖a6#		

Figure 13 shows an histogram which represents the number of moves needed to win each game, starting from metapositions with greatest uncertainty, that is from metapositions where each square not controlled by White may contain a black King. The number of matches won is on the ordinate and the number of moves needed is on the abscissa. The graph depicts the result of all the possible 28,000 matches, which correspond to the 28,000 possibilities for the referee's board or to the 28,000 possible metapositions with greatest uncertainty. We notice that the program wins the whole games with 25 moves in the average.

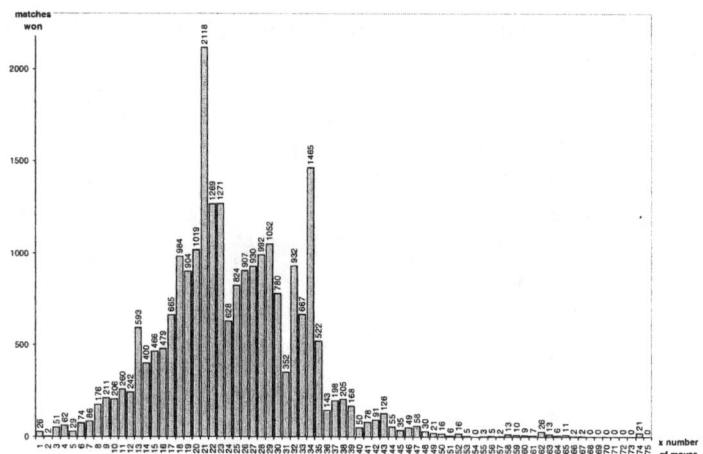

Fig. 13. Detailed histogram of ♔♖♚ ending game

4.3 The Queen Ending (♔♕♚)

The evaluation function is similar to the one described in Subsection 4.2 but we have to consider the Queen instead of the Rook.

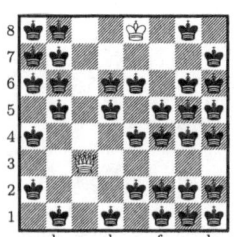

Fig. 14. The Queen cannot move

In some initial experiments we noticed a problem in metapositions with the Queen far from the King and with more than one black King between them. This problem was caused by the choice of bringing the Queen closer to the King. For example, Figure 14 shows a metaposition where the white Queen cannot move without risking to be captured. Thus we introduced three more features with respect to the evaluation function in Subsection 4.2. The first feature aims to avoid this problem and the other two intend to speed up the game by exploiting the power of Queen. So $n = 9$ and in the initial six rules the function is the same as in the Rook case[4], while in the last three:

7. it avoids metapositions where the Queen risks to be captured: $w_7 = -100$ and f_7 is a boolean function that returns true if the Queen is under attack;
8. it penalizes those metapositions with a big number of black Kings: $w_8 = -1$ and f_8 is equal to the number of black Kings on White's reference board;
9. it reduces the number of black Kings in the areas traced by the Queen's diagonals: $w_9 = -1$ and $f_9 = evalRhombArea(S)$ where

$$evalRhombArea(S) = c \cdot (a_0 + a_1 + a_2 + a_3) \qquad (3)$$

---

[4] Notice that rule 7 differs only in weight from rule 1 used for rook ending; these rules could be combined into a single rule, but for the moment we keep them separated in order to maintain strategies for different endings divided.

and $c \in \{1, 2, 3, 4\}$ is a constant which counts the areas that possibly contains a black King and a_i ($i = 0, ..., 3$) counts the number of possible black Kings in i^{th} area.

Figure 15 shows a graphic description of *evalRhombArea()* function. For the miniature on the left the function returns $4 \cdot (12+12+3+6) = 132$.

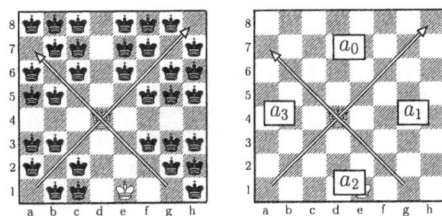

Fig. 15. Graphic description of evalRhombArea()

Now we show some examples of games played by the program. From a starting metaposition where ♕ is on d7, ♔ on c7, and ♛ on a6, ♛ on a7, ♛ on a8, the program correctly plays the move ♕a4#.

From a starting metaposition where ♕ is on h7, ♔ is on d7, and ♛ on a7, ♛ on a8, ♛ on b8, the game goes in accordance with the initial positions of Black as follows.

Assume ♛ on a8:	Assume ♛ on a7:	Assume ♛ on b8:
♔c7; ♛a7	♔c7; ♛a6	♔c7 I, ♔c6; ♛c8
♔b6 I, ♕d3; ♛a8	♔b6 I, ♕d3+; ♛a5	♕h6; ♛d8
♕a6#	♕b3; ♛a6	♔c7 I, ♕f8#
	♕a4#	

Figure 16 shows an histogram analogous to the one for the rook ending. It represents the number of moves needed to win each game, starting from metapositions with greatest uncertainty. The number of matches won is on the ordinate and the number of moves needed is on the abscissa.

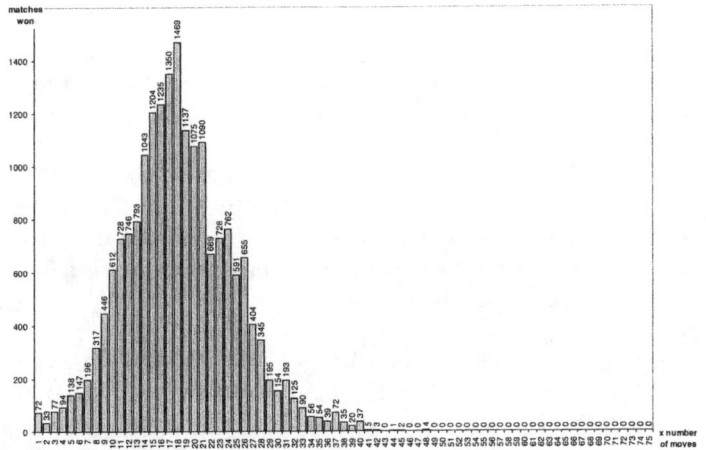

Fig. 16. Detailed histogram of ♔♕♛ ending game

4.4 The Ending with Two Bishops (♔♗♗♚)

In this ending we have to deal with two white pieces besides the King. The evaluation function exploits the same subfunctions previously analyzed, but it assigns different weights.

1. it avoids jeopardizing the Bishop: $w_1 = -1000$ and f_1 is a boolean function which is true if the white Bishop is under attack;
2. it brings the two Kings closer: $w_2 = -1$ and f_2 returns the distance (number of squares) between the two Kings;
3. it avoids the black King to "pass through" the border controlled by the Bishops: $w_3 = -500$ and f_3 is a boolean function which returns true if the black King is inside the rectangle formed by King and Bishop row or King and Bishop column;
4. it keeps close the white Bishops: $w_4 = +2$ and f_4 is a boolean function which returns true if the Bishops are adjacent to each other;
5. it pushes the black King toward the corner of the board: $w_5 = +1$ and $f_5 = \sum_{i=0}^{63} b[i]$, where b is a numerical 64-element vector, shown in figure 17, that returns a grade for each square which possibly holds the black King or returns 0 otherwise.

$$\begin{pmatrix} 0 & -10 & -50 & -100 & -100 & -50 & -10 & 0 \\ -10 & -10 & -40 & -40 & -40 & -40 & -10 & -10 \\ -50 & -40 & -40 & -40 & -40 & -40 & -40 & -50 \\ -100 & -40 & -40 & -50 & -50 & -40 & -40 & -100 \\ -100 & -40 & -40 & -50 & -50 & -40 & -40 & -100 \\ -50 & -40 & -40 & -40 & -40 & -40 & -40 & -50 \\ -10 & -10 & -40 & -40 & -40 & -40 & -10 & -10 \\ 0 & -10 & -50 & -100 & -100 & -50 & -10 & 0 \end{pmatrix}$$

Fig. 17. The numerical matrix $b[]$

6. it keeps the white King on the Bishop's row or column: $w_6 = +1$ and f_6 is a boolean function which returns true if the King and the Bishop are on the same row or column;
7. it penalizes the metapositions where the Bishop risks to be captured: $w_7 = -100$ and f_7 is a boolean function that returns true if Bishops are under attack;
8. it penalizes the metapositions with a big number of black Kings: $w_8 = -1$ and f_8 is equal to the number of black Kings on White's reference board;
9. it reduces the number of black Kings on the areas traced by the Bishop's diagonals: $f_9 = evalRhombArea(m)$, described with equation 3, and
if $evalRhombArea(m) < -600$ $w_9 = -4$;
otherwise $w_9 = \frac{1}{6}$
10. it prefers some particular positioning (we will refer to with the term *key Bishops' positions*) for the white King and Bishops, highlighted in figure 18; for example ♔c7, ♗c4 and ♗c5. Therefore $w_10 = +30$ and f_10 is a boolean function which is true if the Bishops and the King are arranged in one of the key positions.

Below we show some example endings in order to indicate the behavior of the program with two Bishops.

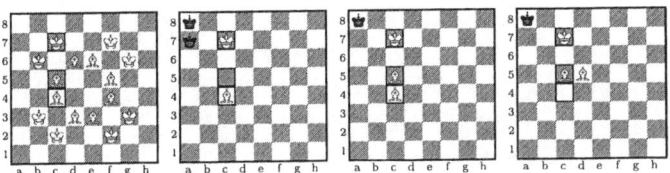

Fig. 18. Key Bishops' positions

Starting from the metaposition depicted in Figure 19, where ♗ is on c4 and ♗ is on h8 and ♚ is on c7, the game continues as follows.

Assume ♛ on a8:
1. ♔b6; ♛b8 2. ♗e6; ♛a8
3. ♗f6; ♛b8 4. ♗e5+; ♛a8
5. ♗d5#.

Assume ♛ is on a7:
1. ♔b6 I, ♗d4+; ♛a8 2. ♗d5#.

Starting from the second metaposition depicted in Figure 19, where ♗ is on c4 and ♗ is on d6 and ♚ is on c7, the game continues as follows.

Assume ♛ on a8:
1. ♗b4; ♛a7 2. ♔b6 I, ♗c5+; ♛a8
3. ♗d5#;

Assume ♛ on a7:
1. ♗b4; ♛a8 2. ♔b6; ♛b8
3. ♗e6; ♛a8 4. ♗e7; ♛b8
5. ♗d6+; ♛a8 6. ♗d5#.

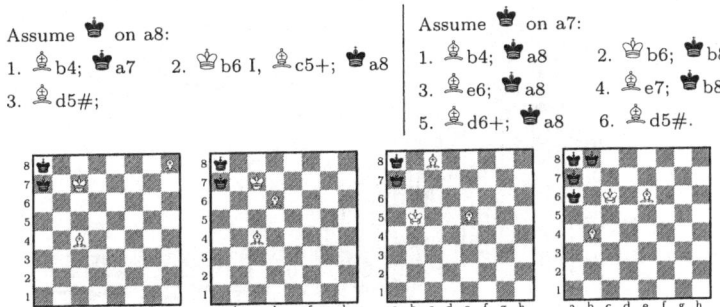

Fig. 19. Bishops ending examples

Starting from the third metaposition depicted in Figure 19, where ♗ is on c8 and ♗ is on e5 and ♚ is on b5, the game continues as follows.

Assume ♛ on a8:
1. ♔c6; ♛a7 2. ♔c7; ♛a8
3. ♔c6; ♛a7 4. ♔c7; ♛a8
5. ♗f6; ♛a7 6. ♔b6 I, ♗d4+; ♛a8
7. ♗b7#.

Assume ♛ is on a7:
1. ♔c6; ♛a8 2. ♔c7; ♛a7
3. ♔c6; ♛a8 4. ♔c7; ♛a7
5. ♗f6; ♛a8 6. ♔b6; ♛b8
7. ♗e6; ♛a8 8. ♗e7; ♛b8
9. ♗d6+; ♛a8 10. ♗d5#.

Starting from the fourth metaposition depicted in Figure 19, where ♗ is on b4 and ♗ is on e6 and ♚ is on c6, the game continues as follows.

Assume ♛ on b8:
1. ♔b6; ♛a8
2. ♔c7; ♛a7
3. ♗b4; ♛a8
4. ♗e7; ♛a7
5. ♗d5+; ♛a8
6. ♗d5#.

Assume ♛ is on a8:
1. ♔b6; ♛b8
2. ♔c7 I, ♗d6+; ♛a8
3. ♗d5#.

Assume ♛ is on a7:
1. ♔b6 I, ♗c4; ♛a8
2. ♔c7; ♛a7
3. ♗c5+; ♛a8
4. ♗d5#.

Assume ♛ is on a6:
1. ♔b6 I, ♗c4+; ♛a7
2. ♔c7; ♛a8
3. ♗e7; ♛a7
4. ♗c5+; ♛a8
5. ♗d5#.

Figure 20 shows an histogram analogous to the one for the rook ending. It represents the number of moves needed to win each game, starting from metapositions with greatest uncertainty. We can notice that for the ♔♗♗♛ ending the game is won in a larger number of moves than those required to win for the ♔♕♛ or the the ♔♖♛ ending. Sometimes the program wins with more than 80 moves.

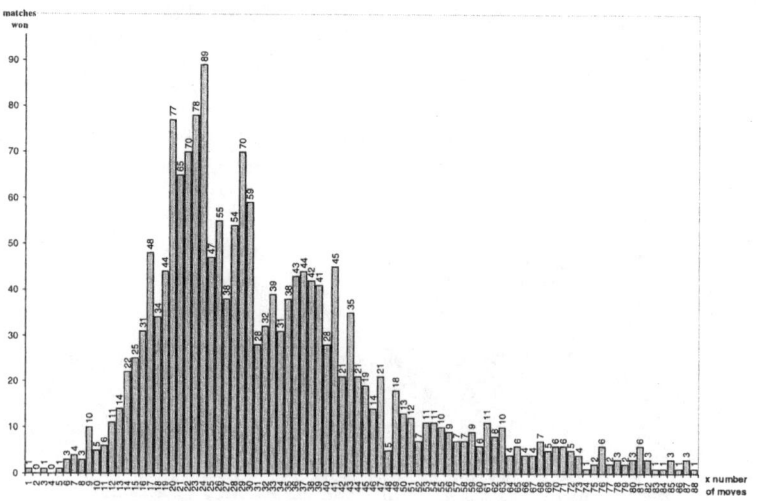

Fig. 20. Detailed histogram of ♔♗♗♛ ending game

4.5 The Pawn Ending (♔♙♛)

The evaluation function for the ♔♙♛ ending takes the discussion in Subsection 3.2 as starting point. It considers the following $n = 5$ different features:

1. it brings the Pawn adjacent to the King: $w_1 = -1$ and f_1 calculates the distance between King and Pawn;
2. it pushes the Pawn: $w_2 = +1$ and $f_2 = 2 \cdot (Pawn's\ row)$;
3. it let the King above the Pawn: $w_3 = +1$ and
   $f_3 = (King's\ row) - (Pawn's\ row)$;
4. **if** $(Pawn's\ row == seventh\ row)$
   if $(Pawn's\ row) > (King's\ row)$ $w_4 = -1000$;
   otherwise $w_4 = +100$

5. **if** (*Pawn's row* == *sixth row*)
   if (*King is on the right of the Pawn*) $w_5 = +5 + rand()$;
   if (*King is on the left of the Pawn*) $w_5 = +5 + rand()$;

The fifth condition implements the stochastic choice and forbids the use of hash techniques.

In order to implement a 1/2 likelihood, it uses a random number generator indicated here with the function $rand()$. Figure 21 shows an histogram which represents the number of moves needed to win each game, starting from random metapositions.

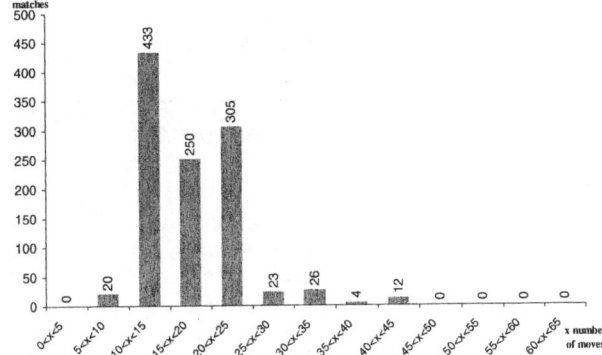

Fig. 21. Histogram of ♔ ♗ ♚ ending game

5 Conclusions

In our knowledge this is the first time that an evaluation function including a notion of progress has been defined for Kriegspiel. We have devoted special care to implement progress inside such an evaluation function. We have tested such a function on some simple endings, with good results except for the KBN vs K case. Future work will lead us to adapt the program to more complex endings, where both players have a larger number of pieces on the board. Our aim consists in writing a complete program for the whole game of Kriegspiel.

References

1. A. Bolognesi and P. Ciancarini. Computer Programming of Kriegspiel Endings: the case of KR vs K. In H.J. van den Herik, H. Iida, and E.A. Heinz, editors, *Advances in Computer Games 10*, pages 325–342. Kluwer, 2003.
2. J. Boyce. A Kriegspiel Endgame. In D. Klarner, editor, *The Mathematical Gardner*, pages 28–36. Prindle, Weber & Smith, 1981.
3. A. Bud, D. Albrecht, A. Nicholson, and I. Zukerman. Information-theoretic Advisors in Invisible Chess. In *Proc. Artificial Intelligence and Statistics 2001 (AISTATS 2001)*, pages 157–162, Florida, USA, 2001. Morgan Kaufman Publishers.
4. P. Ciancarini, F. Dalla Libera, and F. Maran. Decision Making under Uncertainty: A Rational Approach to Kriegspiel. In H.J. van den Herik and J.W.H.M. Uiterwijk, editors, *Advances in Computer Chess 8*, pages 277–298. University of Limburg, Maastricht, The Netherlands, 1997.
5. T. Ferguson. Mate with Bishop and Knight in Kriegspiel. *Theoretical Computer Science*, 96:389–403, 1992.
6. T. Ferguson. Mate with two Bishops in Kriegspiel. Technical report, UCLA, 1995.
7. M. Sakuta and H. Iida. Solving Kriegspiel-like Problems: Exploiting a Transposition Table. *ICCA Journal*, 23(4):218–229, 2000.

The Relative History Heuristic

Mark H.M. Winands, Erik C.D. van der Werf,
H. Jaap van den Herik, and Jos W.H.M. Uiterwijk

Department of Computer Science,
Institute for Knowledge and Agent Technology,
Universiteit Maastricht, Maastricht, The Netherlands
{m.winands, e.vanderwerf, herik, uiterwijk}@cs.unimaas.nl

Abstract. In this paper a new method is described for move ordering, called the relative history heuristic. It is a combination of the history heuristic and the butterfly heuristic. Instead of only recording moves which are the best move in a node, we also record the moves which are applied in the search tree. Both scores are taken into account in the relative history heuristic. In this way we favour moves which on average are good over moves which are sometimes best. Experiments in LOA show that our method gives a reduction between 10 and 15 per cent of the number of nodes searched. Preliminary experiments in Go confirm this result. The relative history heuristic seems to be a valuable element in move ordering.

1 Introduction

Most modern game-playing computer programs successfully use $\alpha\beta$ search [10]. The efficiency of $\alpha\beta$ search is dependent on the enhancements used [4]. Move ordering is one of the main techniques to reduce the size of the search tree. There exist several move-ordering techniques, which can be qualified by their dependency on the search algorithm [11]. *Static* move ordering is independent on the search. These techniques rely on game-dependent knowledge. The ordering can be acquired by using expert knowledge (e.g., favouring capture moves in chess) or by learning techniques (e.g., the Neural MoveMap Heuristic [12]). *Dynamic* move ordering is dependent on the search. These techniques rely on information gained during the search. The transposition-table move [3], the killer moves [1], and the history heuristic [15] are well-known examples.

The history heuristic is a popular choice for ordering moves dynamically, in particular when other techniques are not applicable. In the past the butterfly heuristic [8] was proposed to replace the history heuristic, but it did not succeed. In this paper we propose a new dynamic move-ordering variant, called the relative history heuristic, to replace the history heuristic. The idea is that besides the number of times a move was chosen as a best move, we also record the number of times a particular move was explored.

The paper is organised as follows. In Section 2 we review the history heuristic and the butterfly heuristic. Next, the relative history heuristic is introduced in

Section 3. The test environment is described in Section 4. Subsequently, the results of the experiments are given in Section 5. Finally, in Section 6 we present our conclusion and propose topics for future research.

2 The History Heuristic and the Butterfly Heuristic

The history heuristic is a simple, inexpensive way to reorder moves dynamically at interior nodes of search trees. It was invented by Schaeffer [15] and has been adopted in several game-playing programs. Unlike the killer heuristic, which only maintains a history of the one or two best killer moves at each ply, the history heuristic maintains a history for every legal move seen in the game tree. Since there is only a limited number of legal moves, it is possible to maintain a score for each move in two (black and white) tables. At every interior node in the search tree the history-table entry for the best move found is incremented by a value (e.g., 2^d, where d is the depth of the subtree searched under the node). The best move is in this case defined as the move which either causes an alpha-beta cut-off, or which causes the best score. When a new interior node is examined, moves are re-ordered by descending order of their scores. The scores in the tables can be maintained during the whole game. Each time a new search is started the scores are decremented by a factor (e.g., divided by 2). They are only reset to zero or to some default values at the beginning of a complete new game. Details on the effectiveness or the strategy to maintain history scores during the whole game are dependent on the domain or game program.

The history heuristic does not cost much memory. The history tables are defined as two tables with 4096 entries (64 from_squares × 64 to_squares), where each entry is 4 or 8 bytes large. These tables can be easily indexed by a 12-bit key representing the origin and destination. In the history table we have also defined moves which are illegal.

Hartmann [8] called attention to two disadvantages of the history heuristic. (A) Quite some space for the history table is wasted, because space for illegal moves is reserved too. For instance, in the game of chess 44 per cent of the possible moves are legal. In LOA, this number is even lower because knight moves are not allowed. This gives that 1456 of the 4096 moves are legal, meaning that only 36 per cent of the entries in the table are used. Although this waste of memory is not a problem for games with a small dimensionality of moves, it can be a problem for games with a larger dimensionality of moves (for instance Amazons). (B) Moreover, Hartmann pointed out that some moves are played less frequently than others. There are two reasons for this. (B1) The moves are less frequently considered as good moves. (B2) The moves occur less frequently as legal moves in a game. The disadvantage of the history heuristic is that it is biased towards moves that occur more often in a game than others. However, the history heuristic has as implicit assumption that all the legal moves occur roughly with the same frequency in the game (tree). So, in games where this condition approximately holds an absolute measure seems appropriate. But in other games where some moves occur more frequently than other moves, we should resort to other criteria. For instance, assume we have a move which is quite successful when applicable (e.g., it then causes a cut-off) but it does not occur so often as a legal move in the game tree. This move will not obtain a high history score and is therefore ranked quite low in the move ordering. Therefore a different valuation of such a move may be considered.

To counter some elements of the two disadvantages Hartmann [8] proposed an alternative for the history heuristic, the butterfly heuristic. This heuristic takes the move frequencies in search trees into account. Two tables are needed (one for Black and one for White), called *butterfly boards*. They are defined in the same way as in the history heuristic (i.e., 64 from_squares × 64 to_squares). Any move that is not cut is recorded. Each time a move is executed in the search tree, its corresponding entry in the butterfly board (for each side) is also incremented by a value. Moves are now reordered by their butterfly scores. The butterfly heuristic was denied implementation by its inventor, since he expected that it would be far less effective than the history heuristic.

3 The Relative History Heuristic

We believe that we can considerably improve the performance of the history heuristic in some games by making it *relative* instead of absolute. The score used to order the moves (*movescore*) is given by the following formula:

$$movescore = \frac{hhscore}{bfscore} \quad (1)$$

where *hhscore* is the score found in the history table and *bfscore* is the score found in the butterfly board. We call this move ordering the relative history heuristic. We remark that we only *update* the entries of moves seen in the regular search, not in the quiescence search, because some (maybe better) moves are not investigated in the quiescence search. We *apply* the relative history heuristic everywhere in the tree (including in the quiescence search) for move ordering.

In some sense this heuristic is related to the realization-probability search method [19]. In that scheme the move frequencies gathered *offline* are used to limit or extend the search.

4 Test Environment

In this section the test environment is described which is used for the experiments. First, we briefly explain the game of Lines of Action (LOA). Next, we give some details about the search engine MIA.

4.1 Lines of Action

Lines of Action (LOA) [14] is a two-person zero-sum chess-like connection game with perfect information. It is played on an 8 × 8 board by two sides, Black and White. Each side has twelve pieces at its disposal. The black pieces are placed in two rows along the top and bottom of the board, while the white pieces are placed in two files at the left and right edge of the board. The players alternately move a piece, starting with Black. A move takes place in a straight line, exactly as many squares as there are pieces of either colour anywhere along the line of movement. A player may jump over its own pieces. A player may not jump over the opponent's pieces, but can capture them by landing on them. The goal of a player is to be the first to create a configuration on the board in which all own pieces are connected in one unit. In the case of simultaneous connection, the game is drawn. The connections within the unit may be either orthogonal or diagonal. If a player cannot move, this player has to pass. If a position with the same player to move occurs for the third time, the game is drawn.

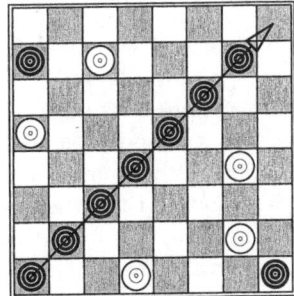 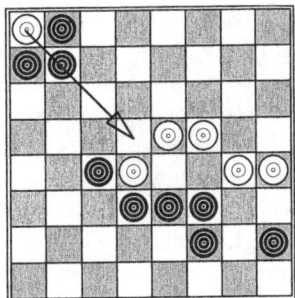

Fig. 1. (a) Rare move; (b) Blocked move

LOA is a nice test bed for the relative history heuristic. Dependent on the position some moves occur more often than others in the search tree. For example, moves going seven squares far are possible if and only if there are seven pieces of the same colour side by side on that line. In Figure 1a it is possible to move **a1-h8**, but it is very rare that in a real game a position occurs where seven pieces of the same colour occupy a diagonal. In contrast, consider a move like **a8-d5** which occurs regularly in a game, but in the position depicted in Figure 1b it will not often be applied in the corresponding search tree, since most of the time Black will not consider to move its piece on **b7**.

4.2 MIA IV

MIA IV is a LOA-playing tournament program, which won the 8th Computer Olympiad. It performs an $\alpha\beta$ depth-first iterative-deepening search. Several techniques are implemented to make the search efficient. The program uses PVS (Principal Variation Search) to narrow the $\alpha\beta$ window as much as possible [13]. A transposition table with 2^{21} double entries (using the *two-deep* replacement scheme [3]) is applied to prune a subtree or to narrow the $\alpha\beta$ window. At all interior nodes which are more than 2 ply away from the leaves, the program generates all the moves to perform the Enhanced Transposition Cutoffs (ETC) scheme [18]. Next, a null move [7] is performed before any other move and it is searched to a lower depth (reduced by R) than other moves. In the search tree we distinguish three types of nodes, namely PV nodes, CUT nodes, and ALL nodes [10,13]. The null move is done at CUT nodes *and* at ALL nodes. At a CUT node a variable scheme, called adaptive null move [9], is used to set R. If the remaining depth is more than 6, R is set to 3. When the number of pieces of the side to move is lower than 5 the remaining depth has to be more than 8 to set R to 3. In all other cases R is set to 2. For ALL nodes $R = 3$ is used. If the null move does not cause a β cut-off, multi-cut [2] is performed. Experiments showed that using multi-cut is not only beneficial at CUT nodes but also at ALL nodes [22]. For move ordering, the move stored in the transposition table, if applicable, is always tried first. Next, two killer moves [1] are tried. These are the last two moves, which were best or at least caused a cut-off at the given depth. Thereafter follow: (1) capture moves going to the inner area (the central 4×4 board) and (2) capture moves going to the middle area (the 6×6 rim). All the other moves are ordered decreasingly according to their scores in the (relative) history table [15]. If a cut-off occurs because of multi-cut, transposition table or ETC, its corresponding entry in the history table or butterfly board is also updated with the depth used for exploration. In the leaf nodes of the tree a quiescence search is performed. This quiescence search looks at capture moves that form or destroy connections [21] and at capture moves going to the central 4×4 board.

5 Experiments

In this section we show the results of various experiments with the relative history heuristic. This is done on a test set of 171 LOA positions.[1] We performed six series of experiments. In the first and second series, we tested the standard history heuristic and the relative history heuristic with different increments, respectively (Subsection 5.1). In the third, fourth, and fifth series, we compared the performance of the relative history heuristic with the standard history heuristic under different configurations (Subsection 5.2). Finally, in the sixth series of experiments, we tested the performance of the relative history heuristic in another domain, namely for 24 test positions for 6×6 Go (Subsection 5.3).

5.1 Increment Settings

In the following two series of experiments we tried to find the optimal increment setting for the history table and the butterfly board, which gave the largest node reduction. Using our set of 171 LOA positions, the program was tested for depth 14 using its normal enhancements as described in the previous subsection.

In the first series of experiments we tested the increments of the history table in a configuration where we used the standard history heuristic. Initially the history heuristic was developed for programs that were searching to a depth

---

[1] The test set can be found at http://www.cs.unimaas.nl/m.winands/loa/TMP.zip.

Table 1. Performance of the history heuristic with different increments on a test set of 171 positions

History Increment	Total Nodes		
	depth 5	depth 9	depth 14
0	2,480,001	188,717,928	30,997,625,767
1	1,901,956	113,163,113	14,478,291,866
d	1,896,429	111,283,177	14,064,388,392
d^2	1,900,055	111,673,124	13,915,673,199
2^d	1,878,114	111,471,652	13,925,222,389

much less than 14. Considering the nature of a search tree, it might be that the best increment to be used depends on the search depth. Therefore we performed experiments for the original depths 5 and 9 used as test depths by Schaeffer [15,17]. The following increments were used: 1, d, d^2, and 2^d, where d is the explored depth of the move causing the cut-off. The increment 2^d is the standard increment of the history table. The result of the search without history heuristic (increment 0) is given for comparison. Table 1 shows that there is not much difference between the size of the search tree using increments d, d^2 and 2^d. For depth 14, the history heuristic gives in all the cases a reduction of approximately 55 per cent of the number of nodes searched. Unlike data for chess, reported in [17], we see a steady growth of the reduction with increasing search depth in LOA. Surprisingly, the increment of 1 is generating for the various depths a search tree that is only slightly larger than other increments. Apparently, the depth of the move explored is not so important in the history heuristic. Hence we may conclude that, unlike some data so far available for chess [16], the choice of the increment is of little value in the test environment we studied.

In the second series of experiments we looked at various increment parameter settings of the history table and the butterfly board (h,b) in our engine using the relative history heuristic and using a search depth of 14 ply. In Table 2 the total number of nodes searched for each combination of parameters is given. In the process of parameter tuning we found that $(d^2, 2^d)$ is the most efficient. However, the difference with several other parameter configurations is not significant (e.g., (1,1), (d^2, d) or $(2^d,1)$). The difference between the best and worst parameter setting is 6 per cent in nodes searched. Hence, we may conclude that the exact choice of parameters seems to be not very critical.

5.2 Performance in LOA

In the third series of experiments we tested the added value of the relative history heuristic, using the optimal parameter setting of the previous subsection, against the same set of 171 LOA positions for several depths in our original search engine under different conditions.

In Figure 2 we plot the relative performance of the two heuristics defined as the size of the search tree investigated using the relative history heuristic divided by that of the standard history heuristic, as a function of the search depth. We

Table 2. Performance of the relative history heuristic with different increments on a test set of 171 positions

History Increment	Butterfly Increment	Total Nodes
1	1	12,261,241,807
1	d	12,544,923,748
1	d^2	12,936,458,114
1	2^d	12,741,580,747
d	1	12,654,462,037
d	d	12,433,892,630
d	d^2	12,914,014,566
d	2^d	13,075,535,903
d^2	1	12,501,473,310
d^2	d	12,238,059,952
d^2	d^2	12,509,830,417
d^2	2^d	12,234,575,562
2^d	1	12,354,762,028
2^d	d	13,081,065,785
2^d	d^2	12,928,253,841
2^d	2^d	12,954,976,931

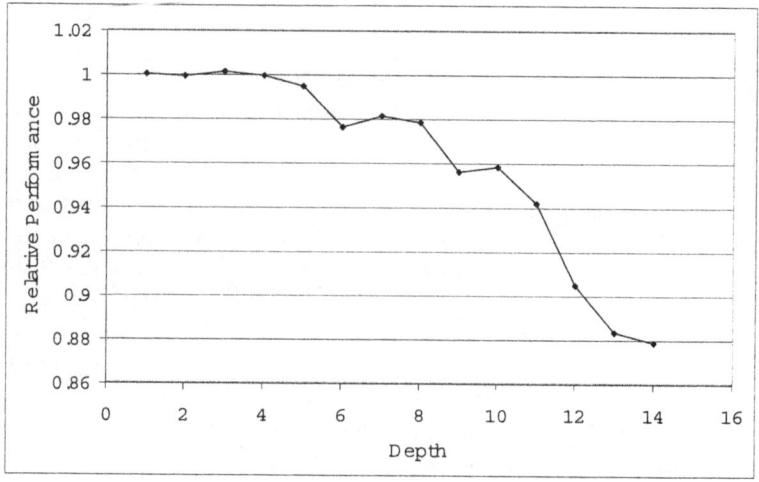

Fig. 2. Performance of the Relative History Heuristic

observe that until depth 8 there is no significant difference between the two move-ordering schemes. From depth 9 onwards the difference increases with the depth to some 12 per cent at depth 14. We see that the search enhanced with the relative history heuristic searches fewer nodes than the one enhanced with the standard setting.

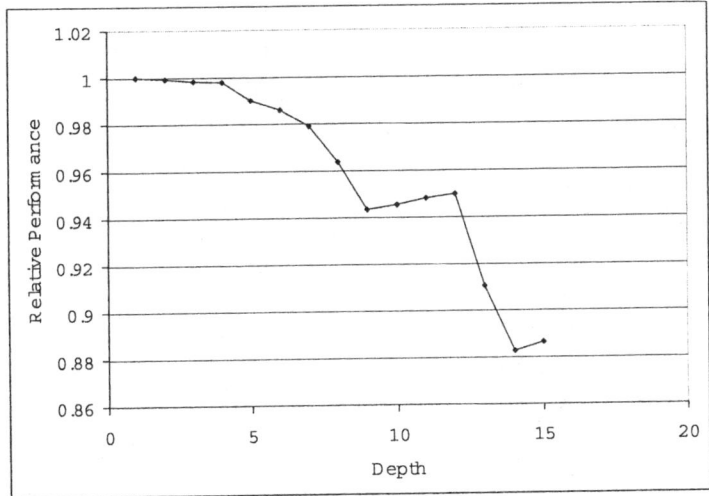

Fig. 3. Validating the Relative History Heuristic

Since we have tuned our heuristic with this particular test set (Subsection 5.1), we performed a fourth series of experiments on a set consisting of 156 different positions[2] to validate the result. The positions were searched up to depth 15. In Figure 3 we see the relative performance of the two heuristics on the validation set. If we compare Figure 2 with Figure 3, we see that similar results are achieved.

Since the performance of many search enhancements may to some extent depend on the search engine, we modified the search engine in the fifth series of experiments by switching the multi-cut forward-pruning mechanism off. Because of the diminished forward pruning the sizes of our search trees increased and we were not able to conduct experiments at depth 13 and further. Looking at Figure 4 we see that the relative history heuristic decreases the search with 11 per cent at depth 12. Hence, we may conclude that the same pattern as in the previous experiments has started. We expect that this pattern will continue, which is to be confirmed in the future by more powerful machines.

5.3 Performance in Go

The relative history heuristic was designed for LOA. To investigate whether the relative history heuristic would be interesting for other domains too, we tested its performance on the small-board games of Go in the sixth series of experiments, for which we used the program MIGOS that recently had solved Go on the 5×5 board [20].

MIGOS uses an iterative-deepening PVS with a transposition table with 2^{24} double entries (using the *two-deep* replacement scheme), enhanced transposi-

---

[2] The test set can be found at http://www.cs.unimaas.nl/m.winands/loa/VMP.zip.

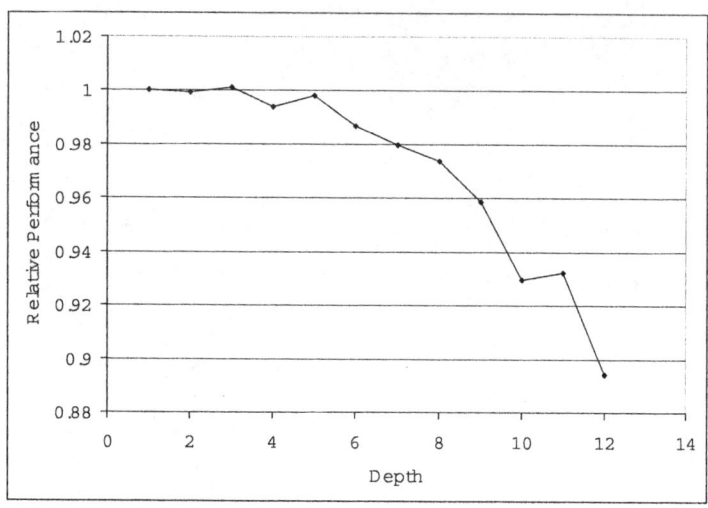

Fig. 4. Relative History Heuristic without using multi-cut

tion cut-offs, symmetry lookups in the transposition table, internal unconditional bounds, and an enhanced move ordering in which the history heuristic is an important component. The implementation of the history heuristic employs one shared table for both the black and white moves which exploits the game-dependent property in Go that moves on the same intersection are often good for both sides. After some parameter tuning for the relative history heuristic increments, which we optimised for solving the empty 5×5 board, we found that using d^3 for both the history and the butterfly board gave quite promising results[3].

The current challenge in small-board Go is solving the 6×6 board (5×5 is the largest square board solved by a computer). Therefore we decided to test the performance of the relative history heuristic on a set of 24 problems for the 6×6 board published in *Go World* by James Davies [5,6]. Figure 5 shows the average relative performance of the relative history heuristic compared to the standard settings without a butterfly table. Since we only used a small number of test positions we also plotted the standard deviations. They tend to increase with the search depth. The reasons for this are (1) the exponential effect of changes in the move ordering, and (2) a reduction in the number of positions because some positions are already solved at smaller depths. The results indicate again that for shallow searches not much should be expected of using the relative history heuristic. However, after about 10 ply the first improvements become noticeable and at about 15 ply the relative history heuristic achieves a reduction of roughly

---

[3] We tested this combination on the LOA test set, too. Our experiments showed that this combination belongs to the better ones.

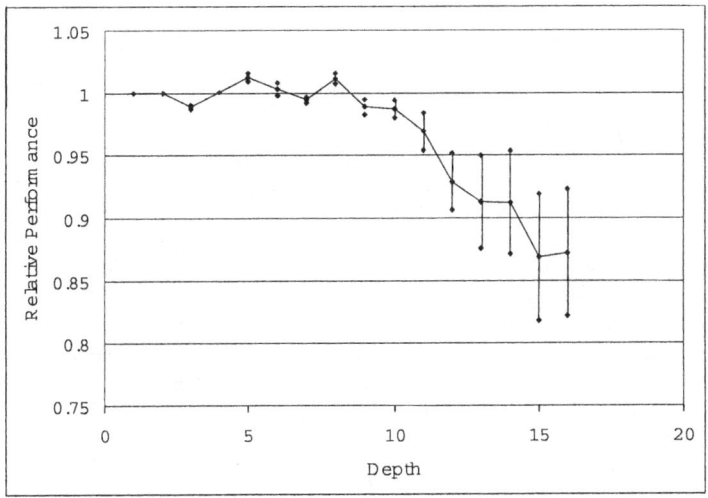

Fig. 5. Performance of the Relative History Heuristic in 6 × 6 Go

13 per cent. However, we remark that the test set is too small to draw strong conclusions. So far the results are favourable for the relative history heuristic and they indicate that the relative history heuristic is worth investigating in other domains as well.

6 Conclusion and Future Research

Combining the ideas of the history heuristic and the butterfly heuristic resulted in the relative history heuristic. This heuristic does not suffer from underestimating less frequently occurring moves in the search tree as the history heuristic does. We favour moves which are the good moves on average instead of moves which are the best move in absolute terms. Both the history heuristic and the relative history heuristic show a steady growth of the reduction with increasing search depth. Using the relative history heuristic our LOA program MIA searches even between 10 and 15 per cent fewer nodes (see Subsection 5.2). The results were confirmed by the Go program MIGOS. Hence, we may conclude that the relative history heuristic is a valuable technique to order the moves in a game tree of considerable depth (more than 12 plies).

It is remarkable that the utility of increments other than 1 does not show much better performance in the (relative) history heuristic for our LOA program MIA. The good performance of the increment of 1 could be the result of some domain-dependent properties.

Finally, it would be interesting for future research to test our heuristic in still more different games, especially in chess, since the original history heuristic was developed for chess.

Acknowledgement. We gratefully acknowledge financial support by the Universiteitsfonds Limburg / SWOL.

References

1. S.G. Akl and M.M. Newborn. The principal continuation and the killer heuristic. In *1977 ACM Annual Conference Proceedings*, pages 466–473. ACM, Seattle, USA, 1977.
2. Y. Björnsson and T.A. Marsland. Multi-cut alpha-beta pruning. In H.J. van den Herik and H. Iida, editors, *Computers and Games, Lecture Notes in Computing Science 1558*, pages 15–24. Springer-Verlag, Berlin, Germany, 1999.
3. D.M. Breuker, J.W.H.M. Uiterwijk, and H.J. van den Herik. Replacement schemes and two-level tables. *ICCA Journal*, 19(3):175–180, 1996.
4. M. Campbell, A.J. Hoane Jr., and F.-h. Hsu. Deep Blue. *Artificial Intelligence*, 134(1-2):57–83, 2002.
5. J. Davies. Small-board problems. *Go World*, 14-16:55–56, 1979.
6. J. Davies. Go in lilliput. *Go World*, 17:55–56, 1980.
7. C. Donninger. Null move and deep search: Selective-search heuristics for obtuse chess programs. *ICCA Journal*, 16(3):137–143, 1993.
8. D. Hartmann. Butterfly boards. *ICCA Journal*, 11(2-3):64–71, 1988.
9. E.A. Heinz. Adaptive null-move pruning. *ICCA Journal*, 22(3):123–132, 1999.
10. D.E. Knuth and R.W. Moore. An analysis of alpha-beta pruning. *Artificial Intelligence*, 6(4):293–326, 1975.
11. L. Kocsis. *Learning Search Decisions*. PhD thesis, Universiteit Maastricht, Maastricht, The Netherlands, 2003.
12. L. Kocsis, J.W.H.M. Uiterwijk, and H.J. van den Herik. Move ordering using neural networks. In L. Montosori, J. Váncza, and M. Ali, editors, *Engineering of Intelligent Systems, Lecture Notes in Artificial Intelligence 2070*, pages 45–50. Springer-Verlag, Berlin, Germany, 2001.
13. T.A. Marsland and M. Campbell. Parallel search of strongly ordered game trees. *Computing Surveys*, 14(4):533–551, 1982.
14. S. Sackson. *A Gamut of Games*. Random House, New York, NY, USA, 1969.
15. J. Schaeffer. The history heuristic. *ICCA Journal*, 6(3):16–19, 1983.
16. J. Schaeffer. *Experiments in Search and Knowledge*. PhD thesis, Department of Computing Science, University of Waterloo, Canada, 1986.
17. J. Schaeffer. The history heuristic and the performance of alpha-beta enhancements. *IEEE Transactions on Pattern Analysis and Machine Intelligence*, 11(11):1203–1212, 1989.
18. J. Schaeffer and A. Plaat. New advances in alpha-beta searching. In *Proceedings of the 1996 ACM 24th Annual Conference on Computer Science*, pages 124–130. ACM Press, New York, NY, USA, 1996.
19. Y. Tsuruoka, D. Yokoyama, and T. Chikayama. Game-tree search algorithm based on realization probability. *ICGA Journal*, 25(3):132–144, 2002.
20. E.C.D. van der Werf, H.J. van den Herik, and J.W.H.M. Uiterwijk. Solving Go on small boards. *ICGA Journal*, 26(2):92–107, 2003.
21. M.H.M. Winands, J.W.H.M. Uiterwijk, and H.J. van den Herik. The quad heuristic in Lines of Action. *ICGA Journal*, 24(1):3–15, 2001.
22. M.H.M. Winands, H.J. van den Herik, J.W.H.M. Uiterwijk, and E.C.D. van der Werf. Enhanced forward pruning. *Information Sciences*, 175(4):258–272, 2005.

Locally Informed Global Search for Sums of Combinatorial Games

Martin Müller and Zhichao Li

Department of Computing Science,
University of Alberta, Edmonton, Canada
{mmueller, zhichao}@cs.ualberta.ca

Abstract. There are two complementary approaches to playing sums of combinatorial games. They can be characterized as local analysis and global search. Algorithms from combinatorial game theory such as *Hotstrat* and *Thermostrat* [2] exploit summary information about each subgame such as its *temperature* or its *thermograph*. These algorithms can achieve good play, with a bounded error. Their runtime depends mainly on the complexity of analyzing individual subgames. Most importantly, it is not exponential in global parameters such as the branching factor and the number of moves in the sum game. One problem of these classic combinatorial game algorithms is that they cannot exploit extra available computation time.

A global minimax search method such as $\alpha\beta$ can determine the optimal result of a sum game. However, the solution cost grows exponentially with the total size of the sum game, even if each subgame by itself is simple. Such an approach does not exploit the independence of subgames.

This paper explores combinations of both local and global-level analysis in order to develop *locally informed global search algorithms* for sum games. The algorithms utilize the subgame structure in order to reduce the runtime of a global $\alpha\beta$ search by orders of magnitude. In contrast to methods such as Hotstrat and Thermostrat, the new algorithms exhibit improving solution quality with increasing time limits.

1 Introduction

Combinatorial game theory studies games that can be decomposed into a sum of independent subgames [2]. Two of many examples are endgames in the games of Go and Amazons. Classical approaches to game tree search work on the global state, and can not profit from the structure of local subgames. Methods for playing sums of games using concepts from combinatorial game theory, that are based purely on local search, have limited accuracy and cannot achieve globally optimal play in general. This paper studies global-search approaches for both optimal and approximate play of sum games that are able to effectively utilize local information. It is organized as follows. Section 2 surveys previous work on sum game algorithms. Section 3 describes several methods for using local information in a global-search framework. Section 4 describes a simple model for generating random abstract hot combinatorial games, and uses it to experimentally evaluate and compare seven methods for playing sums of games. Section 5 concludes the paper and lists future work.

2 Sum Game Algorithms

Algorithms for playing sum games usually contain both a local and a global component. On the local level, individual subgames are analyzed. On the global level, the results of local analyses are combined into an overall strategy. In terms of running time, usually the cost of game-tree search dominates all other factors. Therefore it makes sense to distinguish between algorithms that use only local search, only global search, or a combination of both.

2.1 Approximate Combinatorial Game Algorithms That Use Only Local Search

Computing properties of a combinatorial game position such as its canonical form, its thermograph or its temperature, can be viewed as a local search. Typically, the entire game tree must be expanded. Classical combinatorial game algorithms use only local analyses of this form. On the global level, they perform a static analysis. Examples are Thermostrat, Hotstrat, and Sentestrat [2,10]. All these algorithms play well, but not perfectly. The errors of Thermostrat and Sentestrat can be bounded by the temperature of the hottest or second-hottest subgame [10], while no such bound is known for Hotstrat.

Hotstrat+ – Hotstrat is one of the simplest sum game strategies. It computes the temperature of all subgames, and plays a move in the hottest subgame. A small modification, *Hotstrat+*, performs slightly but consistently better in the experiments reported in this paper. This variant computes a *pseudo-temperature* for each subgame and a given player, defined as the largest t for which the thermograph is equal to its taxed option. For example, in the *sente* position $10|0|| - 1$, the temperature is 1 but Left can already play at $t = 5$, since Left's move to $10|0$ is a threat that raises the temperature of the game to 5. As Left, Hotstrat+ plays in this position at the pseudo-temperature $t = 5$, while plain Hotstrat waits until $t = 1$. As Right, both strategies wait until $t = 1$.

Intuitively, the advantage of Hotstrat+ is that it cashes in free sente moves earlier and reduces the opportunities for the opponent to play "reverse sente" moves such as the move from $10|0|| - 1$ to -1 in the example above.

In this paper, when the meaning is clear from the context, pseudo-temperatures computed by Hotstrat+ will simply be called temperatures.

2.2 Decomposition Search

An exact method for solving sum games that uses a local-search approach is *decomposition search* [6]. In this method, local searches are used to compute the canonical form of each subgame. From the canonical forms, *incentives* of all moves are derived. An incentive measures the difference between the position before and after a move. The incentive of a move by Left from game G to option G^L is defined as $G^L - G$, while the incentive of a Right move from G to G^R is $G - G^R$. The asymmetric definition implements the idea that larger incentives are always better, for either player. In combinatorial games, by convention Left wants to maximize the result while Right wants to minimize it.

Incentives of moves are combinatorial games and as such are partially ordered [1]. If a single move with dominating incentive exists, it is proven to be optimal. In this case the algorithm is very efficient - it found a move using only local search and comparison of locally computed incentives of moves. The other case is when no single move with dominating incentive exists. In this case, a globally optimal play can be found by a combinatorial summation of all subgames. This approach is feasible if the complexity of computing the sum does not lead to a combinatorial explosion. For example, the cooled values of late stage Go endgames investigated by Berlekamp et al. [3,8] often add well and sums can be computed in a reasonable time. However, this approach is no longer practical for sums of "hotter" Go positions that occur earlier in the game. The complexity of computing the sum of such games quickly explodes. This is shown experimentally in Subsection 4.2. Fortunately, finding an optimal move in a sum game can be accomplished much faster than computing the sum.

2.3 Local Move Pruning for Global Minimax Search

A different approach to solving sum games is investigated in [7]. Local enhancements for speeding up full board $\alpha\beta$ search are used for solving the same kind of Go endgames as in decomposition search. Several methods for local move pruning are developed and shown to improve the search performance by orders of magnitude, compared to plain global $\alpha\beta$ search. However, for these late endgames, where usually a single dominated move exists, even a global search method with many local search improvements is completely dominated by decomposition search.

2.4 Complexity Results

The complexity of solving sums of hot games grows quickly. Wolfe [9], building on previous work by Yedwab [10] and Moews [5], showed that playing sums of simple Go endgames of the form $a||b|c$ is PSPACE-hard. In contrast, the experimental results of decomposition search indicate that if dominating moves exist, the complexity of solving sums does not need to grow exponentially with the size of the sum. Since there appears to be a large gap between the theoretical worst case and the typical case in games such as Go, it is important to develop algorithms that will perform well in practice.

3 Using Local Information in Global Search

This paper investigates practical algorithms for playing sums of games in situations where direct summation is impractical. In this sense, the work reported here can be seen as a natural extension of decomposition search to sums of hotter games. The most important goals are to avoid the explicit summation of combinatorial games, and to find fast algorithms for both optimal and approximately optimal play. The algorithms are designed to play a sum of games where each subgame is simple enough such that a complete analysis is possible, and local properties such as incentives, means or temperatures can be computed quickly.

3.1 Exact Search for Optimal Play

In exact search, the minimax value of a sum game for a given first player is determined by a global $\alpha\beta$ search. The effect of using locally computed information for move ordering and for move pruning is studied and compared against standard search techniques.

Move Ordering – In order to maximize the number of cutoffs in the search tree in $\alpha\beta$ search, good move ordering is essential. One of the most effective standard move ordering techniques is to use iterative deepening, and try the best move from the previous iteration first. This is a generic technique that does not use any game-specific features, or the structure of sum games.

A natural way of using information about subgames for move ordering is to compute the temperatures of subgames and order moves starting from the hottest subgame. In the experiments all combinations of these two move-ordering techniques are compared.

Move Pruning – Local analysis allows local move pruning by computing incentives and removing moves with dominated incentives. The tradeoff between the overhead of computing incentives and the gain from pruning dominated moves cannot easily be determined analytically. In this paper it is evaluated empirically.

3.2 Approximate Search

Approximate search methods provide a bridge between local-only analysis methods such as Hotstrat, which have limited accuracy, and exact global search, which may be prohibitively expensive. Approximate search requires a heuristic evaluation function to evaluate non-terminal leaf nodes in the global search.

Heuristic Evaluation of Sum Game Positions – A heuristic evaluation function should approximate the minimax value of a given sum game with a given first player. For example, in the sum game **** the minimax score for left to play is ****, and the evaluation function should predict this score as accurately as possible. Two types of heuristic evaluation functions were investigated.

1. **Static evaluation** – The static evaluation uses a locally computed property, the mean of each subgame. The overall evaluation is the sum of the means of all subgames.
2. **Hotstrat+ rollout** – A leaf node is evaluated by the final result of the game that was completed using the Hotstrat+ strategy for both players.

The two methods have very different characteristics. Static evaluation is fast but quite inaccurate. Hotstrat+ rollouts are usually more precise, but are much slower to compute, especially for long games, because games have to be played out.

Many variations on these basic evaluation functions are possible, and more research is necessary to find possible improvements. One such improvement could be an evaluation bias for the player to move. Let t_{max} be the temperature of the hottest subgame. If a sum game is an "enriched environment" [1], then the first player can achieve a minimax score that is $t_{max}/2$ larger than the sum of the means. However, preliminary experiments with using such a modified static evaluation function in the sum games described below were inconclusive.

Resource-bounded $\alpha\beta$ Search – In order to use a heuristic evaluation function, a control policy must decide when to stop searching and evaluate a position statically. Two such policies were investigated. Both are independent of the specific choice of evaluation function.

1. **Depth-bounded $\alpha\beta$ search** – A simple way of reducing search cost is to limit the maximum search depth. With increasing depth, the heuristic evaluation is applied in closer-to-terminal positions, which should result in increased accuracy. In the limit, if the search depth is sufficient to reach the end of the game along each searched branch, the method computes the optimal result.
2. **Temperature-bounded $\alpha\beta$ search** – This method uses a variable depth search with a temperature bound b. Only hot positions p with $t_{max}(p) > b$ are searched deeper. All positions with $t_{max}(p) \leq b$ are evaluated statically. For $b = 0$, this method is equivalent to a full search. If $b \geq t_{max}(g)$ for the starting position g, only a static evaluation of g is done without any search.

4 Experiments

A series of experiments was performed for both exact and approximate search, using two parameterized test domains with sums of abstract combinatorial games. The tests vary the complexity of individual subgames as well as the number of subgames in a sum. 100 different randomly generated problem instances were used to generate each data point. For exact algorithms, the total runtime (in seconds) for solving all instances is given. For approximate algorithms, both the runtime and the aggregated loss against an optimal player are reported.

4.1 Test Domain for Sum Games

The following simple method, similar to the one in [4], is used to generate random instances from a restricted class of combinatorial games. Let $rnd(n)$ be a function generating a random integer uniformly distributed in the interval $[0, n-1]$. Then a random combinatorial game $g = rcg(k, n)$ of *level* $k > 0$ with *size parameter* $n > 0$ can be generated as follows.

1. Build a complete binary tree with k levels below a root node. The first out-edge of each node corresponds to a Left move, the other to a move by Right.
2. Enumerate the 2^k leaf nodes from right to left, such that the right-most node is node 1 and the left-most node is node 2^k.
3. Assign integers to all leaf nodes. Value v_i is assigned to the i-th leaf node as follows: $v_1 = 0$, $v_{i+1} = v_i + rnd(n)$.

Two properties of this particular generator, that do not hold for all hot combinatorial games, are that values of leaf nodes are monotonically increasing from right to left, and that in each nonterminal position each player has exactly one move.

The 2-level random games used in [4] correspond to games $g = rcg(2, 50) + rnd(50)$ in this framework. The experiments in this paper use 2-level games generated by $rcg(2, 50)$ and 3-level games $rcg(3, 50)$. An example of a $rcg(2, 50)$ game is 114|66||49|0 and an example of a $rcg(3, 50)$ game is 237|191||145|124|||97|57||32|0. Given a subgame generator G, a random *sum game* with s subgames is created simply as the sum of s random subgames generated by G.

4.2 Preliminary Experiment: Combinatorial Summation

This preliminary experiment motivates the need to develop a practical algorithm for playing sums of hot games. It confirms the claim from Subsection 2.2 of this paper that the combinatorial summation of subgames is not a viable algorithm for playing sums of hot games. Table 1 summarizes the results. Even for very small sums of five 2-level or three 3-level games, the algorithm fails due to the quickly growing complexity of computing the canonical form of these sums, and runs out of memory after a few minutes of computation.

4.3 Experiment 1: Move Ordering

Experiments 1 and 2 use a global $\alpha\beta$ search to solve sum games exactly. The first experiment investigates the performance of two move-ordering schemes. *Sort by temperature (TEMP)* sorts all moves according to their temperature. *Best previous first (BEST-PREV)* implements the standard technique of trying the best move from a previous iteration of iterative deepening search before any other move. A basic $\alpha\beta$ search without any move ordering is also included for reference.

Table 1. Time (in seconds) for solving sum games by combinatorial summation. N/C = not completed, out of memory.

Subgames	2	3	4	5
2-level	0.543	9.90	148	N/C
3-level		132	N/C	-	-

Tables 2 and 3 show the performance for 2-level and 3-level games respectively. For very small sums, the overhead of move ordering is greater than the gain. For larger sums, as expected, both move-ordering schemes either alone or in combination perform much better than no ordering. It is very interesting that TEMP alone outperforms both variants where BEST-PREV is used. Even the combination of both methods, which first orders moves by temperature, then moves the previously best move to the front, is slightly but consistently worse than TEMP alone. This may be very surprising to practitioners from other games, where the BEST-PREV enhancement is considered indispensable. It provides a strong indication of how well temperature works as a measure of move urgency in hot games.

To further study the behavior of these two move ordering heuristics, detailed statistics were collected for a series of 100 2-level games with 12 subgames, using the fastest engine including incentive pruning as in Experiment 2. In 80.2% of all searched positions, both heuristics ordered the best move first. In 10.9% of positions, the first-ranked TEMP move was optimal but the first-ranked BEST-PREV move was not. The opposite occurred in 4.3% of all positions, where BEST-PREV had a right move but TEMP did not. Finally, in 4.6% of all positions both heuristics favored a suboptimal move. This result is consistent with the observed better performance of TEMP.

Table 2. Performance of move ordering for 2-level games. Best result in bold.

Method	Subgames							
	2	3	4	5	6	7	8	9
No ordering	**0.172**	**0.254**	**0.339**	0.803	2.77	13.6	85.9	1862
BEST-PREV	0.231	0.288	0.393	**0.709**	2.13	9.09	45.7	293
TEMP	0.204	0.260	0.448	0.766	**1.90**	**6.81**	**29.5**	**131**
BEST-PREV and TEMP	0.262	0.412	0.491	1.09	2.31	10.8	34.5	182

Table 3. Move ordering performance for 3-level games

Method	Subgames				
	2	3	4	5	6
No ordering	**0.347**	1.55	7.80	59.8	956
BEST-PREV	0.452	1.21	2.88	18.4	152
TEMP	0.398	**1.01**	**1.85**	**11.9**	**105**
BEST-PREV and TEMP	0.502	1.11	1.92	13.0	129

4.4 Experiment 2: Move Pruning

Experiment 2 demonstrates the effect of pruning using incentives of moves. Tables 4 and 5 show the performance for 2-level and 3-level games. For comparison, the first row contains the result obtained with TEMP but without using incentive pruning, from line 3 in Tables 2 and 3. In the remaining rows, incentive pruning (INC) is turned on, leading to a huge reduction in search time in all combinations. For all but the smallest instances, the version using TEMP move ordering and INC pruning is consistently the fastest.

Table 4. Incentive pruning for 2-level games

Method	Subgames											
	2	3	4	5	6	7	8	9	10	11	12	13
TEMP	**0.204**	**0.260**	0.448	0.766	1.90	6.81	29.5	131	812	N/A	N/A	N/A
TEMP + INC	0.229	0.296	**0.353**	**0.403**	**0.630**	**1.93**	**4.29**	**8.44**	**19.9**	**62.4**	**162**	**772**
PREV + INC	0.230	0.293	0.446	0.563	0.819	2.36	5.23	11.4	26.7	88.5	221	1082
TEMP + PREV + INC	0.227	0.275	0.433	0.494	0.780	2.02	4.89	10.3	23.1	78.8	192	892

Table 5. Incentive pruning for 3-level games

Method	Subgames							
	2	3	4	5	6	7	8	9
TEMP	**0.398**	1.01	1.85	11.9	105	1276	N/A	N/A
TEMP + INC	0.424	**0.782**	**1.22**	**4.15**	**12.2**	**20.4**	**161**	**791**
PREV + INC	0.431	0.823	1.31	5.95	15.8	29.4	218	1210
TEMP + PREV + INC	0.440	0.815	1.30	5.34	14.0	27.0	196	901

4.5 Experiment 3: Approximation Algorithms

Experiment 3 compares the game-playing strength of different *approximation* algorithms when pitted against a perfect opponent, namely the player using TEMP + INC that performed best in Experiment 2. The results are presented in Tables 6 and 7. The errors given in the tables are the total loss in points incurred by the tested algorithm against the optimal player in a series of 100 games. The smaller the error, the closer the algorithm is to playing perfectly.

This experiment measures only the quality of the players, not the time they need. See Experiment 4 in the next section for an investigation of tradeoffs between execution time and errors. The approximation algorithms compared in this experiment are as follows.

1. **Hotstrat / Hotstrat+** – Compute temperatures / pseudo-temperatures of each subgame, and play in the hottest game as explained in Subection 2.1.
2. **Thermostrat** – Uses a well-known sum-playing strategy based on thermographs [2].
3. **Depth-bounded $\alpha\beta$ search, static evaluation** – Uses a fixed-depth 3 ply search with static sum-of-means evaluation.

4. **Depth-bounded $\alpha\beta$ search, *Hotstrat+* rollouts** – Uses a fixed-depth 3 ply search but with rollouts as evaluation.
5. **Temperature-bounded $\alpha\beta$ search, static evaluation** – The temperature bound for a search of position p is set to $b = 0.8 \times t_{max}(p)$.
6. **Temperature-bounded $\alpha\beta$ search, *Hotstrat+* rollouts** – Uses the same search control as the previous method with a rollout evaluation.

The following abbreviations are used in the tables:

TEMP-$\alpha\beta$ – Temperature-bounded $\alpha\beta$ search;
HR – *Hotstrat+* rollout evaluation;
SE – Static evaluation by sum of means.

Table 6. Total loss over 100 2-level games against perfect opponent. Best result in bold. Best static method in italic. (1st) = first player uses method, (2nd) = second player uses method.

Method	Number of Subgames											
	2	3	4	5	6	7	8	9	10	11	12	13
Hotstrat (1st)	66	68	100	94	151	207	194	205	211	186	195	194
Hotstrat (2nd)	79	111	123	185	137	148	166	176	190	251	169	208
Hotstrat+ (1st)	*68*	*65*	*93*	*84*	*136*	*180*	*158*	*171*	*199*	*176*	*178*	*184*
Hotstrat+ (2nd)	*68*	*82*	*88*	*147*	*115*	*118*	*126*	*138*	*158*	*215*	*147*	*188*
Thermostrat (1st)	*44*	104	121	106	165	181	190	224	217	193	183	192
Thermostrat (2nd)	79	110	120	183	135	146	157	174	188	248	161	192
3-ply $\alpha\beta$, SE (1st)	26	62	68	62	122	142	148	167	189	201	183	212
3-ply $\alpha\beta$, SE (2nd)	0	69	42	103	102	110	130	134	143	178	202	189
3-ply $\alpha\beta$, HR (1st)	0	0	8	21	22	36	47	55	70	89	120	99
3-ply $\alpha\beta$, HR (2nd)	0	0	1	13	18	20	51	**33**	67	86	79	92
TEMP-$\alpha\beta$, SE (1st)	46	87	103	150	325	348	322	378	501	552	538	601
TEMP-$\alpha\beta$, SE (2nd)	33	33	93	164	202	212	288	368	432	512	542	621
TEMP-$\alpha\beta$, HR (1st)	**0**	**0**	**5**	**17**	**19**	**25**	**45**	43	**53**	**61**	**78**	**81**
TEMP-$\alpha\beta$, HR (2nd)	**0**	2	17	**11**	**10**	**19**	39	36	54	62	59	71

Among the static methods, *Hotstrat+* is consistently the best. The differences to plain Hotstrat and Thermostrat are rather small. Overall, *Hotstrat+* rollouts are more precise than static evaluation for the same search depth. 3-ply search with static evaluation is better than static *Hotstrat+* up to about ten 2-level subgames, then it becomes worse. 3-ply search with *Hotstrat+* rollouts is always better than *Hotstrat+* without search. The two results for temperature-bounded search are very different. With static evaluation, this search method does not perform well. However, with rollouts it has the smallest total error in most tests. As the first player in a large number of 3-level subgames, 3-ply search with rollouts also does well.

Table 7. Total loss over 100 3-level games

Method	\multicolumn{8}{c}{Number of Subgames}
	2	3	4	5	6	7	8	9
Hotstrat (1st)	56	135	183	251	195	315	330	366
Hotstrat (2nd)	71	187	262	260	268	274	258	291
Hotstrat+ (1st)	*55*	*111*	*152*	*232*	*182*	*281*	*295*	*331*
Hotstrat+ (2nd)	*67*	*162*	*218*	*229*	*232*	*212*	*215*	*248*
Thermostrat (1st)	70	285	282	334	231	386	334	349
Thermostrat (2nd)	93	*159*	246	248	241	247	252	280
3-ply $\alpha\beta$, SE (1st)	54	111	120	212	202	265	278	294
3-ply $\alpha\beta$, SE (2nd)	4	89	135	168	181	221	234	262
3-ply $\alpha\beta$, HR (1st)	0	**7**	31	88	93	142	**160**	148
3-ply $\alpha\beta$, HR (2nd)	0	23	30	**63**	88	174	202	195
TEMP-$\alpha\beta$, SE (1st)	57	192	262	387	531	662	780	902
TEMP-$\alpha\beta$, SE (2nd)	26	121	202	248	431	601	735	886
TEMP-$\alpha\beta$, HR (1st)	0	8	**30**	85	72	**132**	189	148
TEMP-$\alpha\beta$, HR (2nd)	0	**22**	15	72	56	118	**168**	158

This experiment should be considered as a first exploration of the potential of the different methods. However, the comparison is not fair since the running times of the methods are very different. Static methods are much faster than search-based methods, and search using static evaluation is much faster than search using rollouts. The choice of the search parameters depth and temperature bound in this experiment is also rather arbitrary. The next experiment compares tradeoffs between running time and error of the same set of algorithms.

4.6 Experiment 4: Time-Error Tradeoffs

Figure 1 plots error versus time used for the three static and four search-based algorithms. The test runs used sum games consisting of twelve 2-level games. The search depth of fixed-depth searches varies from 0 to 24 in increments of 1. Depth 24 represents a complete search, since each of the 12 subgames lasts at most 2 moves. For temperature-bounded search, the bound was set to $c \times t_{max}$, where c varied from 0.0 to 1.0 in increments of 0.1.

The clear winner in these experiments is the simplest method: fixed-depth search with static evaluation. Even though the accuracy of rollouts is better for the same fixed search depth, as shown in Experiment 3 above, that evaluation is much too slow to be competitive. It remains unclear whether a different method can be found that achieves some of the precision of rollouts but is much faster. For example, depth-limited partial rollouts are an option.

Temperature-bounded search, which did very well combined with rollouts in the untimed experiment above, was also inferior in the timed experiment. The initial error of this method is very high. However, the slope of the error curve seems to be slanted more than for fixed-depth search, so there is hope that this method will perform well for more complex sums. More research is required.

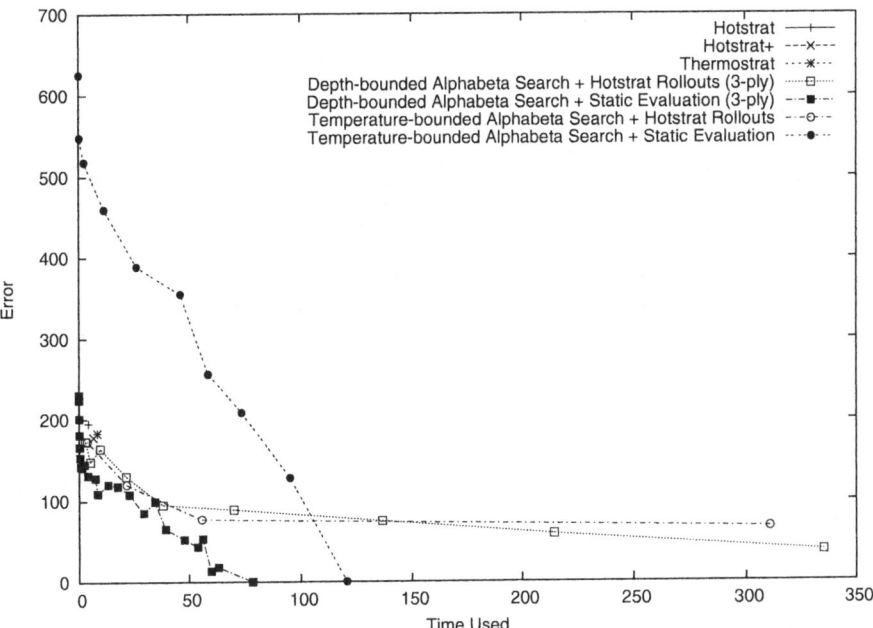

Fig. 1. Decrease of errors (points lost against optimal player) over time (in seconds) for the four search-based algorithms. For the three static methods, their single data point is shown.

5 Conclusions and Future Work

To the best of the authors' knowledge, this paper presents the first systematic experimental study of algorithms for efficiently solving sum games by search. For both exact and heuristic search in sum games, using local information in a global search is very effective. Move ordering by temperature works very well, and pruning of dominated incentives leads to a huge reduction in the search space. Both of these methods greatly improve a search-based sum game solver.

For heuristic search, a simple fixed-depth search with a sum of means static evaluation function performed best in the experiments. The two other techniques investigated, temperature-bounded search and *Hotstrat+* rollout evaluation, showed promise in untimed trials but were not competitive in their current form.

The main intended application of this method is for endgames in combinatorial games with many hot subgames, such as Amazons and Go. Based on the results with artificial games reported here, it seems feasible to evaluate such endgames with high accuracy. If the local games analysis itself is only approximate, as can be expected in complex endgame positions, the error of approximate sum game evaluation may be smaller than the local evaluation error in practice.

One important open question is whether real endgames behave similarly to the sums of abstract games explored in this paper. For example, the relative speed of position evaluation compared to the speed of sum game evaluation will certainly influence the results. As an intermediate step, before applying the method to full-scale Go or Amazons, a database of local endgame positions from such games could be built, and similar experiments repeated with sums of local positions from real games.

Finally, the question of a better random generator for hot combinatorial games should also be investigated. Maybe, specialized generators for more realistic "Go-like" or "Amazons-like" random games could be developed as well.

Acknowledgements

The authors wish to thank NSERC (the Natural Sciences and Engineering Research Council of Canada) for financial support. Tristan Cazenave and Jean Mehat exchanged information about their implementation of classical sum game playing strategies.

References

1. E. Berlekamp. The economist's view of combinatorial games. In R. Nowakowski, editor, *Games of No Chance: Combinatorial Games at MSRI*, pages 365–405. Cambridge University Press, 1996.
2. E. Berlekamp, J. Conway, and R. Guy. *Winning Ways*. Academic Press, London, 1982. Revised version published 2001-2004 by AK Peters.
3. E. Berlekamp and D. Wolfe. *Mathematical Go: Chilling Gets the Last Point*. A K Peters, Wellesley, 1994.
4. T. Cazenave. Comparative evaluation of strategies based on the values of direct threats. In *Board Games in Academia V*, Barcelona, 2002.
5. D.J. Moews. *On Some Combinatorial Games Connected with Go*. PhD thesis, University of California at Berkeley, 1993.
6. M. Müller. Decomposition search: A combinatorial games approach to game tree search, with applications to solving Go endgames. In *IJCAI-99*, pages 578–583, 1999.
7. M. Müller. Global and local game tree search. *Information Sciences*, 135(3–4):187–206, 2001.
8. W. Spight. Go thermography - the 4/21/98 Jiang-Rui endgame. In R. Nowakowski, editor, *More Games of No Chance*, pages 89–105. Cambridge University Press, 2002.
9. D. Wolfe. Go endgames are PSPACE-hard. In R. Nowakowski, editor, *More Games of No Chance*, pages 125–136. Cambridge University Press, 2002.
10. L. Yedwab. On playing well in a sum of games. Master's thesis, MIT, 1985. MIT/LCS/TR-348.

Current Challenges in Multi-player Game Search

Nathan Sturtevant

Department of Computing Science,
University of Alberta, Edmonton, Alberta, Canada
nathanst@cs.ualberta.ca

Abstract. Years of work have gone into algorithms and optimizations for two-player perfect-information games such as Chess and Checkers. It is only more recently that serious research has gone into games with imperfect information, such as Bridge, or game with more than two players or teams of players, such as Poker. This work focuses on multi-player game search in the card games Hearts and Spades, providing an overview of past research in multi-player game search and then presents new research results regarding the optimality of current search techniques and the need for good opponent modeling in multi-player game search. We show that we are already achieving near-optimal pruning in the games Hearts and Spades.

1 Introduction

Artificial Intelligence research in the field of two-player games has been quite successful, with well-publicized victories in games such as chess with DEEP BLUE [1] and checkers with CHINOOK [2]. While there are still interesting challenges in these games, there are many other domains with new challenges yet to be explored. Recent work for programs such as POKI in poker [3] and GIB in bridge [4] has started to shift focus to the difficulties associated with imperfect information and opponent modeling.

Another area that has recently received further attention is the area of multi-player game search. Specifically, this is search applied to domains with three or more players or teams of players. While one might expect that most work from two-player games would extend directly to multi-player games, there are many subtleties that have to be considered. This means that the lessons and techniques derived from work in two-player games must be reconsidered for multi-player games. Techniques of prominence from two-player games may no longer be dominant, while good ideas in two-player games that were not effective enough in practice may find their place in multi-player games.

Without question the most prominent technique from two-player games is minimax search with alpha-beta pruning [5], although this is not the only search algorithm used in two-player games. For instance, there have been attempts to do more selective searches using algorithms like MGSS* [6], or attempts to use an opponent model to increase quality of play, such as in M* [7]. But, in general, these techniques have not found widespread usage, as they cannot compete well with the simplicity and search depth available from minimax with alpha-beta. For instance, while M* was able to out-perform minimax at fixed depth searches, it was unable to do so when the search was limited by node expansions.

The goal of this paper is to provide an overview of the work that has been done in multi-player games and suggest new challenges for future work in the field. The biggest question we wish to address is whether work in multi-player games will follow the route of two-player games, being dominated by algorithms and techniques for deeper search, or whether techniques based on things like selective search or higher-quality shallow search will be more important. As part of that process we present a brief history of multi-player games, new results on the relative size of multi-player game trees in relation to optimally pruned trees, and an overview of some of the techniques and ideas related to opponent modeling that have yet to be explored in multi-player games.

2 Multi-player Game Search Research History

Work on multi-player games traces back, in some sense, to Nash's original work defining equilibrium points, showing that non-cooperative games have at least mixed equilibriums [8]. More importantly, it was later shown that there is a pure equilibrium for every perfect information finite game [9, 10]. These theorems were first realized in the field of Artificial Intelligence by Luckhardt and Irani in the max^n algorithm [11]. In her PhD thesis, Carol Redfield (Luckhardt) also did work on formalizing the max^n algorithm, coalition formation, and pruning [12]. However, pruning was only presented within the context of selectively evaluating the components of each leaf value in the game tree, as opposed to pruning away branches of the game tree, as is done in alpha-beta.

Around the same time David Mutchler showed that max^n displays game-tree pathologies [13], as was originally observed in minimax [14, 15]. That is, searching deeper into the game tree can produce less accurate results. To date, however, no one has shown that these pathologies occur outside of well-structured artificial game trees.

In 1991, Rich Korf published the first thorough analysis of alpha-beta style pruning in max^n [16]. Specifically, without any bounds on the range of leaf values in a max^n game tree, no pruning is generally possible. But, assuming there is an upper bound on the sum of all players' scores, and a lower bound on each player's score, pruning is possible. This pruning can be divided into three classes, immediate pruning, shallow pruning, and deep pruning. Immediate pruning corresponds roughly to a win or loss in a two-player game. Shallow pruning uses the bounds from two consecutive players to prune a game tree. Deep pruning attempts to use the bounds from two non-consecutive players in a game tree, however, is not guaranteed to preserve the max^n value of a game tree, and cannot be applied in general. In the best case, as the branching factor gets large, shallow pruning can reduce the branching factor of a game tree from b to $O(\sqrt{b})$ as b gets large.

No further work in multi-player game search was published until 2000, when Nathan Sturtevant, working with Rich Korf, showed that in some games shallow pruning could never be applied [17]. Specifically, the effectiveness of shallow pruning relies on both the order of nodes in the tree and the range of leaf-values in the tree, as opposed to alpha-beta, whose effectiveness just relies on node ordering. However, this work also showed that information from monotonic heuristics could be used to prune $\max^n$ game trees. What is more, the bounds from shallow pruning can be combined with the bounds from monotonic heuristics to prune even more effectively.

In this same work, the idea of the paranoid algorithm was formally analyzed for the first time. The idea of the paranoid algorithm is to reduce a multi-player game to a two-player game by assuming that all of one's opponents have formed a coalition against you. While this idea had been touched upon in both Mutchler and Luckhardt's work, it had not been seriously considered as a plausible search algorithm, and its properties had not been explored. The paranoid algorithm has the attractive feature that it inherits the ability to use any technique from two-player games, at the cost of using unrealistic opponent models. In an n-player paranoid game tree, alpha-beta pruning will reduce the branching factor from b to $O(b^{n-1/n})$.

The properties of $\max^n$ were further explored in [18]. This work brought to light some of the issues with using the $\max^n$ algorithm, specifically that tie-breaking is a major issue which limits the applicability of some techniques such as zero-window search [19] from two-player games. It also showed that the paranoid algorithm could be competitive in some games, outperforming $\max^n$ in both fixed-depth and node-limited searches in the game of Chinese Checkers.

In 2003 there was a significant advance in pruning algorithms for $\max^n$ game trees with the introduction of speculative pruning [20]. This is the first pruning algorithm for $\max^n$ that, given a constant-sum game, relies only on the ordering of nodes, as opposed to the actual leaf values in the game tree for effective pruning. Asymptotically, it offers the same reduction as the paranoid algorithm, $O(b^{n-1/n})$, however in practice the paranoid algorithm will still dominate speculative pruning.

There is a discrepancy between the best case branching factor of shallow pruning, $O(\sqrt{b})$, and the best case branching factor of paranoid and speculative pruning $O(b^{n-1/n})$. That is, the best case of shallow pruning depends only on the branching factor of a game, not on the number of players in the game. This discrepancy is explained in more detail in [21], but arises from the assumption of independent sub-trees in a $\max^n$ tree. In practice, the best case for shallow pruning actually converges to best-case for immediate pruning, where every leave has a value of win/loss. In this case, only a very precise arrangement of leaf values can achieve the best-case, which will be independent of the number of players in the game, but we do not expect anything close to this in practice.

Given this history, we push to answer the question "What techniques should be used to play multi-player games?" The answer to this question for many two-player games is well known: use minimax with alpha-beta, a high-quality heuristic evaluation function, transposition tables, opening and endgame books. The answer for multi-player games is still being explored. However, a program developed to play Hearts using speculative pruning, transposition tables, and Monte-Carlo simulation for imperfect information was able to outplay one of the better commercial Hearts programs [21], although its play is still weak against humans. We explore this question further in the remainder of this paper.

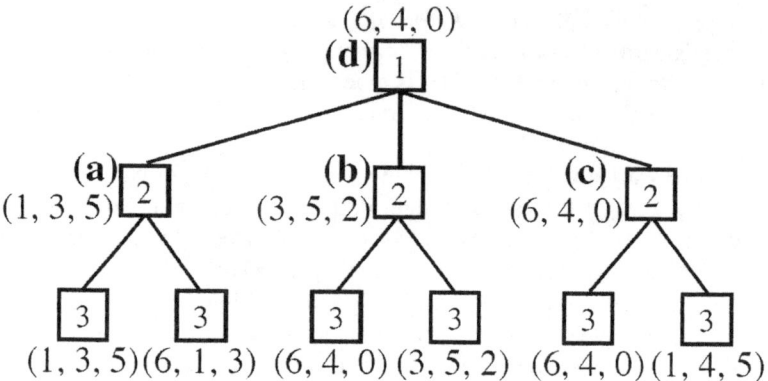

Fig. 1. A 3-player max^n game tree

3 Sample Domains

We use two domains, Hearts and Spades, to illustrate the ideas and concepts in this paper. Hearts and Spades are both trick-based card games. In such games cards are dealt out to each player before the game begins. The first player plays (*leads*) a card face-up on the table, and the other players follow in order, playing the same suit as lead if possible. When all players have played, the player who played the highest card in the suit that was led "wins" or "takes" the trick. He[1] then places the played cards face down in his discard pile, and leads the next trick. This continues until all cards have been played.

Hearts is usually played with four players, but there are variations for playing with two or more players. The goal of Hearts is to take as few points as possible. A player takes points when he takes a trick which contains point cards. Each card in the suit of hearts is worth one point, and the queen of spades is worth 13. At the end of the game, the sum of all scores is always 26, and each player can score between 0 and 26. If a player takes all 26 points, or "shoots the moon," he instead gets 0 points, and the other players all get 26 points each. These fundamental mechanics of the game are unchanged regardless of the number of players.

Spades can be played with 2-4 players. Before the game begins, each player predicts how many tricks they think they are going to take, and they then get a score based on how many tricks they actually do take. With 4 players, the players opposite each other play as a team, collectively trying to make their bids, while in the 3-player version each player plays for themselves. There are, however, several popular games for which the mechanics of play are identical to Spades, except that players do not play in teams no matter how many players there are. These games have various names and rules that differ by region, but we consider them as larger variations on the game of Spades. More in-depth descriptions of these and other multi-player games can be found in Hoyle *et al.* [22]. A much larger database of games and descriptions can currently also be found online at http://www.pagat.com/.

---

[1] In this paper 'he' is used when both 'he' and 'she' are possible.

Although these games are imperfect information games, we will play them as perfect information games, assuming we can see all the cards that our opponents hold. Given this assumption, we can use Monte-Carlo methods in real play to sample possible distributions of opponent cards. These methods have been applied most successfully in Bridge [4], and to a lesser extent in Hearts [21].

4 The Max^n Algorithm

The max^n algorithm [11] is a general form of the minimax algorithm and can be used to play any n-player general-sum game. For two-player games, max^n simply computes the minimax value of a tree. We will generally call what max^n calculates as the max^n value of a game tree, although in a more general sense it is calculating a Nash equilibrium.

In a max^n tree with n players, the leaves of the tree are n-tuples, where the ith element in the tuple is the ith player's score. At the interior nodes in the game tree, the max^n value of a node where player i is to move is the child of that node for which the ith component is maximum. At the leaves of a game tree an exact or heuristic evaluation function can be applied to calculate the n-tuples that are backed up in the game tree.

We demonstrate this in Figure 1. In this tree there are three players. At node (a), Player 2 is to move. Player 2 can get a score of 3 by moving to the left, and a score of 1 by moving to the right. So, Player 2 will choose the left branch, and the max^n value of node (a) is (1, 3, 5). Player 2 acts similarly at node (b) selecting the right branch, and at node (c) breaks the tie to the left, selecting the left branch. At node (d), Player 1 chooses the move at node (c), because 6 is greater than the 1 or 3 available at nodes (a) and (b).

4.1 Speculative Pruning

Speculative pruning [20] is the most general effective pruning technique for multi-player games. It is a non-directional algorithm, meaning that it may not search through a game tree in a strictly left-to-right fashion. The key idea behind this algorithm is that we prune nodes that may still affect the max^n value of the game tree. But, as the search progresses, we can detect whether the pruned nodes could have actually affected the max^n value of the game tree, and then re-explore portions of the tree as necessary. We demonstrate this in Figure 2, however we leave out some of the details for simplicity.

In this example, the sum of all players' scores, *maxsum*, is 10, and each player's score has a lower bound of zero. At the root of the tree, Player 1 can get at least 5 by moving to his left branch. At node (a), Player 2 can get at least 3 points from his left branch. Finally, at node (b) Player 3 can get at least 2 points from his left branch. The sum of consecutive lower bounds at this point in the game tree, 5 + 3 + 2 = 10, is at least as large as *maxsum*, so we can prune the right child of (b). The caveat is that unexplored children of (a) may affect the max^n value of the game. But, we can detect this and re-expand node (b) and its children if necessary. It is possible that poor node

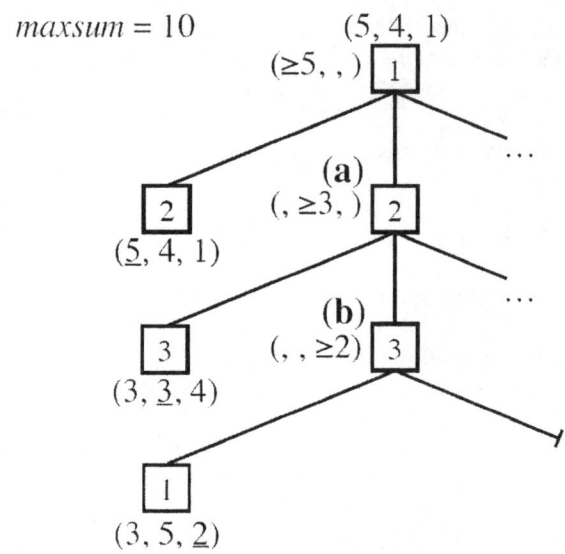

Fig. 2. Simple example of speculative pruning in a $\max^n$ tree

ordering may cause us to expand more nodes using speculative pruning than we would have without it, although we have never seen this in practice.

4.2 Other Properties of Max^n

We cannot cover all the theoretical details of the $\max^n$ algorithm here, but we do need to point out a few issues that strongly affect how the $\max^n$ algorithm can be used. A minimax game tree has a single minimax value, no matter what order the game tree is searched. A $\max^n$ game tree, however, can have many different $\max^n$ values, depending on ties in the game tree and the tie-breaking rule used. This is a property of multi-player games, as each $\max^n$ value corresponds to a different equilibrium point in the game. This means that any time we vary node ordering in the game we have the potential of changing the $\max^n$ value of the corresponding game tree. Furthermore, the change may not be bounded. For instance, the value at the root of a tree can change from a win to a loss based on the difference of how one tie is broken. How large of a problem this is in practice is not yet fully understood, although we address some of the related issues in the following sections.

5 Comparing Max^n Tree Reductions

In this section we begin our presentation of new research. If we wish to understand how various search enhancements will affect the size of $\max^n$ game trees, it is important to have an idea of how current pruning algorithms perform, and whether they are anywhere close to optimal. There are well-known techniques that have been applied to two-player games, such as the history heuristic [23], which have helped to order two-player game trees nearly optimally. If ordering max^n game trees is difficult, we would also like to know if such techniques can help.

5.1 Estimating Minimal Max^n Trees

Computing the minimal size of a game tree is a non-trivial task, as was originally noted by Knuth and Moore in their analysis of alpha-beta pruning [5]. Specifically, they point out that just ordering nodes from best to worst within a game tree may not be optimal. In an unbalanced tree, for instance, if we are likely to get a pruning cut-off, we might want to search the move that leads to the smallest sub-tree first.

This same issues exist in multi-player games, although there are further complexities. For instance, depending on the ties in a multi-player game, each node in the game tree may have not one, but many different possible max^n values, each one corresponding to a different way of breaking ties in the game tree. Thus, we might try to find the best way to break ties in order to create the smallest game tree possible, although this may lead to a poor max^n value. We might also try to make smaller adjustments that do not change the max^n value of an entire game, but still decrease the portion of the game tree that we must search. We demonstrate this in Figure 3.

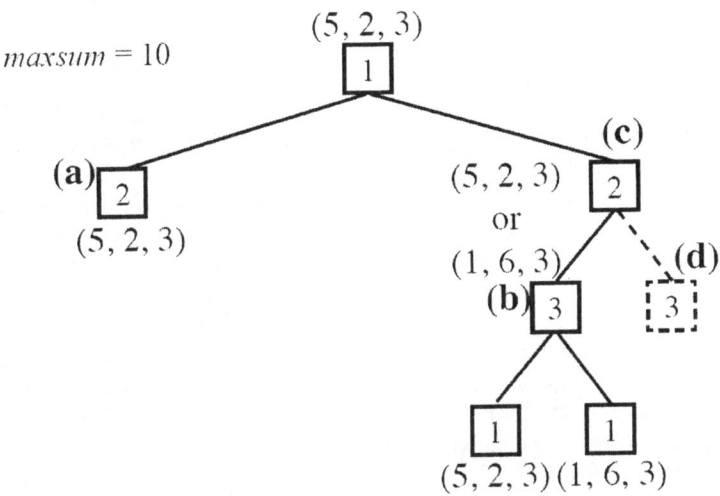

Fig. 3. Increasing pruning by modifying tie-breaking rules

In this figure, Player 1 can get a score of 5 at node (a). When searching the right branch of the tree, Player 3 has a tie at node (b). Normally we break ties to the left, which will not affect the max^n value of this game tree. If we do this, however, Player 1's bound at the root, 5, plus Player 2's bound at (c), 2, will not be adequate to prune, as they do not sum to at least *maxsum*, 10. But, breaking the tie at (b) differently to give Player 2 a higher score at node (c) will allow us to prune node (d) and any of its children, since the best scores for Player 1 and Player 2 would then exceed *maxsum*. Thus, by making subtle changes to a max^n game tree, we could increase pruning. We have not attempted this in practice.

When we prune a n-player max^n game tree, every tree fragment where pruning occurs has a similar structure to Figure 2, meaning we do not usually prune more than n-1 ply away from where any bound originates, unlike in alpha-beta. It is possible to do so correctly, but the bookkeeping costs of making sure such a prune is legal and the potential cost of having to re-search that line under speculative pruning is prohibitive. However, in an optimally ordered tree, such pruning is not only possible, but it will never lead to tree re-expansions, so there will be no additional overhead or bookkeeping costs.

Comparing the effect of such pruning, however, would be misleading, because we would still have to pay such costs in practice. The only time we might consider using such pruning in practice is when we are willing to give up the guarantee that we calculate a correct max^n value, which we will discuss further in Subsection 5.4.

Table 1. Tree sizes in Hearts and Spades given a variety of search enhancements

Game	Full Tree	TT	TT + P	Ordered
Spades	112M	711k	296k	232k
Reduction	-	157x	2.4x	1.3x
Hearts	52.7M	2.04M	1.01M	580k
Reduction	-	26x	2.0x	1.7x

5.2 Current Best Max^n Tree Reductions

Given a best approximation of an optimally ordered game tree to compare against, we can then compare to the best techniques for generating a similar game tree, without omniscient ordering. A good overview of such an attempt in two-player games can be found in [23], which showed that a combination of transposition tables and the history heuristic were able to account for 99% of the possible reductions in a game tree.

We ran four experiments in the games of Hearts and Spades to see how tree sizes compare when using standard techniques. We compared the number of nodes analyzed for the first move in each of the card games. For simplicity, we used a 3-player version of both games, and limited the number of cards in each player's hand so that we could save the game trees to disk. We played a total of 100 hands in each game. Each game tree was saved to disk, with transpositions recorded, to minimize tree size. We then measured the average number of nodes under various combinations of techniques. In Spades, we gave each player 8 cards, for a total search depth of 24 ply, while we gave each player 7 cards in Hearts, for a total search depth of 21 ply. To ensure correctness we used a total ordering on leaf values in the game trees to break ties, which guarantees that each game tree will only have a single max^n value.

Our first experiment was to measure the full size of the game tree, with no search enhancements, besides those inherent in the domain itself. For instance, we only generate one move when a player holds a sequence of consecutive cards in their hand. In Spades these trees were, on average, 112 million nodes, while in Hearts they were 52.7 million nodes. This is the total count of nodes in the tree, not just leaf nodes.

Next, we enabled transposition tables. In Spades, this reduced the size of the game tree to 711k nodes on average, 157x times smaller than the previous tree. In Hearts, transposition tables reduced the average tree size to 2.04 million nodes, a reduction factor of 26x. The reduction is smaller in Hearts because of the significance of the AKQ♠, where in Spades we only care about relative rankings of cards. The next enhancement we added was speculative pruning (which includes shallow pruning). In Spades, this reduced the average tree size to 296k, while in Hearts it reduced it to 1.01 million nodes. Again, this is the total number of nodes in the tree, including re-expansions from speculative pruning.

Finally, we compared the trees with all of these enhancements to ordered trees using transposition tables and speculative pruning. As discussed previously, we cannot measure the true optimal tree sizes, but we can guarantee that the first line of play is the best at every node. Doing so reduced the average Spades tree size to 232k nodes, and the average Hearts tree to 580k nodes.

One enhancement that we did not measure for either of these games is branch and bound pruning. Although it would have reduced the size of these trees further, the reductions are simply based on the properties of the game tree, and are not affected by node ordering.

The most surprising result here is that the game trees are so close to optimal given just a simple manual ordering of nodes. In Spades, for example, this is just composed of a few rules such as "if you can't win the trick, try low cards first" or "if you can use trump to win, try the lowest winning trump first." Also, in Hearts we sometimes observed many re-expansions from speculative max^n, and so we expected these trees to be significantly larger than optimal. But, it appears that transposition tables are able to compensate well for the cost of node re-expansions, although there is more room for improvement in Hearts than in Spades.

Overall these results also explain why, in separate experiments not shown here, we were unable to get significant gains from using the history heuristic, although it was able to perform as well as our custom ordering. In domains where there are not easy or obvious ordering rules, we would still expect it to be effective in improving the efficiency of search.

Table 2. The effect of different techniques on the calculated max^n value

	% Changed moves	Nodes expanded
Regular ordering	-	413k
Random ordering	58%	618k
Approx. deep pruning	10%	120k

5.3 Implications

These results have significant implications for future directions of multi-player game research. We are currently within a factor of two times optimal, which is on the order that we often give up with techniques like iterative deepening. As we have achieved near optimal pruning relatively easily in practice, this means that instead of focusing our efforts improving node ordering, there are two directions to push future research.

The first direction, which we consider in the next section, is to continue to develop pruning techniques and methods to try to extend search depth even further. The second possible direction is to use a shallower search, but to apply more complex reasoning methods to the search in an attempt to improve performance.

5.4 Further Search Techniques

Speculative pruning was the first pruning technique that is generally applicable to multi-player games. We do not expect that there are any other general pruning techniques that are significantly better than speculative pruning in practice, however there are better pruning techniques in theory. For instance, suppose we want to avoid the bookkeeping costs from speculative pruning, and ignore the need to re-expand portions of the game tree. This would only be marginally faster in practice. But, taking this idea further, if we are going to ignore re-expansions we could re-consider deep pruning. That is, expanding the idea of speculative max^n beyond consecutive nodes in the game tree. Normally the bookkeeping and re-expansions costs would be prohibitive, but if we are not going to re-expand nodes, we might as well get the best gain possible, since we are not guaranteed to preserve the same max^n value in the game tree either way. We call this approximate deep max^n.

This approach may not have much justification in two-player games, but in multi-player games there is some possible justification. First, a max^n game tree may have several max^n values depending on the way ties are broken in the tree. Second, we must model both our opponents leaf utility function and the way they are going to break their ties when they play the game. If we break ties incorrectly, the max^n value we calculate may not be the one our opponents calculate. Thus, since we are prevented in practice from being able to guarantee that our max^n value will coincide with the one our opponents calculate, we can argue that the small amount of noise introduced by deep pruning may be offset by increased search depth. This may be particularly true in card games, where the branching factor decreases the farther we search into the tree. So a small reduction in node expansions may bring a larger increase in search depth.

To investigate this in practice, we generated 100 hands in the three-player variation of the game of Hearts. We then searched these hands to determine which move would be made at the root of the tree using three different methods. The first method was simply to use a predefined node ordering under speculative pruning. The second method was to use a randomized node ordering under speculative pruning. The third method was to use the predefined ordering with approximate deep pruning. The results of these experiments are in Table 2. Using a random move ordering changed the move that would be made 58% of the time, while using approximate deep pruning changed the move only 10% of the time. As expected, the average number of nodes expanded was higher when we used a random ordering. When using approximate deep pruning we were able to reduce the nodes expanded by over a factor of three. These results suggest that it is possible that any mistakes made from approximate deep pruning may be more than compensated for by the fact that we do not have an exact model of how our opponent will order their moves. Results on quality of play when compared to various opponents, not shown here, were inconsistent, partially due to issues of opponent modeling, which we will discuss in the next section.

We note that there is a similar argument to why we should use the paranoid algorithm. Specifically, because it is a form of minimax, any techniques developed for minimax can be applied to the paranoid algorithm, and in practice the paranoid assumption allows you to search deeper than you can with max^n. But, the paranoid algorithm can also lead to extremely pessimistic play. The one exception to this is in the game of Chinese Checkers. Our conjecture, first suggested in [21], is that if we had some measure m of the ability of our opponents to collude, we will only want to use the paranoid algorithm when m is low, that is, when it is difficult for them to collude, otherwise we will want to use something more similar to max^n.

6 Opponent Modeling

While there has been a reasonable amount of work on incorporating opponent modeling into two-player games, this work has not been widely used in practice. For instance, Jansen [24] describes various positions in a game where you would need to consider your opponents strategy, Iida et al. suggested other potential applications of opponent modeling [25,26], and various other algorithms for opponent modeling have been suggested by Carmel and Markovitch [7], Korf [27], and Donkers et al. [28].

Results from Carmel and Markovitch give a strong indication of why this work has not been more widely applied. In their experiments in checkers they showed that while opponent modeling methods were superior given the same search depth, they were not effective when budgeted the same number of node expansions. This is because opponent modeling necessarily reduces the amount of pruning possible in a game tree. So, alpha-beta is able to overcome handicaps in opponent modeling due to its advantage in search depth.

Donkers presents further experiments with opponent modeling in [29], showing that in certain situations gains from opponent modeling are possible, although it is a difficult task, even when given a perfect model of one's opponent.

Most work in opponent modeling has been focused on two-player perfect information games. But, in multi-player games we do not see the same large gains from pruning as in two-player games. This means that if there are any benefits to be obtained by good opponent modeling, they are less likely to be overshadowed by deeper search by a version of max^n that does not do sophisticated opponent modeling.

There are several ways we can add opponent modeling to max^n. One way is without modifying the max^n algorithm itself, but instead just by changing the evaluation function used by our opponents at the leaves of the game tree. This is a limited form of opponent modeling that is inherently assuming that the model we have of our opponent is correct, and that our opponent also has a correct model of our own behavior. This approach will not directly allow us to use a sophisticated model of our opponent to trick them into making sub-optimal plays, because we are assuming that we both have accurate models of each other, but it is a reasonable starting point. We will use this idea in a moment, but we first provide a stronger motivation of why we cannot ignore opponent modeling in multi-player games.

6.1 Opponent Modeling Motivation

A different explanation for why opponent-modeling techniques were not developed earlier and have not been widely applied in two-player game search is that the assumption of an optimal opponent in the minimax strategy is adequate for most play. While this is true in two-player, zero-sum games, it is not true in multi-player games, because there is not necessarily a good definition of optimality for our opponents.

For instance, we may define an optimal opponent as one that plays a Nash equilibrium. But, as we have mentioned multiple times, there can be many different Nash equilibriums in a multi-player game, and we do not know which equilibrium our opponent will be playing. Furthermore, if we are each playing parts of different equilibrium strategies, our combined strategy will not necessarily be part of any equilibrium strategy. We might try to avoid this issue by guaranteeing that a game tree only has one equilibrium point, by eliminating all ties in the game tree.

We define a *pure tie* between two max^n values to be one in which all players have exactly the same scores. Any other tie just assumes that a single player has a tie between their own components of a max^n value. If every tie in a game tree is *pure*, there will be exactly one max^n value of the game tree. For many games, it is fairly easy to modify the evaluation function to avoid ties.

But, while this will avoid the equilibrium selection problem, it leaves an equally difficult opponent modeling problem. We now must have a perfect model of our opponent's evaluation function, which is impossible in practice. For instance, one player might be maximizing their own score while another is maximizing the difference between their score and their opponents' score. Thus, the fundamental problem has not changed. Unlike two-player games, there is no widely applicable "optimal" opponent model for use in multi-player games.

This can be seen from Korf's proof of why deep pruning fails in multi-player games [16]. In this proof he shows that changing the value of any single node in a max^n game tree can potentially change the max^n value of the entire tree. So, if we have modeled our opponent's utility function incorrectly at any node in a game tree, the max^n value of the tree can change, and our expected utility from our calculated max^n value may be incorrect. This means, unless we have some oracle providing perfect information about our opponent, we cannot be guaranteed that the max^n strategy we are playing with is correct. We can now show how this occurs in practice.

6.2 Opponent Modeling Experiments

To understand better the role opponent modeling plays in multi-player games, we ran experiments in the games of Hearts and Spades. For each of these games, we chose two different evaluation functions to compare the effects of correct and incorrect opponent modeling. Players were dealt small enough hands (8 cards in Spades, 7 in Hearts) so that they could search entire game trees, meaning the evaluation function is exact.

In Spades, each player bids on how many tricks they expect to take at the beginning of the game, and is trying to take at least that many tricks, with as few overtricks as possible. Taking 10 overtricks leads to a 100-point penalty. The first evaluation function we used for Spades simply tried to maximize the number of tricks taken, so we will call it 'MT'. The second evaluation function tried to both make the bid, but also minimize overtricks, so we will call it 'mOT'. We then created two variations on each of these players. The first variation was aware of the type of player it was playing against, and could model that player's evaluation function as part of its search. When a player is using a model of their opponent, we add a subscript m to their player type. If they are not modeling their opponent correctly, each player assumes that all other players are the same type that they are. This gives us four player types, MT, MT_m, mOT, and mOT_m.

Players receive a cumulative score throughout a round until they reach a predetermined bound, which we set at 200 points, when a winner is declared. For each of the four variations on player types (<MT, mOT>, <MT_m, mOT>, <MT, mOT_m>, <MT_m, mOT_m>) we played 600 rounds and then measured the average score and number of wins. Within these rounds, we repeated each round six times with the same cards being dealt to each position at the table in order to account for all combinations of players and positions in a 3-player game, as shown in Table 3.

Table 3. The six possible ways to assign MT and mOT player types to a 3-player game

	Player 1	Player 2	Player 3
1	MT	MT	mOT
2	MT	mOT	MT
3	MT	mOT	mOT
4	mOT	MT	MT
5	mOT	MT	mOT
6	mOT	mOT	MT

The results of this experiment are in Table 4. As expected, when both players had correct models of their opponent, the mOT players outplayed the maximizing players, winning 53% of the games and having 10 more points on average. But, when the mOT player did have not a correct opponent model, it played much more poorly, essentially tying the game when the mOT player was the only player with a correct model, and losing badly when the mOT player did not have a correct model of the opponent.

Table 4. Spades Games

	"Maximizing tricks" Player		"Minimizing overtricks" Player	
Comparison	Average score	% wins	Average score	% wins
MT_m v. mOT_m	139.0	47.0	149.2	53.0
MT_m v. mOT	145.0	65.0	78.0	35.0
MT v. mOT_m	144.9	49.3	144.1	50.7
MT v. mOT	147.9	63.2	95.8	36.8

This is not surprising, since, in Spades in particular, having a bad opponent model can be very costly. This is because of the large penalty for missing your bid. If you are also trying to minimize overtricks, you must be certain that you will first get the chance to make your bid. An incorrect assumption of how your opponent is going to play can break this assumption and carry a high penalty.

We performed similar experiments in the game of Hearts. In Hearts we played until one player had a score that exceeded 100, where players are trying to minimize their scores. When a round ends, the player with the lowest score wins. Our first player type simply tried to minimize their score (mS), while the second player tried to maximize their lead (ML) over the next best player of they were ahead, and tried to minimize the amount they trailed the best player if they were behind. The results are in Table 5. We see that the players trying to minimize their score were generally successful at that task, but not as successful in the larger goal of actually winning games. When both players modeled each other correctly, the ML player averaged 92 points per round versus 86.6 for the minimizing player, but won 54.5% of the total games. When the ML player had a model of the minimizing player, but not vice versa, the ML player won 58.1% of the games with a slightly lower average score of 89.3. In contrast, when the ML player did not have a proper model of the opponent, and the mS player did, the ML player only won 45% of the games, and averaged 95.1 per game. Finally, when neither player had a model of each other, they both won roughly the same number of games, although the ML player still had a much higher (worse) average score.

In both of these games, our experiments show that having a good opponent model is important, but if the modeling done is incorrectly, it can have strongly negative repercussions. We note that except for the case in which both players had models of each other, the modeling is incomplete. Specifically, when a player had a model of their opponent, they also assumed that their opponent had a model of them, which was not always true.

Table 5. Hearts Games

Comparison	"Minimizing score" Player		"Maximizing lead" Player	
	Average score	% wins	Average score	% wins
mS_m v. ML_m	86.6	45.5	92.0	54.5
mS v. ML_m	86.5	41.9	89.3	58.1
mS_m v. ML	72.7	54.5	94.4	46.5
mS v. ML	74.4	49.1	93.2	50.9

In order to correct this deficiency in modeling, we need to extend M* or a similar algorithm to n-player games, as max^n cannot do full opponent modeling by only changing the leaf evaluations in the game tree.

7 Conclusions

In this work we have shown that existing pruning and tree reduction techniques are adequate to produce multi-player game trees to within a factor of two of optimal sized trees. This, along with results on opponent modeling, suggest that a key question for multi-player game tree search is whether we better off trying to search as deep as possible into a game tree, or is it be better to spend our resources doing a more intelligent, but shallower search.

We have taken the first steps towards answering this question by suggesting that using an approximate search strategy like approximate deep max^n will allow us to search deeper into a game tree, and the penalty for doing an approximate search may be no more than what we pay for having an incorrect opponent model. We have also demonstrated how opponent models are crucial to the quality of play in a multi-player search algorithm.

There are many areas in which this work can be extended. First, we could consider extending other two-player algorithms besides minimax to multi-player games. Given this, we also need to compare the relationship between quality of play to search depth, as opponent modeling necessarily reduces search depth. Additionally, there is the question of where we can get an opponent model in practice.

In conclusion, there are many open questions in the field of multi-player game-tree search. Experimental and theoretical results in this paper point to the fact that we may not be able to get by with just brute-force search as we often have in two-player games, although further research must be done to give more conclusive results.

Acknowledgements

We wish to thank the referees and Akihiro Kishimoto for their comments and suggestions after reading this paper. This research has also benefited from conversations with Darse Billings, Michael Bowling, Martin Müller, and Jonathan Schaeffer.

References

1. Hu, F. (2002). *Behind DEEP BLUE*. Princeton University Press.
2. Schaeffer, J., Culberson, J., Treloar, N., Knight, B., Lu, P., and Szafron, D. (1992). A world championship caliber checkers program, *Artificial Intelligence*, Vol. 53, pp. 273–290.
3. Billings, D., Peña, L., Schaeffer, J., and Szafron, D. (1999). Using Probabilistic Knowledge and Simulation to Play Poker, *AAAI-99*, pp. 697–703.
4. Ginsberg, M., GIB (2001). Imperfect Information in a Computationally Challenging Game, *Journal of Artificial Intelligence Research*, vol. 14, pp. 303–358.
5. Knuth, D., and Moore, R. (1975). An Analysis of Alpha-Beta Pruning, Artificial Intelligence, vol. 6 no. 4, pp. 293–326.
6. Russell, S.J. and Wefald, E.H. (1989). On optimal game-tree search using rational meta-reasoning. *IJCAI-89*, Detroit, MI, pp. 334–340.
7. Carmel, D. and Markovitch, S. (1996). Incorporating Opponent Models into Adversary Search, *AAAI-96*, Portland, OR.
8. Nash, J.F. (1951). Non-cooperative games, *Annals of Mathematics*, vol. 54, pp. 286–295.
9. Jones, A.J. (1980). *Game Theory: Mathematical Models of Conflict*, Ellis Horwood, West Sussex, England.
10. Kuhn, H.W. (1953). Extensive Games and the Problem of Information, in Kuhn and Tucker (editors), *Contributions to the Theory of Games*, Princeton University Press, New Jersey.
11. Luckhardt, C. and Irani, K. (1986). An algorithmic solution of N-person games, *AAAI-86*, Philadelphia, PA, pp. 158–162.

12. Luckhardt, C. (1989). N-Person Game Playing and Artificial Intelligence, PhD. Thesis, University of Michigan.
13. Mutchler, D. (1993). The Multi-Player Version of Minimax Displays Game-Tree Pathology, *Artificial Intelligence*, vol. 64, pp. 323–336.
14. Nao, D.S. (1980). Pathology on game trees: A summary of results, *AAAI-80*, Stanford, CA, pp. 102–104.
15. Beal, D.F. (1980). An analysis of Minimax, In Clarke, M.R.B. (editor), *Advances in Computer Chess 2*, Edinburgh University Press, Edinburgh, Scotland, pp. 103–109.
16. Korf, R. (1991). Multiplayer Alpha-Beta Pruning. *Artificial Intelligence*, vol. 48 no. 1, pp. 99–111.
17. Sturtevant, N., and Korf, R. (2000). On Pruning Techniques for Multi-Player Games, *AAAI-00*, Austin, TX, pp. 201–207.
18. Sturtevant, N. (2002). A Comparison of Algorithms for Multi-Player Games, in Schaeffer, Müller and Björnsson (editors), *Proceedings of the 3rd International Conference on Computers and Games*, pp. 108–122.
19. Pearl, J. (1984). *Heuristics,* Addison-Wesley, Reading, MA.
20. Sturtevant, N. (2003). Last-Branch and Speculative Pruning Algorithms for Max^n, *IJCAI-2003*, Acapulco, Mexico.
21. Sturtevant, N. (2003). Multi-Player Games: Algorithms and Approaches, PhD Thesis, UCLA.
22. Hoyle, E., and Frey, R.L., Morehead, A.L., and Mott-Smith, G, (1991). *The Authoritative Guide to the Official Rules of All Popular Games of Skill and Chance*, Doubleday.
23. Schaeffer, J. (1989). The History Heuristic and Alpha-Beta Enhancements in Practice, *IEEE Transactions on Pattern Analysis and Machine Intelligence*, vol. 11, no. 11, pp. 1203–1212
24. Jansen, P. (1990). Problematic Positions and Speculative Play, in Marsland and Schaeffer (editors), *Computers, Chess and Cognition,* Springer, New York, pp. 169–182.
25. Iida, H., Uiterwijk, J.W.H.M., van den Herik, H.J, and Herschberg, I.S. (1993). Potential Applications of Opponent-Model Search, Part 1: The domain of applicability, *ICCA Journal*, vol. 16, no. 4, pp. 201–208
26. Iida, H., Uiterwijk, J.W.H.M., van den Herik, H.J, and Herschberg, I.S. (1994). Potential Applications of Opponent-Model Search, Part 2: Risks and Strategies, *ICCA Journal*, vol. 17, no. 1, pp. 10–14
27. Korf, R. (1989). Generalized Game Trees, *IJCAI-89*, pp. 328–333
28. Donkers, H.H.L.M., Uiterwijk, J.W.H.M., and Herik, H.J. van den (2001). Probabilistic Opponent-Model Search, *Information Sciences*, vol. 135, nos. 3–4, pp. 123–149.
29. Donkers, H.H.L.M. (2003). Nosce Hostem: Searching with Opponent Models, PhD Thesis, University of Maastricht.

Preventing Look-Ahead Cheating with Active Objects

Jouni Smed and Harri Hakonen

Turku Centre for Computer Science (TUCS) and
Department of Information Technology,
University of Turku, Turku, Finland
{jouni.smed, harri.hakonen}@cs.utu.fi

Abstract. In a turn-based networked multiplayer computer game, it is possible to cheat by delaying the announcement of one's action for a turn until one has received messages from all the other players. This look-ahead cheating can be prevented with a lockstep protocol, which requires that the player first announces a commitment to an action and later on the action itself, which can be compared with the earlier announced commitment. However, because the lockstep protocol requires separate transmissions for the commitment and the action and a synchronization step before the actions can be announced, it slows down the turns of the game. In this paper, we propose that active objects can be used to prevent look-ahead cheating. Moreover, we can parameterize the probability of catching cheaters: The smaller this probability is, the less bandwidth and transmissions are required. In most cases, the mere threat of getting caught is enough to discourage cheating, and, consequently, this probability can be quite small.

1 Introduction

Online security has recently become a major concern for the entertainment industry. The gaming sites periodically report on attacks and warn the users against misbehaviour and cheating. The cheaters attacking the games are mainly motivated by an appetite for vandalism or dominance. However, only a minority of the cheaters try to create open and immediate havoc, whereas most of them want to achieve a dominating, superhuman position and hold sway over the other players.

As the online gaming is becoming a more lucrative business, potential financial losses, caused directly or indirectly by cheaters, are now a major concern among the online gaming sites and the main motivation to implement counter-measures against cheating. In this respect, cheating prevention has three distinct goals [1,2]:

- protect the sensitive information,
- provide a fair playing field, and
- uphold justice inside the game world.

Each of these goals can be viewed from a technical or social perspective: Sensitive information (e.g., players' accounts) can be gained, for instance, by cracking the passwords or by pretending to be an administrator and asking the players to give their passwords. A fair playing field can be compromised, for instance, by tampering with the network traffic or by colluding with other players. The sense of justice can be violated, for instance, by abusing inexperienced and ill-equipped players or by ganging up and controlling parts of the game world.

Although this paper concentrates on a specific problem concerning the fair playing field, namely preventing look-ahead cheating in turn-based games, we begin with a review of the common cheating methods in Section 2. In Section 3 we examine the lockstep protocol and its variations, which aim at preventing look-ahead cheating. To improve the responsiveness, we introduce a scheme which uses active objects to prevent and detect look-ahead cheating in Section 4. The concluding remarks appear in Section 5.

2 Methods Used in Cheating

In a networked multiplayer game, a cheater can attack the clients, the servers, or the network connecting them. Figure 1 illustrates the typical attacks [3]. On the client side, the attacks focus on compromising the software or game data, and tampering with the network traffic. Game servers are vulnerable to network attacks as well as physical attacks such as theft or vandalism. Third party attacks on clients or servers include IP spoofing and denial-of-service attacks. In the following subsections we briefly review the common cheating methods and their counter-measures.

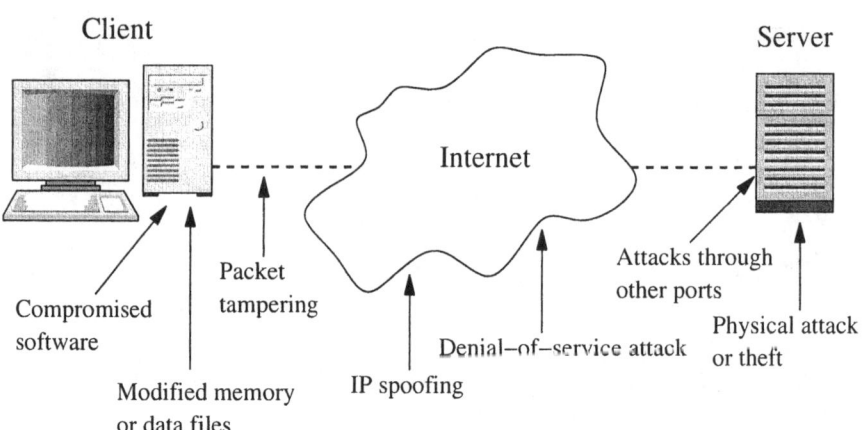

Fig. 1. Typical attacks in a networked multiplayer game

2.1 Tampering with the Network Traffic

In first-person shooter games, a usual way to cheat is to enhance the player's reactions with *reflex augmentation* [4]. For example, an aiming proxy can monitor the network traffic and keep a record of the opponents' positions. When the cheater fires, the proxy uses this information and sends additional rotation and movement control packets before the fire command thus improving the aim. Reversely, in *packet interception* the proxy prevents certain packets from reaching the cheating player. For example, if the packets containing damage information are suppressed, the cheater becomes invulnerable. In *look-ahead cheating*, time-stamped packets are forged so that they seem to be issued before they actually were (see Figure 2). This is usually counteracted with a lockstep protocol, which is discussed in Section 3. In a *packet replay* attack, the same packet is sent repeatedly. For example, if a weapon can be fired only once in a second, the cheater can send the fire command packet hundred times a second to boost its firing rate.

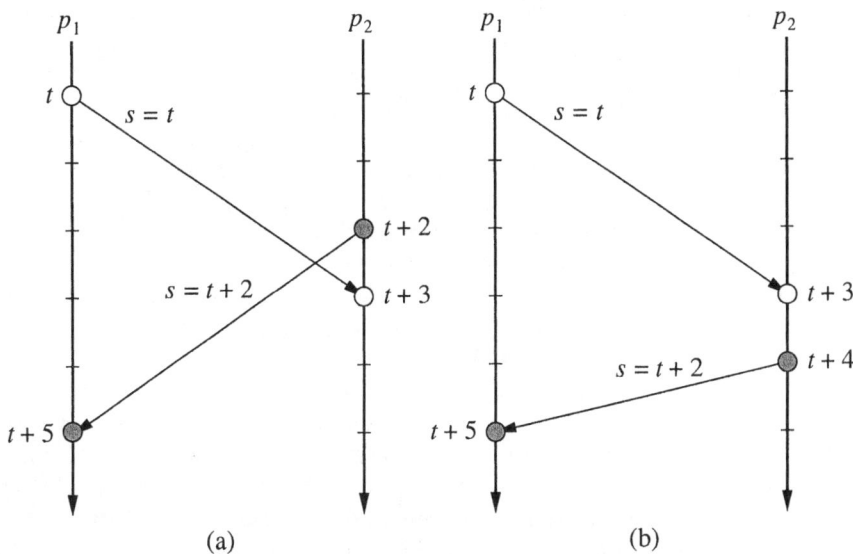

Fig. 2. Assume the players must time-stamp their outgoing messages, and the latency between the players is 3 time units. (a) If both players are fair, p_1 can be sure that the message from p_2, which has the time-stamp $t+2$, was sent before the message issued at t has arrived. (b) If p_1 has latency of 3 time units and p_2 has a latency of 1 time unit but pretends that it is 3, look-ahead cheating using forged time-stamps allows p_2 to base decisions on information that it should not have.

A common method for breaking the control protocol is to change bytes in a packet and observe the effects. A straightforward way to prevent this is to use checksums. For this purpose, the MD5 algorithm [5] is widely used (e.g., in the PunkBuster system [6]), because it is well tested, publicly available and fast enough for real-time computer games. However, there are two weaknesses that cannot be prevented with checksums alone: (1) the cheaters can reverse engineer the checksum algorithm and (2) they can attack with packet replay.

By encrypting the command packets, the cheaters have a lesser chance to record and forge information. However, to prevent a packet replay attack requires that the packets carry some state information so that even the packets with a similar payload appear to be different. Instead of serial numbering, pseudo-random numbers provide a better alternative. Random numbers can also be used to modify the packets so that even identical packets do not appear the same. Dissimilarity can be further induced by adding a variable amount of junk data to the packets, which eliminates the possibility of analysing their contents by the size.

2.2 Illicit Information

A cracked client software may allow the cheater to gain access to the replicated, hidden game data (e.g., the status of other players) [7]. On the surface, this kind of passive cheating does not tamper with the network traffic, but the cheaters can base their decisions on more accurate knowledge than they are supposed to have. For example, typical exposed data in real-time strategy games are the variables controlling the visible area on the screen (i.e., the fog of war). This problem is common also in first-person shooters, where compromised graphics rendering drivers may allow the player to see through walls.

Strictly speaking, information exposure problems stem from the software and cannot be prevented with networking alone. Clearly, the sensitive data should be encoded and its location in the memory should be hard to detect. Nevertheless, it is always susceptible to ingenious hackers and, therefore, requires some additional counter-measures. In a centralized architecture, an obvious solution is to utilize the server, which can check whether a client issuing a command is actually aware of the object with which it is operating. For example, if a player has not seen the opponent's base, he cannot give an order to attack it—unless he is cheating. When the server detects cheating, it can drop out the cheating client. A democratized version of the same method can be applied in a replicated architecture: every node checks the validity of each other's commands, and if some discrepancy is detected, the nodes vote whether its source should be debarred from participating in the game. Moreover, the game world can be partially replicated so that the server sends data to each participating client on a need-to-know basis, which prevents the clients from having—and hacking—complete information about the game world [8].

2.3 Exploiting Design Defects

Network traffic and software are not the only vulnerable places in a computer game, but design defects can create loop-holes, which the cheaters are apt to exploit. For example, if the clients are designed to trust each other, the game is unshielded from *client authority abuse*. In that case, a compromised client can exaggerate the damage caused by a cheater, and the rest accept this information as such. Although this problem can be tackled by using checksums to ensure that each client has the same binaries, it is more advisable to alter the design so that the clients can issue command requests, which the server puts into operation.

In addition to poor design, distribution—especially the heterogeneity of network environments—can be the source of unexpected behaviour. For instance, there may be features that become eminent only when the latency is extremely high or when the server is under a denial-of-service attack (i.e., an attacker sends it a large number of spurious requests to slow down or to block communication).

2.4 Collusion

The basic assumption of *imperfect information games* is that each player has access only to a limited amount of information. A typical example of such a game is poker, where the judgements are based on the player's ability to infer information from the bets thus outwitting the opponents. A usual method of cheating in individual imperfect information games is *collusion*, where two or more players play together without informing the rest of the participants. Normally this would not pose a problem, since the players are physically present and can (at least in theory) detect any attempts of collusion (e.g., coughs, hand signals, or coded language). However, when the game is played online, the players cannot be sure whether there are colluding players present. This means a serious threat to the e-casinos and other online gaming sites, because they cannot guarantee a fair playing field [9].

Only the organizer of an online game, who has the full information on the game, can take counter-measures against collusion. These counter-measures fall into two categories: *tracking* (determining who the players are) and *styling* (analysing how the players play the game). Unfortunately, there are no preemptive nor real-time counter-measures against collusion. Although tracking can be done in real-time, it alone is not sufficient. Physical identity does not reflect who is actually playing the game, and a cheater can always avoid network location tracking with rerouting techniques. Styling allows to find out if there are players who participate often in the same games and, over a long period, profit more than they should. However, this analysis requires a sufficient amount of game data, and collusion can be detected only afterwards.

The situation becomes even worse, when we look at the types of collusion in which the cheating players can engage. In active collusion, cheating players play more aggressively than they normally would. In poker, for example, the colluding players can outbet the non-colluding ones. In passive collusion, cheating players play more cautiously that they normally would. In poker, for example, the colluding players can let only the one with strongest hand to continue the play whilst the rest of them fold. Although styling can manage to detect active collusion, it is hard—if not impossible—to discern passive collusion from a normal play.

2.5 Offending Other Players

Although players may act in accordance with the rules of a game, they can cheat by acting against the spirit of the game. For example, in online role-playing games, killing and stealing from other players are surprisingly common problems [10]. The players committing these "crimes" are not necessarily cheating, because they can operate well within the rules of the game. For example, in the online version of *Terminus* (Vicarious Visions, 2001) different gangs have ended up owning different parts of the game world, where they assault all trespassers. Nonetheless, we may consider an ambush by a more experienced and better-equipped player on a beginner cheating, because it is not fair nor justified.

There are different approaches to handle this problem. *Ultima Online* (Origin, 1997) originally left the policing to the players, but eventually this led to gangs of player killers which terrorized the whole game. This was counteracted with a rating system, where everybody is initially innocent, but any misconduct against other players (including the synthetic ones) brands the player as a criminal. Each crime increases the bounty on their head ultimately preventing them from entering shops. The only way to clear one's name is not to commit crimes for a given time. *EverQuest* (Verant Interactive, 1999) uses a different approach, where the players can mark themselves able to attack and be attacked by other players, or completely unable to engage in such activities. However, killing and stealing are not the only ways to harm another player but there are other, non-violent ways to offend such as blocking exits, interfering with fights, and verbal abuse.

3 Lockstep Protocol

In multiplayer games based on a peer-to-peer architecture, all nodes uphold the game state, and the players' time-stamped actions must be conveyed to all nodes. This opens a possibility to use look-ahead cheating (see Figure 2), where the cheater gains an unfair advantage by delaying his actions—as if he had a high latency—to see what the other players do before choosing his own action.

The *lockstep protocol* tackles this problem by requiring that each player announces first a commitment to an action; when everyone has received the commitments, the players reveal their actions, which can be then verified by comparing them with the original commitments [11]. The commitment must meet two requirements: (1) it cannot be used to infer the action, but (2) it should be easy to compare whether an action corresponds to a commitment. An obvious choice for constructing the commitments is to calculate a hash value of the action. Figure 3 describes an implementation for the lockstep protocol. Although the details of the auxiliary functions, including the function HASH($\cdot$), are omitted here, their intention should be apparent.

We can readily see that the game progresses in the terms of the slowest player because of the synchronization. Nobody announces their action, until everybody has received all commitments—otherwise, if a player announces his action too early, those that have not yet announced their commitments could still change

LOCKSTEP(a_p, P)
    in: action a_p of the local player p; set of remote players P
    out: set of players' actions A
    local: commitment c; action a; set of commitments C
1: $c \leftarrow \text{HASH}(a_p)$
2: SEND-ALL(P, c)   ▷ Announce commitment to other players.
3: $C \leftarrow \{c\}$
4: $C \leftarrow C \cup$ RECEIVE-ALL(P)   ▷ Get other players' commitments.
5: SYNCHRONIZE(P)   ▷ Wait until everyone is ready.
6: SEND-ALL(P, a_p)   ▷ Announce the action.
7: $A \leftarrow \{a_p\}$
8: $A \leftarrow A \cup$ RECEIVE-ALL(P)   ▷ Get other players' actions.
9: **for all** $a \in A$ **do**
10:     find $c \in C$ for which $sender(c) = sender(a)$
11:     **if** $c \neq \text{HASH}(a)$ **then**   ▷ Are commitment and action different?
12:         **error** action a does not comply with commitment c
13:     **end if**
14: **end for**
15: **return** A

Fig. 3. Pseudocode for the lockstep protocol

their mind. This may suit a turn-based game, which is not time critical, but if we want to use the lockstep protocol in a real-time game, the turns have to be short or there has to be a time limit inside which a player must announce the action or lose that turn altogether.

To overcome this drawback, we can use an *asynchronous lockstep protocol*, where each player advances in time asynchronously from the other players but enters into a lockstep mode whenever interaction is required. The mode is defined by a sphere of influence surrounding each player, which outlines the game world that can possibly be affected by a player in the next turn (or subsequent turns). If two players' spheres of influence do not intersect, they cannot affect each other in the next turn, and therefore their decisions will not affect each other when the next the game state is computed and they can proceed asynchronously.

In the *pipelined lockstep protocol*, synchronization is loosened by having a pipe of size ℓ where the incoming commitments are stored (i.e., in basic lockstep $\ell = 1$) [12]. Instead of synchronizing at each turn, the players can send several commitments, which are pipelined, before the corresponding opponents' commitments are received. In other words, when player p has received all other players r commitments c_r^t for the turn t, it announces its action a_p^t (see Figure 4). The pipeline may include commitments for the turns $(t + 1), \ldots, (t + \ell)$, and player p can announce commitments $c_p^{t+1}, \ldots, c_p^{t+\ell}$ before it has to announce action a_p^t. However, this opens a possibility to reintroduce look-ahead cheating: if a player announces its action earlier than required by the protocol, other players can change both their commitments and actions based on that knowledge. This can be counteracted with an *adaptive pipeline protocol*, where the idea is to measure

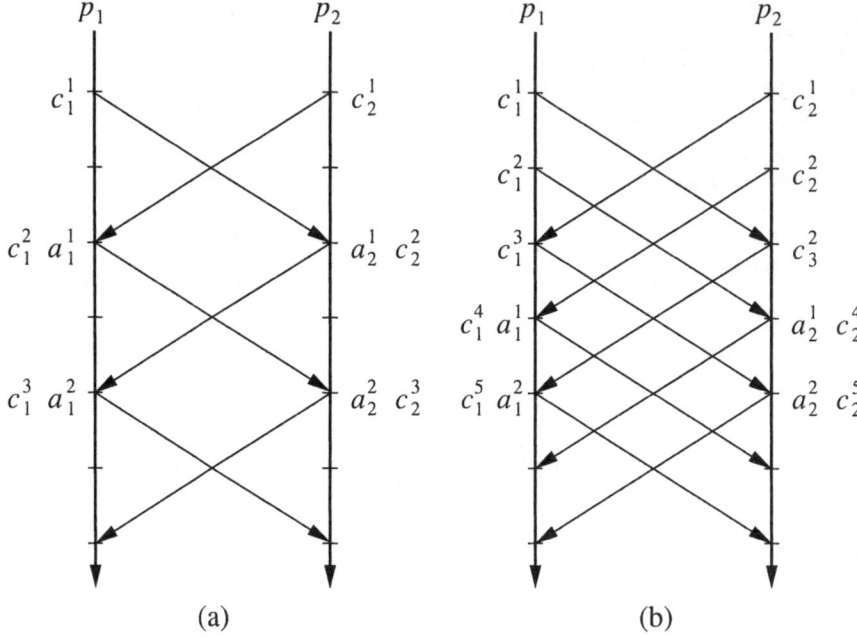

Fig. 4. (a) The lockstep protocol synchronizes after each turn and waits until everybody has received all commitments. Hence, player p_1 can announce only commitment c_1^1 for the turn 1 before it must announce action a_1^1, but of course that happens only after it has received commitment c_2^1 from player p_2. (b) The pipelined lockstep protocol has a fixed size pipe (here $\ell = 3$), which holds a sequence of commitments. Now, player p_1 can announce commitments c_1^1, c_1^2 and c_1^3 before it must announce action a_1^1.

the actual latencies between the players and to grow or shrink the pipeline size accordingly [13].

4 Active Objects

The lockstep protocol requires that the players send two transmissions—one for the commitment and one for the action—in each turn. Let us now address the question, whether we can use only one transmission and still detect look-ahead cheating. Single transmission means that the action must be included in the outgoing message, but the receiver is allowed to view it only after it has replied with its own action. But this leaves open the question how a player can make sure that the exchange of messages in another player's computer has not been compromised. It is possible that he is a cheater who intercepts and alters the outgoing messages or has hacked the communication system.

We can use active objects to secure the exchange of messages which happens in a possibly "hostile" environment. Now, the player (or the originator) provides an active object, a *delegate*, which includes a program code to be run by the other player (or the host). The delegate acts then as a trusted party for the originator by guaranteeing the message exchange in the host's system.

Let us illustrate the idea using the game rock-paper-scissors as an example. Player p goes through the following stages.

1. Player p decides the action "paper", puts this message inside a box and locks it. The key to the box can be generated by the delegate of player p, which has been sent beforehand to player r.
2. Player p gives the box to the delegate of player r, which closes it inside another box before sending it to player r. Thus, when the message comes out from the delegate, player p cannot tamper with its contents.
3. Once the double-boxed message has been sent, the delegate of player r generates a key and gives it to player p. This key will open the box enclosing the incoming message from player r.
4. When player p receives a double-boxed message originating from player r, it can open the outer box, closed by its own delegate, and the inner box using the key it received from the delegate of player r.
5. Player p can now view the action of player r.

At the same time, player r goes through the following stages.

1. Player r receives a box from player p. It can open the outer box, closed by its own delegate, but not the inner box.
2. To get the key to the inner box, player r must inform its action to the delegate of player p. Player r chooses "rock", puts it in a box and passes it to the delegate.
3. When the message has been sent, player r receives the key to the inner box from the delegate of player p.
4. Player r can now view the action of player p.

Although we can trust, at least to some extent, to our delegates, there still remains two problems to be solved. First, the delegate must ensure that it really has a connection to its originator, which seems to incur extra talk-back communication. Second, although we have secured one-to-one exchange of messages, there is no guarantee that the player does not alter its action when it sends a message to another player.

Let us tackle first the problem of ensuring the communication channel. Ideally, the delegate, once started, should contact the originator and convey a unique identification of itself. This identification should be a combination of dynamic information (e.g., the memory address where the delegate is located or the system time when the delegate was created) and static information (e.g., built-in identification number or the Internet address of the host's computer where the delegate is being run). Dynamic information is needed to prevent a cheating host from creating a copy of the delegate and using that as surrogate to work out how it operates. Static information allows to ensure that the delegate has not been moved to somewhere else or replaced after the communication check.

If we could trust the run environment where the delegate resides, there would be no need to do any check-ups at all. In contrast, in a completely hostile environment we would have to ensure the communication channel every time, and there would be no improvement over the lockstep protocol. To reduce the number of check-up messages the delegate can initiate them randomly with some parameterized probability. In practice, this probability can be relatively low—especially if the number of turns in the game is high. Rather than detecting the cheats this imposes a threat of being detected: Although a player can get away with a cheat, in the long run attempts to cheat are likely to be noticed. Moreover, as the number of participating players increases, it also increases the possibility of getting caught.

A similar approach helps us to solve the problem of preventing a player from sending differing actions to the other players. Rather than detecting an inconsistent action in the current turn, the players can "gossip" among themselves about the actions made in the previous turns. These gossips can then be compared with the recorded actions from the previous turns, and any discrepancy indicates that somebody has cheated. Although the gossip can comprise all earlier actions, it is enough to include only a small, randomly chosen subset of them—especially if the number of participants is high. This gossiping does not require any extra transmissions because it can be piggybacked in the ordinary messages.

In the following subsections we give a more detailed description of the delegate and the communication turn.

4.1 Delegate

Pseudocodes for the routines in the delegate are listed in Figure 5. The delegate receives from the host a sender object S and a receiver object R, which provide communication routines for networking, and message e to be encoded. The message exchange can begin with a connection check, where the delegate encodes its identification and sends it to the originator. Because connection checks are done randomly, the hosting system cannot tell whether the outgoing message is a connection check or the actual transmission. If the decoded reply complies with the identification, the delegate presumes that a connection has been established and proceeds to encode and send the message. After that, it constructs a decoder object, which will open the message from the originator in the current turn. We omit the details of the encoding and decoding routines ENCODE-MESSAGE($\cdot$), ENCODE-ID($\cdot$), and DECODE-ID($\cdot$).

The idea of the delegate is that it is hard for the host to find out analytically how the delegate and its originator communicate and how to generate an appropriate decoder—and to make it even more difficult the originator can change the structure and functionality of the delegate at each turn. An easy solution for creating varying delegates is to have a predefined set of delegates from which to choose randomly one for the next turn. Another approach is to have a parameterized generator for creating delegates and to select the parameters randomly. A neurotic originator can even add spurious decoy data and scramble the execution flow to make it even harder to follow the construction of the decoder.

A delegate poses a possible security threat to the system which hosts it: the delegate could spy on the host system, tamper with its network traffic or waste its resources. Therefore, the delegate should be run in a sandbox, which limits

EXCHANGE$(e, \mathcal{S}, \mathcal{R})$
    in: encoded message e to be sent; sender $\mathcal{S}$; receiver $\mathcal{R}$
    out: originator's decoder $\mathcal{C}_p^t$ for the current turn
    local: encoded outgoing message f
    static: originator p; unique and dynamically created identification i
1: CHECK-CONNECTION$(\mathcal{S}, \mathcal{R}, p, i)$
2: $f \leftarrow$ ENCODE-MESSAGE$(\langle e, i \rangle)$
3: $\mathcal{S}$.SEND(p, f)
4: construct $\mathcal{C}_p^t$
5: **return** $\mathcal{C}_p^t$

CHECK-CONNECTION$(\mathcal{S}, \mathcal{R}, p, i)$
    in: sender $\mathcal{S}$; receiver $\mathcal{R}$; player p; unique identification i
    local: encoded response c from the originator
    static: probability b ($0 \leq b \leq 1$) of checking the connection
1: **if** RANDOM$() < b$ **then**
2:     $\mathcal{S}$.SEND$(p,$ ENCODE-ID$(i))$
3:     $c \leftarrow \mathcal{R}$.RECEIVE(p) ▷ Waits until a message arrives.
4:     **if** $i \neq$ DECODE-ID(c) **then**
5:         **error** connection to the player p is uncertain
6:     **end if**
7: **end if**

Fig. 5. Pseudocodes for the routines in a delegate

the allowed operations to the bare necessities. For example, the Java Runtime Environment allows to set a strict security policy for the executed code limiting its access to the files and networking capabilities [14].

4.2 Communication Turn

Figure 6 lists the pseudocode for the communication turn, which is run by each participating player. The message to be encoded includes three pieces of data: (1) the action to be carried out in the current turn, (2) the delegate to be used in the next turn, and (3) gossip about the actions made in the previous turn. The implementation of the routine ENCODE-MESSAGE$(\cdot)$ is omitted, but it should comply with the decoder that was sent along delegate in the previous turn. The presented approach does not prevent from using any cryptographic method (e.g., shared key, public key, or one-time pads). However, the chosen method does not have to be too complex, because the encoding function can be changed at every turn and it is enough that it can withhold the information for a few minutes.

Once the message is encoded it is exchanged with all the delegates, which were received in the previous turn. The resulting set of decoders is then used to decode the incoming messages. The actions, delegates for the next turn, and gossip from the previous turn are extracted from the decoded messages. Finally, the received gossip is compared with the known actions from the previous turn. Although the method given in Figure 6 is synchronized (i.e., all players communicate with

COMMUNICATE(M, S, R, P, D^t, A^{t-1})
  in: message $M = \langle a_p^t, \mathcal{D}_p^{t+1}, G_p^{t-1} \rangle$ of player's p action a_p^t for the current turn, delegate $\mathcal{D}_p^{t+1}$ for the next turn, and gossip G_p^{t-1} from the previous turn; sender S; receiver R; set of remote players P; set of delegates D^t for the current turn; set of all actions A^{t-1} from the previous turn
  out: set of players' actions A^t for the current turn; set of delegates D^{t+1} for the next turn
  local: set of decoders C; encoded message e; delegate $\mathcal{D}$; decoder $\mathcal{C}$; set G^{t-1} of gossiped actions of the previous turn; set of encoded messages E; encoded message f; decoded message N

1: $C \leftarrow \emptyset$
2: $e \leftarrow$ ENCODE-MESSAGE(M)
3: **for all** $\mathcal{D} \in D^t$ **do**   ▷ Get decoders from all delegates.
4:     $\mathcal{C} \leftarrow \mathcal{D}$.EXCHAGE($e, S, R$)
5:     $C \leftarrow C \cup \mathcal{C}$
6: **end for**
7: $A^t \leftarrow \{action(M)\}$ ▷ Store the player's own action.
8: $D^{t+1} \leftarrow \emptyset$
9: $G^{t-1} \leftarrow \emptyset$
10: $E \leftarrow R$.RECEIVE(P, t)   ▷ Get all messages for the current turn.
11: **for all** $e \in E$ **do**
12:    $f \leftarrow$ DECODE-MESSAGE(e)
13:    find decoder $\mathcal{C} \in C$ for which $origin(\mathcal{C}) = sender(f)$
14:    $N \leftarrow \mathcal{C}$.DECODE(f)
15:    $A^t \leftarrow A^t \cup \{action(N)\}$
16:    $D^{t+1} \leftarrow D^{t+1} \cup \{delegate(N)\}$
17:    $G^{t-1} \leftarrow G^{t-1} \cup gossip(N)$
18: **end for**
19: **if** $G^{t-1} \not\subseteq A^{t-1}$ **then**
20:    **error** actions in $(G^{t-1} \setminus A^{t-1})$ are conflicting
21: **end if**
22: **return** A^t, D^{t+1}

Fig. 6. Pseudocode for the communication turn

each other in every turn), we need to make only slight modifications to it to allow asynchronous communication; in particular, the delegates in D^t that have not been used in the current turn t (i.e., whose originator has not send a message) can be retained to the next turn $t + 1$.

Figure 7 illustrates the situation from a single player's point of view. The game process comprises the application and delegates from the other players. The application communicates with the current delegates in D^t by passing encoded messages e and receiving decoder objects $\mathcal{C}$. The sender object S and receiver object R provide a communication layer to the network. The game process of each player connects to network as illustrated in Figure 8.

Because all players provide their own delegates, in a game of n players each player hosts $n - 1$ delegates. If the execution of the delegates consumes too much processing power on the hosting system, it incurs an increase in the communi-

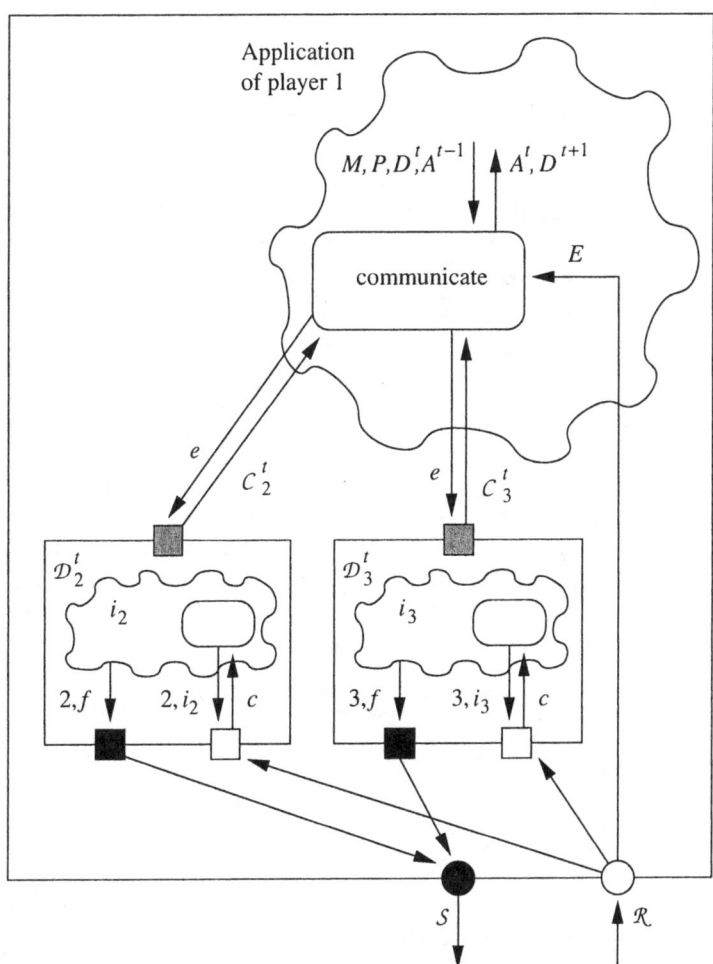

Fig. 7. The game process of player 1 comprises the application and the delegates of the other two players. Grey square represents message exchange, black square connection to the sender service (black circle), and white square connection to the receiver service (white circle). The oval inside the delegate cloud indicates the connection check.

cation delays. However, there is rarely a need to execute all delegates in a single turn, because the player usually communicates only with a few other players at a time. Moreover, the delegates can be executed in separate threads and prioritized so that the slow-down on the hosting system is as small as possible. Naturally, it is possible to attack against the host by sending delegates that waste processing power. This kind of malevolent behaviour rebounds immediately, because by slowing down the host's processing the malefactor also slows downs its own communication: The next delegate, which is required in the next exchange of messages, is not available until the current delegate returns a decoder, and when

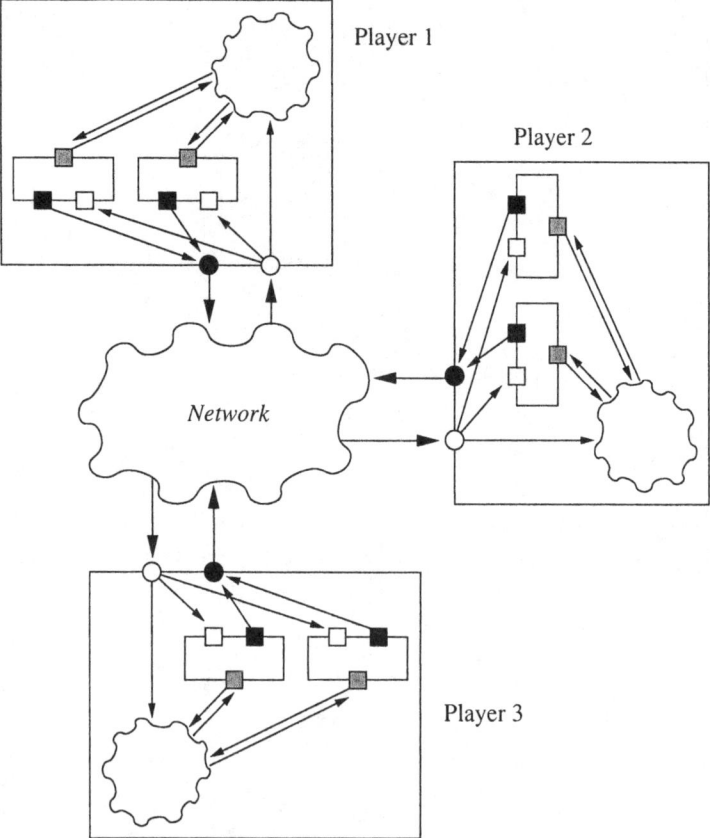

Fig. 8. Delegates in a three-player peer-to-peer network. Each player has received delegates to the other players' game processes.

the host has received the decoder, it can terminate and delete the delegate freely. We can also set limits on the time and memory consumption for the delegates and monitor the resource usages to spot abnormal behaviour. Moreover, we can add notifications on possible resource wasters to the gossip information, which can be then used to ban them from the game.

5 Final Remarks

In this paper we described how to prevent look-ahead cheating by using active objects. Each player has a delegate which is run in the other player's game process and which acts there as a trusted party in the message exchange. Since the delegate comprises executable code, its behaviour cannot be worked out by analytical methods alone. Moreover, by establishing the existence of a communication channel to its originator, the delegate becomes unique, which ensures that it cannot be replaced. The approach allows to parameterize the need for network transmissions, because we can select how frequently the delegates do the connection checks. Inconsistencies in the announced actions are controlled by including gossip about the previous actions in the messages.

Further research on the effect of different parameter values is still required. We gave only the bare bones of the delegate and decoder objects, and a more rigorous technical study on their implementation is needed. Especially, it would be interesting to know how hard they are to reverse engineer and whether they are vulnerable to other kinds of attacks than those mentioned in this paper.

References

1. Smed, J., Kaukoranta, T., Hakonen, H.: Aspects of networking in multiplayer computer games. The Electronic Library **20** (2002) 87–97
2. Yan, J.J., Choi, H.J.: Security issues in online games. The Electronic Library **20** (2002) 125–33
3. Kirmse, A., Kirmse, C.: Security in online games. Game Developer, July (1997) 20–8
4. Kirmse, A.: A network protocol for online games. In DeLoura, M., ed.: Game Programming Gems. Charles River Media (2000) 104–8
5. Rivest, R.: The MD5 message digest algorithm. Internet RFC 1321 (1992) Available at ⟨http://theory.lcs.mit.edu/~rivest/Rivest-MD5.txt⟩.
6. Ray, T.: PunkBuster User Manual. (2001) Available at ⟨http://www.punkbuster.com/pbmanual/userman.htm⟩.
7. Pritchard, M.: How to hurt hackers: The scoop on Internet cheating and how you can combat it. Gamasutra, July 24 (2000) Available at ⟨http://www.gamasutra.com/features/20000724/pritchard_01.htm⟩.
8. Buro, M.: ORTS: A hack-free RTS game environment. In Schaeffer, J., Müller, M., Björnsson, Y., eds.: Proceedings of the Third International Conference on Computers and Games, Edmonton, Canada. Volume 2883 of Lecture Notes in Computer Science. Springer-Verlag (2002) 280–91
9. Johansson, U., Sönströd, C., König, R.: Cheating by sharing information—the doom of online poker? In Sing, L.W., Man, W.H., Wai, W., eds.: Proceedings of the 2nd International Conference on Application and Development of Computer Games, Hong Kong SAR, China (2003) 16–22
10. Sanderson, D.: Online justice systems. Game Developer, April (1999) 42–9
11. Baughman, N.E., Levine, B.N.: Cheat-proof playout for centralized and distributed online games. In: Proceedings of the Twentieth IEEE Computer and Communication Society INFOCOM Conference, Anchorage, AK (2001)
12. Lee, H., Kozlowski, E., Lenker, S., Jamin, S.: Multiplayer game cheating prevention with pipelined lockstep protocol. In Nakatsu, R., Hoshino, J., eds.: Entertainment Computing: Technologies and Applications, IFIP First International Workshop on Entertainment Computing, Makuhari, Japan (2002) 31–9
13. Cronin, E., Filstrup, B., Jamin, S.: Cheat-proofing dead reckoned multiplayer games. In Sing, L.W., Man, W.H., Wai, W., eds.: Proceedings of the 2nd International Conference on Application and Development of Computer Games, Hong Kong SAR, China (2003) 23–9
14. Gong, L.: Java 2 platform security architecture. Web page (2002) Available at ⟨http://java.sun.com/j2se/1.5.0/docs/guide/security/spec/security-spec.doc.html⟩.

Strategic Interactions in the TAC 2003 Supply Chain Tournament

Joshua Estelle, Yevgeniy Vorobeychik, Michael P. Wellman, Satinder Singh, Christopher Kiekintveld, and Vishal Soni

Artificial Intelligence Laboratory,
University of Michigan, Ann Arbor, MI, USA
{jestelle, yvorobey, wellman, baveja, ckiekint, soniv}@umich.edu

Abstract. The TAC 2003 supply-chain game presented automated trading agents with a challenging strategic problem. Embedded within a complex stochastic environment was a pivotal strategic decision about initial procurement of components. Early evidence suggested that the entrant field was headed toward a self-destructive, mutually unprofitable equilibrium. Our agent, DEEP MAIZE, introduced a preemptive strategy designed to neutralize aggressive procurement, perturbing the field to a more profitable equilibrium. It worked. Not only did preemption improve DEEP MAIZE's profitability, it improved profitability for the whole field. Whereas it is perhaps counterintuitive that action designed to prevent others from achieving their goals actually helps them, strategic analysis employing an empirical game-theoretic methodology verifies and provides insight into the reasons of this outcome.

1 Introduction

Like classic computer games, multiagent research competitions [11] present well-defined problems for testing and comparing AI techniques and systems. The annual Trading Agent Competition (TAC) series provides a forum for research on strategic market behavior, and has led to several promising concepts and methods for implementing strategies in such domains [12].

The TAC Supply Chain Management (TAC/SCM) scenario [10] defines a complex six-player game with severely incomplete and imperfect information, and high-dimensional strategy spaces. Like the real supply-chain environments it is intended to model, the TAC/SCM game presents participants with challenging decision problems in a context of great strategic uncertainty. This paper is a case study of a strategic issue that arose in the first TAC/SCM tournament. We present our reasoning about the issue, and our effort to perturb the environment from an "equilibrium" we considered undesirable, to another more profitable domain of operation. We recount the experience as it played out in the competition, and analyze the outcome of this naturalistic experiment. We then perform a more controlled experimental analysis of the issue, applying empirical game-theoretic methods to produce compelling results, narrow in scope but arguably accounting well for strategic interactions.

2 TAC/SCM Game

In the TAC/SCM scenario,[1] six agents representing PC (personal computer) assemblers operate in a common market environment, over a simulated year. The environment constitutes a *supply chain*, in that agents trade simultaneously in markets for supplies (PC components) and the market for finished PCs. Agents may assemble for sale 16 different models of PCs, defined by the compatible combinations of the four component types: CPU, motherboard, memory, and hard disk.

Figure 1 diagrams the basic configuration of the supply chain. The six agents (arrayed vertically in the middle of the figure) procure components from the eight suppliers on the left, and sell PCs to the entity representing customers, on the right. Trades at both levels are negotiated through a *request-for-quote* (RFQ) mechanism, which proceeds in three steps.

1. Buyer issues RFQs to one or more sellers.
2. Sellers respond to RFQs with *offers*.
3. Buyers accept or reject offers. An accepted offer becomes an *order*.

The suppliers and customer implement fixed negotiation policies, defined in the game specification, and discussed in detail below where applicable.

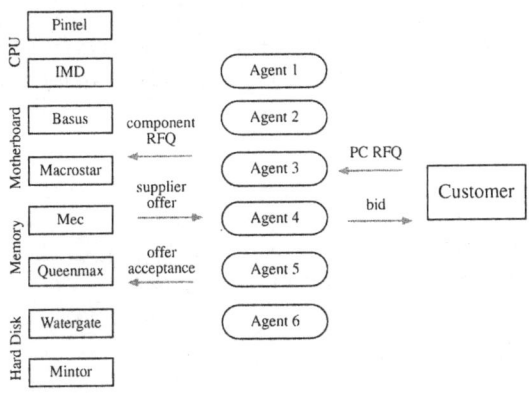

Fig. 1. TAC/SCM supply chain

The game runs for 220 simulated days. On each day, the agent may receive offers and component delivery notices from suppliers, and RFQs and offer acceptance notifications from customers. It then must make several decisions.

---

[1] For complete details of the game rules, see the specification document [1]. This is available at http://www.sics.se/tac, as is much additional information about TAC/SCM and TAC in general.

1. What RFQs to issue to component suppliers.
2. Given offers from suppliers (based on the previous day's RFQs), which to accept.
3. Given component inventory and factory capacity, what PCs to manufacture.
4. Given inventory of finished PCs, which customer orders to ship.
5. Given RFQs from customers, to which to respond and with what offers.

In the simulation, the agent has 15 seconds to compute and communicate its daily decisions to the game server. At the end of the game, agents are evaluated by total profit, with any outstanding component or PC inventory valued at zero.

As we describe below, a key stochastic feature of the game environment is level of demand for PCs. The underlying demand level is defined by an integer parameter Q (called RFQ_{avg} in the specification document [1, Section 6]). Each day, the customer issues a set of $\hat{Q}$ RFQs, where $\hat{Q}$ is drawn from a Poisson distribution with mean value defined by the parameter Q for that day. Since the order quantity, PC model, and reserve price are set independently for each customer RFQ, the number of RFQs serves as a sufficient statistic for the overall demand, which in turn is a major determinant of the potential profits available to the agents.

The demand parameter Q evolves according to a given stochastic process. In each game instance, an initial value, Q_0, is drawn uniformly from [80,320]. If Q_d is the value of Q on day d, then its value on the next day is given by [1, Section 6]:

$$Q_{d+1} = \min(320, \max(80, \tau_d Q_d)), \tag{1}$$

where τ is a trend parameter that also evolves stochastically. The initial trend is neutral, $\tau_0 = 1$, with subsequent trends updated by a perturbation $\epsilon \sim U[-0.01, 0.01]$:

$$\tau_{d+1} = \max(0.95, \min(1/0.95, \tau_d + \epsilon)). \tag{2}$$

In a given game, the demand may stay at predominantly high or low levels, or oscillate back and forth. The probabilistic behavior of Q figures importantly in our analysis, as presented in Subsection 5.5 below.

3 Deep Maize

The University of Michigan's entry in TAC-03/SCM is an agent called DEEP MAIZE [6,7]. The agent is organized in modular functional units controlling procurement, manufacturing, and sales. Its behavior is based on distributed feedback control, in that it acts to maintain a reference zone of profitable operation. To coordinate the distributed modules, DEEP MAIZE employs aggregate price signals, derived from a market equilibrium analysis and continual Bayesian demand projection. The design of DEEP MAIZE optimizes for performance in the steady-state, with little explicit attention to transient or end-game behaviors.

In the present study we focus on one pivotal feature of DEEP MAIZE's strategy, described in full detail below. We thus defer specifics of the rest of our agent's strategy to our other reports (which in turn do not address the strategic analysis presented here).

4 Day-0 Procurement Strategies

A close examination of the game rules suggests that procurement of components at the very beginning of the game (*day-0 procurement*) may be a pivotal strategic issue. This was indeed borne out by the behavior observed in preliminary rounds of the tournament, as discussed below. In this section, we explain the reason for expecting day-0 procurement to be so significant, and its ramifications for DEEP MAIZE and other agents.

4.1 Supplier Pricing

In the TAC/SCM market, suppliers set prices for components based on an analysis of their available capacity. Conceptually, there exist separate prices for each type of component, from each supplier. Moreover, these prices vary over time: both the time that the deal is struck, and time that the component is promised for delivery.

The TAC/SCM component catalog [1, Figure 3] associates every component c with a *base price*, b_c. The correspondence between price and quantity for component supplies is defined by the suppliers' pricing formula [1, Subsection 5.5]. The price offered by a supplier at day d for an order to be delivered on day $d+i$ is

$$p_c(d+i) = b_c - 0.5 b_c \frac{\kappa_c(d+i)}{500i}, \qquad (3)$$

where $\kappa_c(j)$ denotes the cumulative capacity for c the supplier projects to have available from the current day through day j. The denominator, $500i$, represents the *nominal capacity* controlled by the supplier over i days, not accounting for any capacity committed to existing orders.

Supplier prices according to Eq. (3) are date-specific, depending on the particular pattern of capacity commitments in place at the time the supplier evaluates the given RFQ. A key observation is that component prices are never lower than at the start of the game ($d = 0$), when $\kappa_c(i) = 500i$ and therefore $p_c(i) = 0.5 b_c$, for all c and i.[2] As the supplier approaches fully committed capacity ($\kappa_c(d+i) \to 0$), $p_c(d+i)$ approaches b_c.

In general, one would expect that procuring components at half their base price would be profitable, up to the limits of production capacity. Customer reserve prices range between 0.75 and 1.25 the base price of PCs, defined as the sum of base prices of components. Therefore, unless there is a significant oversupply, prices for PCs should easily exceed the component cost, based on day-0 prices.

---

[2] As discussed below, this creates a powerful incentive for early procurement, with significant consequences for game balance. In retrospect, the supplier pricing rule was generally considered a design flaw in the game, and has been substantially revised for the 2004 TAC/SCM tournament.

An agent's procurement strategy must also take into account the specific TAC/SCM RFQ process. Each day, agents may submit up to 10 RFQs, ordered by priority, to each supplier. The suppliers then repeatedly execute the following, until all RFQs are exhausted: (1) randomly choose an agent, (2) take the highest-priority RFQ remaining on its list, (3) generate a corresponding offer, if possible. In responding to an RFQ, if the supplier has sufficient available capacity to meet the requested quantity and due date, it offers to do so according to its pricing function. If it does not, the supplier instead offers a partial quantity at the requested date and/or the full quantity at a later date, to the best of its ability given its existing commitments. In all cases, the supplier quotes prices based on Eq. (3), and reserves sufficient capacity to meet the quantity and date offered.

4.2 Implications of Aggressive Day-0 Procurement

From the discussion above, it would appear advantageous to any agent that it attempt to procure a large number of components on day 0. We call this strategy *aggressive day-0 procurement*, or simply *aggressive*. From each agent's perspective, the main effect of being aggressive is on its own component procurement profile. If every agent is aggressive, however, it can significantly change the character of the game environment.

An aggressive day-0 procurement commits to large component orders before overall demand over the game horizon is known. This leaves agents with little flexibility to respond to cases of low demand, except by lowering PC prices to customers. Since component costs are sunk at the beginning, there is little to keep prices from dropping below (ex ante) profitable levels.

As more agents procure aggressively, several factors make aggressiveness even more compelling. The aggressive agents reserve significant fractions of supplier capacity, thus reducing subsequent availability and raising prices, according to their pricing function (3). A natural response might induce a "race" dynamic, where agents issue day-0 RFQs in increasingly large chunks, ultimately requesting all components they expect to be able to use over the entire game horizon. Not only does this exacerbate the risk of locking in aggregate oversupply, it also produces a less interleaved and more unbalanced distribution of components, especially at the beginning of the game. This in turn can prevent many agents from being able to acquire key components needed for particular PC models until relatively far into the production year.

For all these reasons, the aggressive strategy is appealing to individual agents, yet potentially quite damaging for the agent pool overall. We considered this situation particularly bad for our agent, given that it was designed for high performance in the steady state [6]. DEEP MAIZE devotes a considerable effort toward developing accurate demand projections, and thus is quite responsive to actual demand conditions. If most of the game's component procurement is up front, we never reach a steady state, and the ability to respond to demand conditions is much less relevant.

The DEEP MAIZE development team therefore decided not to employ aggressive day-0 procurement in the preliminary rounds, instead treating it just like

Table 1. TAC-03/SCM tournament participants, and their performance in preliminary rounds. Results from the qualifying rounds are weighted, seeding rounds are unweighted.

Agent	Affiliation	Average Profit ($M)		
		Qualifying	Seeding 1	Seeding 2
TacTex	U Texas	33.65	32.66	32.97
RedAgent	McGill U	15.09	24.57	29.52
Botticelli	Brown U	13.88	17.29	28.03
Jackaroo	U Western Sydney	14.89	35.55	19.23
WhiteBear	Cornell U	−3.17	13.57	16.50
PSUTAC	Pennsylvania State U	−120.0	15.52	15.25
HarTAC	Harvard U	12.41	4.19	10.72
UMBCTAC	U Maryland Baltimore Cty	−13.94	30.16	10.23
Sirish		−109.4	−0.17	8.27
Deep Maize	U Michigan	1.85	0.45	7.49
TAC-o-matic	Uppsala U	0.22	1.79	7.07
RonaX	Xonar GmbH	−0.92	9.24	4.29
MinneTAC	U Minnesota	10.88	6.56	−0.32
Mertacor	Aristotle U Thessaloniki	9.29	−0.38	−3.53
zepp	Poli Bucharest	−24.83	−7.80	−5.46
PackaTAC	N Carolina State U	−5.11	−25.67	−5.71
Socrates	U Essex	−48.94	−3.31	−6.84
Argos	Bogazici U	3.65	−4.24	−8.43
DummerAgent		−8.08	−20.56	—
DAI_hard	U Tulsa	−11.36	−39.05	—

any other day. We did not really expect that others would miss the opportunity, but did not want to encourage or accelerate it.

5 TAC-03 Tournament

The twenty agents who participated in the TAC-03/SCM tournament are listed in Table 1. The table presents average scores from each of three preliminary rounds, measured in millions of dollars of profit. Results from the semifinal and final rounds are presented in Subsection 5.3.

Two seeding rounds were held during the periods 7–11 and 14–18 July,[3] with each agent playing 60 and 66 games, respectively. Two agents were eliminated based on scores and/or inactivity after Seeding Round 1. The remaining 18 agents advanced to the semifinals, with assignment to heats based on standing in Seeding Round 2. The semifinals and finals were held live at IJCAI-03, 11–13 August in Acapulco, Mexico, each round consisting of nine games in one day. Semifinal 1 heat 1 (S1H1) comprised agents seeded 1–6 and 16–18, and

---

[3] An earlier "qualifying" round spanned 16–27 June, but this was mainly for debugging and no agents were eliminated.

the 7-15 seeds played in S1H2. The top six teams from each S1 heat (9 games) proceeded to the second semifinal round. S2H1 comprised teams ranked 1-3 in S1H1, and those ranked 4-6 in S1H2. The top three in S1H2 played, along with the second three in S1H1, in S2H2. The top three from each of S2H1 and S2H2 then entered the finals on 13 August. Further details about the TAC-03 tournament are available at http://www.sics.se/tac.

5.1 Evolution of Day-0 Policies in Preliminary Rounds

As we expected, competition entrants noticed the individual advantages of aggressive day-0 procurement. Early in the qualifying rounds we noticed JACKAROO's distinct saw-tooth shaped profits, indicating a steady increase in wealth with large periodic drops corresponding to large deliveries of supplies. This pattern was the result of large supply orders placed early in the game (over the first seven days, not just day 0) for delivery at regular intervals [14].

Based on our subsequent analysis of early game logs,[4] we can identify TACTEX [8] as the first to employ an aggressive day-0 strategy in competition. In their very first qualifying round game, TACTEX requested 8000 of each component from each supplier. Although we have found many agents performed mild day-0 procurement during the qualifying rounds, TACTEX was more aggressive, earlier—likely a factor in their supremacy this first round.

Throughout the first seeding round, more agents began using increasingly aggressive day-0 procurement strategies. In particular we noticed the successful agents TACTEX, BOTTICELLI, REDAGENT, UMBCTAC, and JACKAROO ordering very large quantities on day 0 and very little later in the game. Interestingly, there was no discussion of the issue on the TAC/SCM message boards, possibly because entrants recognized its strategic sensitivity. By the second seeding round it was obvious that the majority of agents were using aggressive strategies. In particular, we verified that all the agents that placed higher than DEEP MAIZE in the second seeding round (see Table 1) employed aggressive day-0 procurement.

While observing the increase in aggressiveness, we compiled detailed dossiers describing the day-0 strategies of other agents. We hoped to use this data to understand how widespread the use of day-0 procurement had become, and to understand how it was affecting the dynamics of the game.

5.2 Deep Maize Preemptive Strategy

After much deliberation, we decided that the only way to prevent the disastrous rush toward all-aggressive equilibrium was to *preempt* the other agents' day-0 RFQs. By requesting an extremely large quantity of a particular component, we would prevent the supplier from making reasonable offers to subsequent agents, at least in response to their requests on that day. Our premise was that it would

---

[4] The TAC/SCM game server records all agent actions (e.g., RFQs, manufacturing, bids) along with supplier and customer behavior, and releases the log files after each game instance is complete.

be sufficient to preempt only day-0 RFQs, since after day 0 prices are not so especially attractive.

The DEEP MAIZE preemptive strategy operates by submitting a large RFQ to each supplier for each component produced. The preemptive RFQ requests 85000 units—representing 170 days' worth of supplier capacity—to be delivered by day 30. See Figure 2. It is of course impossible for the supplier to actually fulfill this request. Instead, the supplier will offer us both a partial delivery on day 30 of the components they can offer by that date (if any), and an earliest-complete offer fulfilling the entire quantity (unless the supplier has already committed 50 days of capacity). With these offers, the supplier reserves necessary capacity. This has the effect of preempting subsequent RFQs, since we can be sure that the supplier will have committed capacity at least through day 172. (The extra two days account for negotiation and shipment time.) We will accept the partial-delivery offer, if any (and thereby reject the earliest-complete), giving us at most 14000 component units to be delivered on day 30, a large but feasible number of components to use up by the end of the game.

Fig. 2. DEEP MAIZE's preemptive RFQ

The TAC/SCM designers anticipated the possibility of preemptive RFQ generation, (there was much discussion about it in the original design correspondence), and took steps to inhibit it. The designers instated a reputation mechanism, in which refusing offers from suppliers reduces the priority of an agent's RFQs being considered in the future. Even with this deterrent, we felt our preemptive strategy would be worthwhile. Since most agents were focusing strongly on day 0, priority for RFQ selection on subsequent days might not turn out to be crucial.

5.3 Tournament Story

Having developed the preemptive strategy, we still faced the question of when to deploy it. Based on our performance in preliminaries, we were reasonably confident that we could make the top six out of nine in S1H2 without resorting to preemption, and instead chose to implement a moderate form of aggressive day-0 procurement. As expected, other agents actually scaled up their day-0 procurement, and consequently, DEEP MAIZE did not put on a very strong showing in this round. Fortunately, fourth place was sufficient to advance to the next round.

Table 2 presents results for the top twelve agents after Semifinal 1. Network problems at the competition venue caused difficulties for agents running locally—JACKAROO and HARTAC, in particular.[5]

---

[5] The problems did not affect the majority of agents communicating over the Internet from entrants' home institutions to the servers in Sweden.

Table 2. Results for twelve agents participating in the second semifinal and final rounds

Agent	Average Profit ($M)		
	Semifinal 1	Semifinal 2	Final
RedAgent	12.75 (H1)	25.09 (H1)	11.61
Deep Maize	10.51 (H2)	15.28 (H1)	9.47
TacTex	1.85 (H1)	−15.54 (H2)	5.02
Botticelli	5.69 (H1)	−4.83 (H2)	3.33
PackaTAC	18.31 (H1)	8.70 (H1)	−1.68
WhiteBear	5.26 (H1)	−9.58 (H2)	−3.45
PSUTAC	17.81 (H1)	−1.56 (H1)	—
TAC-o-matic	−1.24 (H2)	−13.50 (H1)	—
Sirish	15.86 (H2)	−20.21 (H2)	—
MinneTAC	13.92 (H2)	−24.98 (H2)	—
UMBCTAC	10.78 (H2)	−29.91 (H2)	—
HarTAC[6]	2.59 (H2)	−32.95 (H1)	—

After the first semifinal closed, the next few hours were filled with a great deal of hustle as the team activated the preemptive strategy that would be played the next day. On the one hand, these hours were also filled with anxiety. We had only intuition about the effect of preemptive strategy on DEEP MAIZE and other agents, but had never had a chance to test it against other competitors. On the other hand, we could hardly wait to see the "unexpected" dramatic change in DEEP MAIZE behavior in the arena with presumably the three best agents (since we did not place very highly in the first round, we would play the top three placing agents from the other heat).

In the morning of 12 August, the DEEP MAIZE team stood waiting by the computer screen as the second round of semifinals began. As day 29 rolled around, everyone held their breath, releasing it when the first large delivery of components dropped in. Once we saw distinct manifestations of the preemptive strategy, we began to wonder how other agents would react. Our suspense did not last long: soon after the game's midpoint, a comment emerged in the TAC game chatroom: "why we can't get hard disks? How server handle purchase RFQs? is the administrator around!!!?" Apparently, one agent at least was taking for granted that its day-0 requests would be fulfilled.

At the end of S2H1, DEEP MAIZE came in second behind the eventual tournament winner, RED AGENT [5], followed closely by PACKATAC [2]. These agents, it turned out, were relatively resilient to the preemptive strategy, as they did not excessively rely on day-0 procurement, but adaptively purchased components throughout the game.

Although none had anticipated it explicitly, it turned out that most agents playing in the finals were individually flexible enough to recover from day-0

---

[6] The score of HARTAC in Semifinal 2 was adversely affected by one game in which it experienced connectivity problems and lost $364M. Omitting this game would boost their average profit to $8.46M.

Table 3. Effect of preemption on day 1 component orders and average profits

	S1H1	S1H2	S2H1	S2H2	Finals
(DM?, P?, N)	-,-,9	DM,-,9	DM,P,8	-,-,9	DM,P,16
components	59390	46989	27377	70744	27172
avg profits	2.97	−3.05	7.02	−17.51	4.05

preemption. By preempting, it seemed that DEEP MAIZE had leveled the playing field, but REDAGENT's apparent adaptivity in procurement and sales [5] earned it the top spot in the competition rankings.

5.4 Analysis

Did DEEP MAIZE's preemption strategy work? We can first examine whether it had its intended direct effect, namely, to reduce the number of components ordered at the very beginning of the game. Table 3 presents, for each tournament round, the number of components ordered on day 1 (based on day-0 RFQs). Each value represents a total over delivery dates and agents, averaged over the 16 supplier-component pairs. Above the component numbers we indicate whether DEEP MAIZE played in that round (DM), whether it employed preemption (P), and the number of games. Note that this data includes one game in S2H1 and two in the finals in which DEEP MAIZE failed to preempt due to network problems. It does exclude one anomalous S2H1 game, in which HARTAC experienced connectivity problems, to wildly distorting effect.

From the table, it is clear that the preemptive day-0 strategy had a large effect. The difference is most dramatic in Semifinal 2, where the heat with DEEP MAIZE preempting saw an average of 27377 components committed on day 1, as compared to 70744 in the heat without DEEP MAIZE.

The tournament results also indicate that preemption was successful. The fact that DEEP MAIZE performed well overall is suggestive, though of course there are many other elements of DEEP MAIZE contributing to its behavior. Evidence that the preemptive strategy in particular was helpful can be found in the results from Semifinal 1, where DEEP MAIZE did not preempt and ended up in fourth place. This was sufficient for advancing in the tournament, but clearly not as creditable as its second place showing in the finals, among the (presumably) top agents in the field.

We may conclude, then, that preemption helped DEEP MAIZE. How did it affect the rest of the field? Table 3 also suggests a positive relation between preemption and profits averaged over *all* agents. Again, the contrast is greatest between S2H1 and S2H2. In the heat without DEEP MAIZE, it appears that competition among aggressive agents led to an average *loss* of $17.51M. With DEEP MAIZE preempting in S2H1, profits are a healthy $7.02M per agent. Preemption was also operative in the finals, and profits there were also positive. That it is preemption and not DEEP MAIZE per se is supported by examination of Semifinal 1, in which the heat without our agent appears to be substantially more profitable on average.

Pooling all of these semifinal and final games, we compared average profits for games with and without preemption. Games with preemption averaged $3.97M in profits, compared to a loss of $4.02M in games without preemption. Given the small dataset and large variance, this difference is only marginally statistically significant ($p = .09$).

Drawing inferences from tournament results is complicated by the presence of many varying and interacting factors. These include details of participating agents, and random features of environment, in particular the level of demand. To test the influence of demand, we measured the overall demand level for each game, $\bar{Q}$, defined as the average number of customer RFQs per day. Figure 3 presents a scatterplot of the tournament games, showing $\bar{Q}$ and per-agent profits for each. We distinguish the games with and without preemption, and for each class, fit a line to the points. The linear fit was quite good for the games with preemption ($R^2 = 0.84$), capturing somewhat less of the variance for the games without ($R^2 = 0.66$).

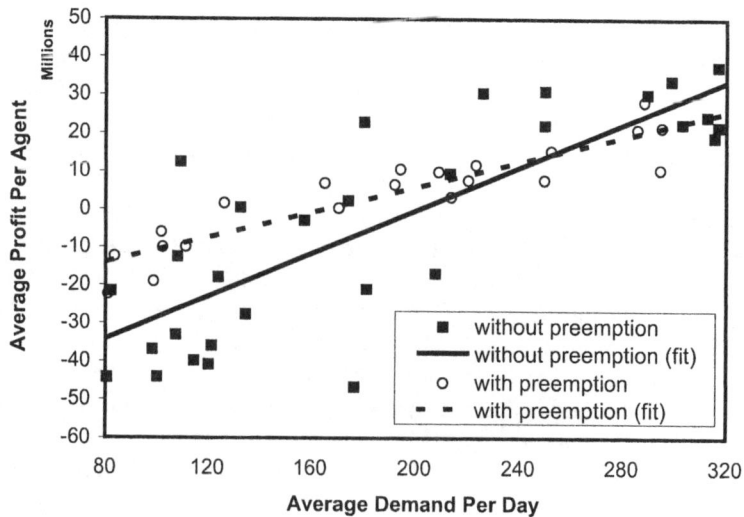

Fig. 3. Profits versus $\bar{Q}$ in TAC-03 tournament games. The lines represent best fits to data from games with and without preemption.

As seen in the figure, with or without preemption, demand clearly exhibits a significant ($p < 10^{-6}$) relation to profits. The relation is attenuated by preemption, and indeed the revealed trend indicates that preemption is beneficial when demand is low, and detrimental in the highest-demand games. This is what we would expect, given that the primary effect of preemption is to inhibit early commitment to large supplies. Given the apparently important influence of demand, we developed a more elaborate mechanism to control for demand in our analysis of tournament games as well as our post-competition experiments.

5.5 Demand Adjustment

Given a sufficient number of random instances, the problem of variance due to stochastic demand would subside, as the sample means for outcomes of interest would converge to their true expectations. However, for TAC/SCM, sample data is quite expensive, as each game instance takes approximately one hour. Therefore, datasets from tournaments and even offline experiments will necessarily reflect only limited sampling from the distribution of demand environments.

To address this issue, we can calibrate a given sample with respect to the known underlying distribution of demand ($\bar{Q}$). Our approach is closely related to the standard method of variance reduction by conditioning [9, Section 11.6.2]. Given a specification for the expectation of some game statistic y as a function of $\bar{Q}$, its overall expectation accounting for demand is given by

$$E[y] = \int_{\bar{Q}} E[y|\bar{Q}] \Pr(\bar{Q}) d\bar{Q}. \qquad (4)$$

Although we do not have a closed-form characterization of the density function $\Pr(\bar{Q})$, we do have a specification of the underlying stochastic demand process. From this, we can generate Monte-Carlo samples of demand trajectories over a simulated game.[7] We then employ a kernel-based density estimation method using Parzen windows [3] to approximate the probability density function for $\bar{Q}$. This distribution is shown in Figure 4. Its mean is 196, with a standard deviation of 77.4. Note that much of the probability is massed at the extremes of demand, with a skew toward the low end. The tendency toward the extremes comes from the combination of trend (τ) momentum and bounding of Q. The skew toward the low end comes from the fact that the trend is multiplicative, so the process tends to transition more rapidly while at the higher levels of demand.

Given this distribution, we define *demand-adjusted profit* (DAP) as the expected profit, adjusted for demand. We calculate this by substituting the per-agent profit for y in Eq. (4). Using this formula requires an estimate for profits as a function of $\bar{Q}$, which we obtain by linear regression from the sample data. The two lines in Figure 3 thus represent our estimates for profits given $\bar{Q}$ for the two sets of TAC-03 tournament games. Although the linearity assumption is not correct, for limited samples this is compensated by the reduction in variance due to adjusting for $\bar{Q}$.

From the linear model of profits given $\bar{Q}$, we can obtain a summary comparison of overall profits with and without preemption. For the TAC-03 games without preemption, DAP was –$1.41M. Preemption increased DAP to $5.20M. Thus, we find that on average, DEEP MAIZE's preemptive strategy improved not only its own profits, but those of the other agents as well. These results are corroborated by controlled experiments described below.

---

[7] We could also use historical game data, but simulating Eqs. (1) and (2) is much faster. The 200,000 data points we generated for our density estimate would take 22.8 years of game simulation time to produce.

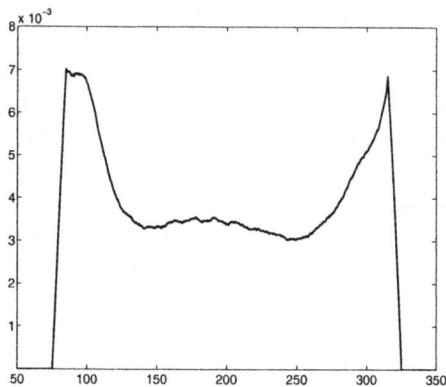

Fig. 4. Probability density for average RFQs per day ($\bar{Q}$)

6 Game-Theoretic Model

Although the tournament results presented above are illuminating, it is difficult to support general conclusions due to the many contributing factors and differences among agents. To isolate the effect of preemption on the key strategic variable (aggressiveness of day-0 procurement), we developed a stylized game-theoretic model, then calibrated it using simulation experiments. Our results are summarized here; see the extended version of this paper [13] for our detailed analysis.

As noted at the outset, TAC/SCM defines a six-player game of incomplete and imperfect information, with an enormous space of available strategies. The game is *symmetric* [4], in that agents have identical action possibilities, and face the same environmental conditions. In our stylized model, we restrict the agents to two strategies, differing only in their approach to day-0 procurement. Both are implemented as variants of DEEP MAIZE. In strategy A (aggressive), the agent requests large quantities of components from every supplier on day 0. In strategy B (baseline), the agent treats day 0 just like any other day, issuing requests according to its usual policy of serving anticipated demand and maintaining a buffer inventory [6].

To calibrate our models, we ran 30 or more simulated games for each strategy profile (i.e., combination of A and B), with and without the presence of an agent playing the preemptive strategy, P. For each sample, we collected the average profits for the As and Bs, as well as the demand level, $\bar{Q}$. We derive the demand-adjusted payoff (DAP) for each strategy, using the method described in Subsection 5.5. Our findings are as follows.

Aggressiveness has a negative effect on total profits – We regressed total DAP for each profile on the number of aggressive agents in that profile. For games without preemption, the linear relationship was quite strong ($p = 0.0018$, $R^2 = 0.88$), with each A in the profile subtracting \$20.9M from total profits, on average. With a preemptive agent, the effect was insignificant ($p = 0.54$, $R^2 = 0.10$).

Preemption neutralizes aggressive procurement – In non-preemptive profiles, the raw difference in average profits between aggressive and baseline agents was on the order of $10M, as compared to $1M for the preemptive profiles. Moreover, the average variance across agents for non-preemptive profiles was an order of magnitude larger than the average variance for preemptive profiles.

The expected behaviors obtain in equilibrium – Our observations about the game's propensity to promote aggressive procurement were consistent with the 2003 tournament results, but does this actually constitute rational behavior? From our empirical payoff function, we can derive pure-strategy Nash equilibria, providing one way to answer this question. As seen in Figure 5, the unique pure Nash equilibrium profile without preemption comprises four As and two Bs. A similar analysis with preemption reveals three equilibria, with zero, two, or four As, respectively. The differences are much smaller in this case, and the statistical differences much less significant. This is consistent with our finding above that preemption neutralizes the difference between A and B. Without preemption, a predominance of As is expected.

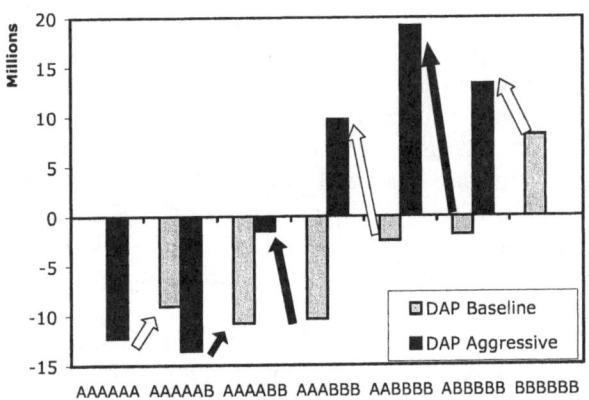

Fig. 5. DAP payoffs for strategy profiles, without preemption. Arrows indicate for each column, whether an agent in that profile would prefer to stay with that strategy (arrow head), or switch (arrow tail). Solid black arrows denote statistically significant comparisons.

Symmetric equilibria confirm these findings – For the game without preemption, the unique symmetric mixed-strategy equilibrium plays strategy A with probability 0.82. With preemption, there are two equilibria, at probabilities 0.03 and 0.99.

Preemption increases average profits for everybody – Analysis of the mixed-strategy equilibrium of the game without preemption reveals that the expected payoff (equal for A and B, by definition) is a *loss* of $9.59M. With preemption, the two equilibria have expected payoffs of $5.92M and $7.01M, respectively.

To evaluate the degree to which preemption neutralizes the difference between A and B, we can identify an ϵ^* for each game such that *any* mixed strategy is a symmetric ϵ-Nash equilibrium at $\epsilon = \epsilon^*$. A profile is ϵ-Nash if no agent can improve its payoff by more than ϵ by deviating from its assigned strategy. For games without preemption, ϵ^* is $10.6M. With preemption, ϵ^* is only $0.97M. This provides a bound on how much it can matter to make the right choice about aggressiveness, given a symmetric set of other agents.

Preemption obtains in equilibrium – When agents are allowed to choose among all three strategies (A, B, and P), some will choose to preempt. Among the 28 distinct strategy profiles, there are four pure-strategy equilibria, which have 1–3 preemptors. We have also identified a symmetric mixed-strategy equilibrium, in which agents preempt with probability 0.58.

7 Conclusion

The TAC supply-chain game presented automated trading agents (and their designers) with a challenging strategic problem. Embedded within a highly-dimensional stochastic environment was a pivotal strategic decision about initial procurement of components. Our reading of the game rules and observation of the preliminary rounds suggested to us that the entrant field was headed toward a self-destructive, mutually unprofitable equilibrium of chronic oversupply. Our agent, DEEP MAIZE, introduced a preemptive strategy designed to neutralize aggressive procurement. It worked. Not only did preemption improve DEEP MAIZE's profitability, it improved profitability for the whole field. Whereas it is perhaps counterintuitive that actions designed to prevent others from achieving their goals actually helps them, strategic analysis explains how that can be the case.

Investigating strategic behavior in the context of a research competition has several distinct advantages. First, the game is designed by someone other than the investigator, avoiding the kinds of bias that often doom research projects to success. Second, the entry pool is uncontrolled, and so we may encounter unanticipated behavior of individual agents and aggregates. Third, the games are complex, avoiding many of the biases following from the need to preserve analytical or computational tractability. Fourth, the environment model is precisely specified and repeatable, thus subject to controlled experimentation. We have exploited all of these features in our study, in the process developing a repertoire of methods for empirical game-theoretic analysis, which we expect to prove useful for a range of problems.

There is no doubt that this form of study also has several limitations, for example in justifying generalizations beyond the particular environment studied. Nevertheless, we believe that the methods developed here provide a useful complement to the kinds of (a priori) stylized modeling most often pursued in game-theoretic analysis, and to the non-strategic analyses typically applied to simulation environments.

Acknowledgements. Thanks to the TAC/SCM organizers and participants. At the University of Michigan, DEEP MAIZE was designed and implemented with the additional help of Shih-Fen Cheng, Thede Loder, Kevin O'Malley, and Matthew Rudary. Daniel Reeves assisted our equilibrium analyses. This research was supported in part by NSF grant IIS-0205435.

References

1. Raghu Arunachalam, Joakim Eriksson, Niclas Finne, Sverker Janson, and Norman Sadeh. The TAC supply chain management game. Technical report, Swedish Institute of Computer Science, 2003. Draft Version 0.62.
2. Erik Dahlgren. PackaTAC: A conservative trading agent. Master's thesis, Lund University, 2003.
3. Richard O. Duda, Peter E. Hart, and David G. Stork. *Pattern Classification*. Wiley-Interscience, second edition, 2000.
4. Herbert Gintis. *Game Theory Evolving*. Princeton University Press, 2000.
5. Philipp W. Keller, Félix-Olivier Duguay, and Doina Precup. RedAgent: Winner of TAC SCM 2003. *SIGecom Exchanges*, 4(3):1–8, 2004.
6. Christopher Kiekintveld, Michael P. Wellman, Satinder Singh, Joshua Estelle, Yevgeniy Vorobeychik, Vishal Soni, and Matthew Rudary. Distributed feedback control for decision making on supply chains. In *Fourteenth International Conference on Automated Planning and Scheduling*, Whistler, BC, 2004a.
7. Christopher Kiekintveld, Michael P. Wellman, Satinder Singh, and Vishal Soni. Value-driven procurement in the TAC supply chain game. *SIGecom Exchanges*, 4(3):9–18, 2004b.
8. David Pardoe and Peter Stone. TacTex-03: A supply chain management agent. *SIGecom Exchanges*, 4(3):19–28, 2004.
9. Sheldon M. Ross. *Introduction to Probability Models*. Academic Press, sixth edition, 1997.
10. Norman Sadeh, Raghu Arunachalam, Joakim Eriksson, Niclas Finne, and Sverker Janson. TAC-03: A supply-chain trading competition. *AI Magazine*, 24(1):92–94, 2003.
11. Peter Stone. Multiagent competitions and research: Lessons from RoboCup and TAC. In *Sixth RoboCup International Symposium*, Fukuoka, Japan, 2002.
12. Michael P. Wellman, Shih-Fen Cheng, Daniel M. Reeves, and Kevin M. Lochner. Trading agents competing: Performance, progress, and market effectiveness. *IEEE Intelligent Systems*, 18(6):48–53, 2003.
13. Michael P. Wellman, Joshua Estelle, Satinder Singh, Yevgeniy Vorobeychik, Christopher Kiekintveld, and Vishal Soni. Strategic interactions in a supply chain game. Technical report, University of Michigan, 2004.
14. Dongmo Zhang, Kanghua Zhao, Chia-Ming Liang, Gonelur Begum Huq, and Tze-Haw Huang. Strategic trading agents via market modelling. *SIGecom Exchanges*, 4(3):46–55, 2004.

Author Index

Billings, Darse 21
Bolognesi, Andrea 246
Bouzy, Bruno 67
Bowling, Michael 21
Burch, Neil 21
Buro, Michael 35, 51

Cazenave, Tristan 220
Ciancarini, Paolo 246
Croonenborghs, Tom 113

Davidson, Aaron 21
Donkers, H. (Jeroen) H.L.M. 202

Estelle, Joshua 316

Fang, Haw-ren 129
Felner, Ariel 187
Fleischer, Rudolf 232
Fotland, David 175

Hakonen, Harri 301
Hauk, Thomas 35, 51
Helmstetter, Bernard 220
Holte, Robert 21
Hsu, Tsan-sheng 145

Karapetyan, Akop 161
Kiekintveld, Christopher 316

Li, Zhichao 273
Liu, Ping-Yi 145
Lorentz, Richard J. 161

Müller, Martin 97, 273

Netanyahu, Nathan S. 187
Niu, Xiaozhen 97

Ramon, Jan 113

Schaeffer, Jonathan 21, 35, 51
Schauenberg, Terence 21
Sheppard, Brian 1
Singh, Satinder 316
Smed, Jouni 301
Soni, Vishal 316
Sturtevant, Nathan 285
Szafron, Duane 21

Tabibi, Omid David 187
Trippen, Gerhard 232

Uiterwijk, Jos W.H.M. 81, 202, 262

van den Herik, H. Jaap 81, 202, 262
van der Werf, Erik C.D. 81, 262
Vorobeychik, Yevgeniy 316

Wellman, Michael P. 316
Winands, Mark H.M. 262
Wu, Ping-hsun 145

Lecture Notes in Computer Science

For information about Vols. 1–3777

please contact your bookseller or Springer

Vol. 3878: A. Gelbukh (Ed.), Computational Linguistics and Intelligent Text Processing. XVII, 589 pages. 2006.

Vol. 3872: H. Bunke, A. L. Spitz (Eds.), Document Analysis Systems VII. XIII, 630 pages. 2006.

Vol. 3870: S. Spaccapietra, P. Atzeni, W.W. Chu, T. Catarci, K.P. Sycara (Eds.), Journal on Data Semantics V. XIII, 237 pages. 2006.

Vol. 3868: K. Römer, H. Karl, F. Mattern (Eds.), Wireless Sensor Networks. XI, 342 pages. 2006.

Vol. 3863: M. Kohlhase (Ed.), Mathematical Knowledge Management. XI, 405 pages. 2006. (Sublibrary LNAI).

Vol. 3861: J. Dix, S.J. Hegner (Eds.), Foundations of Information and Knowledge Systems. X, 331 pages. 2006.

Vol. 3860: D. Pointcheval (Ed.), Topics in Cryptology – CT-RSA 2006. XI, 365 pages. 2006.

Vol. 3858: A. Valdes, D. Zamboni (Eds.), Recent Advances in Intrusion Detection. X, 351 pages. 2006.

Vol. 3857: M. Fossorier, H. Imai, S. Lin, A. Poli (Eds.), Applied Algebra, Algebraic Algorithms and Error-Correcting Codes. XI, 350 pages. 2006.

Vol. 3855: E. A. Emerson, K.S. Namjoshi (Eds.), Verification, Model Checking, and Abstract Interpretation. XI, 443 pages. 2005.

Vol. 3853: A.J. Ijspeert, T. Masuzawa, S. Kusumoto (Eds.), Biologically Inspired Approaches to Advanced Information Technology. XIV, 388 pages. 2006.

Vol. 3852: P.J. Narayanan, S.K. Nayar, H.-Y. Shum (Eds.), Computer Vision - ACCV 2006, Part II. XXXI, 977 pages. 2005.

Vol. 3851: P.J. Narayanan, S.K. Nayar, H.-Y. Shum (Eds.), Computer Vision - ACCV 2006, Part I. XXXI, 973 pages. 2006.

Vol. 3850: R. Freund, G. Păun, G. Rozenberg, A. Salomaa (Eds.), Membrane Computing. IX, 371 pages. 2006.

Vol. 3848: J.-F. Boulicaut, L. De Raedt, H. Mannila (Eds.), Constraint-Based Mining and Inductive Databases. X, 401 pages. 2006. (Sublibrary LNAI).

Vol. 3847: K.P. Jantke, A. Lunzer, N. Spyratos, Y. Tanaka (Eds.), Federation over the Web. X, 215 pages. 2006. (Sublibrary LNAI).

Vol. 3846: H. J. van den Herik, Y. Björnsson, N.S. Netanyahu (Eds.), Computers and Games. XIV, 333 pages. 2006.

Vol. 3844: J.-M. Bruel (Ed.), Satellite Events at the MoD-ELS 2005 Conference. XIII, 360 pages. 2006.

Vol. 3843: P. Healy, N.S. Nikolov (Eds.), Graph Drawing. XVII, 536 pages. 2006.

Vol. 3842: H.T. Shen, J. Li, M. Li, J. Ni, W. Wang (Eds.), Advanced Web and Network Technologies, and Applications. XXVII, 1057 pages. 2006.

Vol. 3841: X. Zhou, J. Li, H.T. Shen, M. Kitsuregawa, Y. Zhang (Eds.), Frontiers of WWW Research and Development - APWeb 2006. XXIV, 1223 pages. 2006.

Vol. 3840: M. Li, B. Boehm, L.J. Osterweil (Eds.), Unifying the Software Process Spectrum. XVI, 522 pages. 2006.

Vol. 3839: J.-C. Filliâtre, C. Paulin-Mohring, B. Werner (Eds.), Types for Proofs and Programs. VIII, 275 pages. 2006.

Vol. 3838: A. Middeldorp, V. van Oostrom, F. van Raamsdonk, R. de Vrijer (Eds.), Processes, Terms and Cycles: Steps on the Road to Infinity. XVIII, 639 pages. 2005.

Vol. 3837: K. Cho, P. Jacquet (Eds.), Technologies for Advanced Heterogeneous Networks. IX, 307 pages. 2005.

Vol. 3836: J.-M. Pierson (Ed.), Data Management in Grids. X, 143 pages. 2006.

Vol. 3835: G. Sutcliffe, A. Voronkov (Eds.), Logic for Programming, Artificial Intelligence, and Reasoning. XIV, 744 pages. 2005. (Sublibrary LNAI).

Vol. 3834: D.G. Feitelson, E. Frachtenberg, L. Rudolph, U. Schwiegelshohn (Eds.), Job Scheduling Strategies for Parallel Processing. VIII, 283 pages. 2005.

Vol. 3833: K.-J. Li, C. Vangenot (Eds.), Web and Wireless Geographical Information Systems. XI, 309 pages. 2005.

Vol. 3832: D. Zhang, A.K. Jain (Eds.), Advances in Biometrics. XX, 796 pages. 2005.

Vol. 3831: J. Wiedermann, G. Tel, J. Pokorný, M. Bieliková, J. Štuller (Eds.), SOFSEM 2006: Theory and Practice of Computer Science. XV, 576 pages. 2006.

Vol. 3829: P. Pettersson, W. Yi (Eds.), Formal Modeling and Analysis of Timed Systems. IX, 305 pages. 2005.

Vol. 3828: X. Deng, Y. Ye (Eds.), Internet and Network Economics. XVII, 1106 pages. 2005.

Vol. 3827: X. Deng, D.-Z. Du (Eds.), Algorithms and Computation. XX, 1190 pages. 2005.

Vol. 3826: B. Benatallah, F. Casati, P. Traverso (Eds.), Service-Oriented Computing - ICSOC 2005. XVIII, 597 pages. 2005.

Vol. 3824: L.T. Yang, M. Amamiya, Z. Liu, M. Guo, F.J. Rammig (Eds.), Embedded and Ubiquitous Computing – EUC 2005. XXIII, 1204 pages. 2005.

Vol. 3823: T. Enokido, L. Yan, B. Xiao, D. Kim, Y. Dai, L.T. Yang (Eds.), Embedded and Ubiquitous Computing – EUC 2005 Workshops. XXXII, 1317 pages. 2005.

Vol. 3822: D. Feng, D. Lin, M. Yung (Eds.), Information Security and Cryptology. XII, 420 pages. 2005.

Vol. 3821: R. Ramanujam, S. Sen (Eds.), FSTTCS 2005: Foundations of Software Technology and Theoretical Computer Science. XIV, 566 pages. 2005.

Vol. 3820: L.T. Yang, X.-s. Zhou, W. Zhao, Z. Wu, Y. Zhu, M. Lin (Eds.), Embedded Software and Systems. XXVIII, 779 pages. 2005.

Vol. 3819: P. Van Hentenryck (Ed.), Practical Aspects of Declarative Languages. X, 231 pages. 2005.

Vol. 3818: S. Grumbach, L. Sui, V. Vianu (Eds.), Advances in Computer Science – ASIAN 2005. XIII, 294 pages. 2005.

Vol. 3817: M. Faundez-Zanuy, L. Janer, A. Esposito, A. Satue-Villar, J. Roure, V. Espinosa-Duro (Eds.), Nonlinear Analyses and Algorithms for Speech Processing. XII, 380 pages. 2006. (Sublibrary LNAI).

Vol. 3816: G. Chakraborty (Ed.), Distributed Computing and Internet Technology. XXI, 606 pages. 2005.

Vol. 3815: E.A. Fox, E.J. Neuhold, P. Premsmit, V. Wuwongse (Eds.), Digital Libraries: Implementing Strategies and Sharing Experiences. XVII, 529 pages. 2005.

Vol. 3814: M. Maybury, O. Stock, W. Wahlster (Eds.), Intelligent Technologies for Interactive Entertainment. XV, 342 pages. 2005. (Sublibrary LNAI).

Vol. 3813: R. Molva, G. Tsudik, D. Westhoff (Eds.), Security and Privacy in Ad-hoc and Sensor Networks. VIII, 219 pages. 2005.

Vol. 3811: C. Bussler, M.-C. Shan (Eds.), Technologies for E-Services. VIII, 127 pages. 2006.

Vol. 3810: Y.G. Desmedt, H. Wang, Y. Mu, Y. Li (Eds.), Cryptology and Network Security. XI, 349 pages. 2005.

Vol. 3809: S. Zhang, R. Jarvis (Eds.), AI 2005: Advances in Artificial Intelligence. XXVII, 1344 pages. 2005. (Sublibrary LNAI).

Vol. 3808: C. Bento, A. Cardoso, G. Dias (Eds.), Progress in Artificial Intelligence. XVIII, 704 pages. 2005. (Sublibrary LNAI).

Vol. 3807: M. Dean, Y. Guo, W. Jun, R. Kaschek, S. Krishnaswamy, Z. Pan, Q.Z. Sheng (Eds.), Web Information Systems Engineering – WISE 2005 Workshops. XV, 275 pages. 2005.

Vol. 3806: A.H. H. Ngu, M. Kitsuregawa, E.J. Neuhold, J.-Y. Chung, Q.Z. Sheng (Eds.), Web Information Systems Engineering – WISE 2005. XXI, 771 pages. 2005.

Vol. 3805: G. Subsol (Ed.), Virtual Storytelling. XII, 289 pages. 2005.

Vol. 3804: G. Bebis, R. Boyle, D. Koracin, B. Parvin (Eds.), Advances in Visual Computing. XX, 755 pages. 2005.

Vol. 3803: S. Jajodia, C. Mazumdar (Eds.), Information Systems Security. XI, 342 pages. 2005.

Vol. 3802: Y. Hao, J. Liu, Y.-P. Wang, Y.-m. Cheung, H. Yin, L. Jiao, J. Ma, Y.-C. Jiao (Eds.), Computational Intelligence and Security, Part II. XLII, 1166 pages. 2005. (Sublibrary LNAI).

Vol. 3801: Y. Hao, J. Liu, Y.-P. Wang, Y.-m. Cheung, H. Yin, L. Jiao, J. Ma, Y.-C. Jiao (Eds.), Computational Intelligence and Security, Part I. XLI, 1122 pages. 2005. (Sublibrary LNAI).

Vol. 3799: M. A. Rodríguez, I.F. Cruz, S. Levashkin, M.J. Egenhofer (Eds.), GeoSpatial Semantics. X, 259 pages. 2005.

Vol. 3798: A. Dearle, S. Eisenbach (Eds.), Component Deployment. X, 197 pages. 2005.

Vol. 3797: S. Maitra, C. E. V. Madhavan, R. Venkatesan (Eds.), Progress in Cryptology - INDOCRYPT 2005. XIV, 417 pages. 2005.

Vol. 3796: N.P. Smart (Ed.), Cryptography and Coding. XI, 461 pages. 2005.

Vol. 3795: H. Zhuge, G.C. Fox (Eds.), Grid and Cooperative Computing - GCC 2005. XXI, 1203 pages. 2005.

Vol. 3794: X. Jia, J. Wu, Y. He (Eds.), Mobile Ad-hoc and Sensor Networks. XX, 1136 pages. 2005.

Vol. 3793: T. Conte, N. Navarro, W.-m.W. Hwu, M. Valero, T. Ungerer (Eds.), High Performance Embedded Architectures and Compilers. XIII, 317 pages. 2005.

Vol. 3792: I. Richardson, P. Abrahamsson, R. Messnarz (Eds.), Software Process Improvement. VIII, 215 pages. 2005.

Vol. 3791: A. Adi, S. Stoutenburg, S. Tabet (Eds.), Rules and Rule Markup Languages for the Semantic Web. X, 225 pages. 2005.

Vol. 3790: G. Alonso (Ed.), Middleware 2005. XIII, 443 pages. 2005.

Vol. 3789: A. Gelbukh, Á. de Albornoz, H. Terashima-Marín (Eds.), MICAI 2005: Advances in Artificial Intelligence. XXVI, 1198 pages. 2005. (Sublibrary LNAI).

Vol. 3788: B. Roy (Ed.), Advances in Cryptology - ASIACRYPT 2005. XIV, 703 pages. 2005.

Vol. 3787: D. Kratsch (Ed.), Graph-Theoretic Concepts in Computer Science. XIV, 470 pages. 2005.

Vol. 3786: J. Song, T. Kwon, M. Yung (Eds.), Information Security Applications. XI, 378 pages. 2006.

Vol. 3785: K.-K. Lau, R. Banach (Eds.), Formal Methods and Software Engineering. XIV, 496 pages. 2005.

Vol. 3784: J. Tao, T. Tan, R.W. Picard (Eds.), Affective Computing and Intelligent Interaction. XIX, 1008 pages. 2005.

Vol. 3783: S. Qing, W. Mao, J. Lopez, G. Wang (Eds.), Information and Communications Security. XIV, 492 pages. 2005.

Vol. 3782: K.-D. Althoff, A. Dengel, R. Bergmann, M. Nick, T.R. Roth-Berghofer (Eds.), Professional Knowledge Management. XXIII, 739 pages. 2005. (Sublibrary LNAI).

Vol. 3781: S.Z. Li, Z. Sun, T. Tan, S. Pankanti, G. Chollet, D. Zhang (Eds.), Advances in Biometric Person Authentication. XI, 250 pages. 2005.

Vol. 3780: K. Yi (Ed.), Programming Languages and Systems. XI, 435 pages. 2005.

Vol. 3779: H. Jin, D. Reed, W. Jiang (Eds.), Network and Parallel Computing. XV, 513 pages. 2005.

Vol. 3778: C. Atkinson, C. Bunse, H.-G. Gross, C. Peper (Eds.), Component-Based Software Development for Embedded Systems. VIII, 345 pages. 2005.